PROPENSITY SCORE ANALYSIS

Advanced Quantitative Techniques
in the Social Sciences

VOLUMES IN THE SERIES

PROPENSITY SCORE ANALYSIS

Statistical Methods and Applications

SHENYANG GUO

University of North Carolina at Chapel Hill

MARK W. FRASER

University of North Carolina at Chapel Hill

Advanced Quantitative Techniques
in the Social Sciences Series **11**

Los Angeles | London | New Delhi
Singapore | Washington DC

For information:

SAGE Publications, Inc.
2455 Teller Road
Thousand Oaks, California 91320
E-mail: order@sagepub.com

SAGE Publications India Pvt. Ltd.
B 1/I 1 Mohan Cooperative
 Industrial Area
Mathura Road, New Delhi 110 044
India

SAGE Publications Ltd.
1 Oliver's Yard
55 City Road
London EC1Y 1SP
United Kingdom

SAGE Publications Asia-Pacific Pte. Ltd.
33 Pekin Street #02-01
Far East Square
Singapore 048763

Printed in the United States of America.

Library of Congress Cataloging-in-Publication Data

Guo, Shenyang.
Propensity score analysis : statistical methods and applications/Shenyang Guo, Mark W. Fraser.
    p. cm.—(Advanced quantitative techniques in the social sciences; v. 11)
Includes bibliographical references and index.
ISBN 978-1-4129-5356-6 (cloth)
    1. Social sciences—Statistical methods. 2. Analysis of variance. I. Fraser, Mark W., 1946- II. Title.

HA29.G91775 2010
519.5'3—dc22   2009012335

This book is printed on acid-free paper.

09   10   11   12   13   10   9   8   7   6   5   4   3   2

Acquisitions Editor:	Vicki Knight
Associate Editor:	Sean Connelly
Editorial Assistant:	Lauren Habib
Production Editor:	Karen Wiley
Copy Editor:	QuADS Prepress (P) Ltd.
Typesetter:	C&M Digitals (P) Ltd.
Proofreader:	Scott Oney
Indexer:	Kathleen Paparchontis
Cover Designer:	Glenn Vogel
Marketing Manager:	Stephanie Adams

Contents

List of Tables

List of Figures

Acknowledgments

We have many people to thank for help in preparing this book. First, we thank the original developers of new methods for observational studies, including James Heckman, Paul Rosenbaum, Donald Rubin, Alberto Abadie, Guido Imbens, Hidehiko Ichimura, and Petra Todd, whose contributions in developing the four statistical methods for observational studies make this book possible.

We thank Richard Berk, the editor-in-chief of the book, for his many innovative ideas and helpful guidance during the whole process. When preparing this book, we received invaluable comments, suggestions, and direct help from Paul Rosenbaum, Ben Hansen, Guido Imbens, Petra Todd, John Fox, Michael Foster, and two anonymous reviewers. Shenyang Guo thanks his mentor William Mason whose innovative and rigorous research in statistics greatly shaped his career and led him to choose quantitative methodology as his main research area.

We thank our former acquisitions editor Lisa Shaw and the current senior acquisitions editor Vicki Knight at Sage Publications for their help in preparing the book.

We thank many of our colleagues at the University of North Carolina at Chapel Hill and in the field of social work research. Dean Jack Richman's vision of promoting research rigor, which was shown by his full support of this project, motivated us to write this book. Richard Barth engaged in the original discussion of designing the book and contributed excellent suggestions for its completion—we cited several studies led by Barth to illustrate the applications of targeted methods in social work research. Diane Wyant provided excellent editorial help for the entire book. Alan Ellis helped with programming in R for some illustrating examples. Jung-Sook Lee helped manage the PSID and CDS data that were employed in several illustrating examples. Carrie Pettus Davis and Jilan Li helped with the search for computing procedures currently available in the field. We thank Cyrette Cotten-Fleming for assistance with copying and other logistics. Part of the financial support was provided by the John A. Tate Distinguished Professorship held by Mark Fraser.

Finally, we thank our families for their support, understanding, and patience. Specifically, Shenyang Guo thanks his wife Shenyan Li and children Han and Hanzhe; and Mark Fraser thanks his wife Mary Fraser and children Alex and Katy. This book is dedicated to you all!

1

Introduction

Propensity score analysis is a relatively new and innovative class of statistical methods that has proven useful for evaluating treatment effects when using nonexperimental or observational data. Specifically, propensity score analysis offers an approach to program evaluation when randomized clinical trials are infeasible or unethical, or when researchers need to assess treatment effects from survey data, census data, administrative data, or other types of data "collected through the observation of systems as they operate in normal practice without any interventions implemented by randomized assignment rules" (Rubin, 1997, p. 757).

As such, this book focuses on four closely related but technically distinct models for estimating treatment effects: (1) Heckman's sample selection model (Heckman, 1976, 1978, 1979) and its revised version (Maddala, 1983); (2) propensity score matching (Rosenbaum & Rubin, 1983) and related models; (3) matching estimators (Abadie & Imbens, 2002, 2006); and (4) propensity score analysis with nonparametric regression (Heckman, Ichimura, & Todd, 1997, 1998).

Although statisticians and econometricians have not reached consensus on the scope and content of propensity score analysis, the statistical models described in this book share several similar characteristics: Each has the objective of assessing treatment effects and controlling for covariates, each represents state-of-the-art analysis in program evaluation, and each can be employed to overcome various kinds of challenges encountered in research.

Although the randomized controlled trial is deemed to be the gold standard in research design, true experimental designs are not always possible, practical, or even desirable in human social science research. Given that social science research continues to rely heavily on quasi-experimental research design, researchers have increasingly sought methods of improved program evaluation.

Over the past 30 years, methods of program evaluation have undergone a significant change as researchers have recognized the need to develop more

1

efficient approaches for assessing treatment effects from studies based on observational data and for evaluations based on quasi-experimental designs. This growing interest in seeking consistent and efficient estimators of program effectiveness led to a surge in work focused on estimating average treatment effects under various sets of assumptions. Statisticians (e.g., Rosenbaum & Rubin, 1983) and econometricians (e.g., Heckman, 1978, 1979) have made substantial contributions by developing and refining new approaches for the estimation of causal effects from observational data; collectively, these approaches are known as propensity score analysis.

Econometricians have integrated propensity score models into other econometric models (i.e., instrumental variable, control function, difference-in-differences estimators) to perform less expensive and less intrusive nonexperimental evaluations of social, educational, and health programs. Furthermore, recent criticism and reformulations of the classical experimental approach in econometrics symbolize an important shift in evaluation methods. The significance of this movement was evidenced by the selection of James Heckman as one of the 2000 Nobel Prize award winners in the field of economics. The Prize recognized his development of theory and methods for data analysis in selective samples.

Representing the interest in—and indeed perceived utility of—these new methods, the propensity score approach has been employed in a variety of disciplines and professions such as education (Morgan, 2001), epidemiology (Normand et al., 2001), medicine (e.g., Earle et al., 2001; Gum, Thamilarasan, Watanabe, Blackstone, & Lauer, 2001), psychology (Jones, D'Agostino, Gondolf, & Heckert, 2004), social work (Barth, Greeson, Guo, & Green, 2007; Barth, Lee, Wildfire, & Guo, 2006; Guo, Barth, & Gibbons, 2006; Weigensberg, Barth, & Guo, 2009), and sociology (Smith, 1997). In social welfare studies, economists and others used propensity score methods in evaluations of the National Job Training Partnership Act program (Heckman, Ichimura, & Todd, 1997), the National Supported Work Demonstration (LaLonde, 1986), and the National Evaluation of Welfare-to-Work Strategies Study (Michalopoulos, Bloom, & Hill, 2004).

In describing these new methods, the preparation and writing of this book was guided by two primary objectives. The first objective was to introduce readers to the origins, main features, and debates centering on the four models of propensity score analysis. We hope this introduction will help accomplish our second objective of illuminating new ideas, concepts, and approaches that social behavioral researchers can apply to their own fields to solve problems they might encounter in their research efforts. In addition, this book has two overarching goals. Our primary goal is to make the past three decades of theoretical and technological advances in analytic methods accessible and available in a less technical and more practical fashion. The second goal is to promote discussions among social behavioral researchers regarding the challenges, strategies, and best methods of estimating causal effects using nonexperimental methods.

The aim of this chapter is to provide an overview of the propensity score approach. Section 1.1 presents a definition of observational study. Section 1.2 reviews the history and development of the methods. Section 1.3 is an overview of the randomized experimental approach, which is the gold standard developed by statisticians, and the model that should serve as a foundation for the nonexperimental approach. Section 1.4 offers examples drawn from literature beyond the fields of econometrics and statistics. These examples are intended to help readers determine the situations in which the propensity score approach may be appropriate. Section 1.5 reviews the computing software packages that are currently available for propensity score analysis and the main features of the package used in the models presented throughout this book. Section 1.6 outlines the organization of the book.

1.1 Observational Studies

The statistical methods we discuss may be generally categorized as methods for *observational studies*. According to Cochran (1965), an observational study is an empirical investigation whose objective is to elucidate causal relationships (i.e., cause and effect), when it is infeasible to use controlled experimentation and to assign participants at random to different procedures.

In the general literature related to program evaluation (i.e., non-statistically oriented literature), researchers use the term *quasi-experimental* more frequently than *observational studies*, with the term defined as experimental studies that compare groups, but which lack the critical element of random assignment. Indeed, quasi-experiments can be used interchangeably with observational studies as described in the following quote from Shadish, Cook, and Campbell (2002):

> Quasi-experiments share with all other experiments a similar purpose—to test descriptive causal hypotheses about manipulable causes—as well as many structural details, such as the frequent presence of control groups and pretest measures, to support a counterfactual inference about what would have happened in the absence of treatment. But, by definition, quasi-experiments lack random assignment. Assignment to conditions is by means of *self-selection,* by which units choose treatment for themselves, or means of *administrator selection,* by which teachers, bureaucrats, legislators, therapists, physicians, or others decide which persons should get which treatment. (pp. 13–14)

Two features about observational studies merit particular emphasis. First, an observational study concerns treatment effects. A study without a treatment— often called an *intervention* or a *program*—is neither an experiment nor an observational study. Most public opinion polls, forecasting efforts, investigations of fairness and discrimination, and many other important empirical studies are neither experiments nor observational studies (Rosenbaum, 2002b). Second, observational studies can employ data from nonexperimental, nonobservational studies as long as the focus is on assessing treatment or the effects of receiving a particular service. By this definition, observational data refer to data that were generated by something other than a randomized experiment and typically include surveys, censuses, or administrative records (Winship & Morgan, 1999).

1.2 History and Development

The term *propensity score* first appeared in a 1983 article by Rosenbaum and Rubin, who described the estimation of causal effects from observational data. Heckman's (1978, 1979) work on dummy endogenous variables using simultaneous equation modeling addressed the same issue of estimating treatment effects when assignment was nonrandom; however, Heckman approached this issue from a perspective of sample selection. Although Heckman's work on the dummy endogenous variable problem employed different terminology, he used the same approach toward estimating a participant's probability of receiving one of two conditions. Both schools of thought (i.e., the econometric tradition of Heckman and the statistical tradition of Rosenbaum and Rubin) have had a significant influence on the direction of the field, although the term *propensity score analysis*, coined by Rosenbaum and Rubin, is used most frequently as a general term for the set of related techniques used to correct for selection bias in observational studies.

The development of the propensity score approach signified a convergence of two traditions in studying causal inferences: the econometric tradition that primarily relies on structural equation modeling and the statistical tradition that primarily relies on randomized experiments (Angrist, Imbens, & Rubin, 1996; Heckman, 2005). The econometric tradition dates back to Trygve Haavelmo (1943, 1944), whose pioneering work developed a system of linear simultaneous equations that allowed analysts to capture interdependence among outcomes, to distinguish between fixing and conditioning on inputs, and to parse out true causal effects and spurious causal effects. The task of estimating *counterfactuals*, a term generally developed and used by statisticians, is explored by econometricians in the form of a *switching regression model* (Maddala, 1983; Quandt, 1958, 1972). Heckman's (1978,

1979) development of a two-step estimator is credited as the field's pioneering work in explicitly modeling the causes of selection in the form of a dummy endogenous variable. As previously mentioned, Heckman's work followed econometric conventions and solved the problem through structural equation modeling.

Historically quite distinct from the econometric tradition, the statistical tradition can be traced back to Fisher (1935/1971), Neyman (1923), and Rubin (1974, 1978). Unlike conventions based on structural equation models, the statistical tradition is fundamentally based on the randomized experiment. The principal notion in this formulation is the study of *potential outcomes*, known as the *Neyman-Rubin counterfactual framework*. Under this framework, the causal effects of treatment on sample participants (already exposed to treatments) are explored by observing outcomes of participants in samples not exposed to the treatments. Rubin extended the counterfactual framework to more complicated situations, such as observational studies without randomization.

For a detailed discussion of these two traditions, readers are referred to a special issue of the *Journal of the American Statistical Association* (1996, Vol. 91, No. 434), which presents an interesting dialogue between statisticians and econometricians. Significant scholars in the field—including Greenland, Heckman, Moffitt, Robins, and Rosenbaum—participated in a discussion of a paper that used instrumental variables to identify causal effects, particularly the local average treatment effect (Angrist et al., 1996).

1.3 Randomized Experiments

The statistical premise of program evaluation is grounded in the tradition of the randomized experiment. Therefore, a natural starting point in a discussion of causal attribution in observational studies is to review key features of the randomized experiment. According to Rosenbaum (2002b), a theory of observational studies must have a clear view of the role of randomization, so it can have an equally clear view of the consequences of its absence. For example, sensitivity analysis is among Rosenbaum's approaches to handling data with hidden selection bias; this approach includes the use of test statistics that were developed primarily for randomized experiments, such as Wilcoxon's signed-rank statistic and Hodges-Lehmann estimates. However, the critiques of social experiments by econometricians (e.g., Heckman & Smith, 1995) frequently include description of the conditions under which randomization is infeasible, particularly under the setting of social behavioral research. Thus, it is important to review principles and types of randomized experiments, randomization tests, and the challenges to this tradition. Each of these topics is addressed in the following sections.

1.3.1 FISHER'S RANDOMIZED EXPERIMENT

The invention of the randomized experiment is generally credited to Sir Ronald Fisher, one of the foremost statisticians of the 20th century. Fisher's book, *The Design of Experiments* (1935/1971), introduced the principles of randomization, demonstrating them with the now-famous example of testing a British woman's tea-tasting ability. This example has been cited repeatedly to illustrate the power of randomization and the logic of hypothesis testing (see, e.g., Maxwell & Delaney, 1990; Rosenbaum, 2002b). In a somewhat less technical fashion, we include this example as an illustration of important concepts in randomized experimentation.

In Fisher's (1935/1971) words, the problem is as follows:

> A lady declares that by tasting a cup of tea made with milk she can discriminate whether the milk or the tea infusion was first added to the cup. We will consider the problem of designing an experiment by means of which this assertion can be tested. (p. 11)

During Fisher's time, the dominant practice in experimentation was to control covariates or confounding factors that might contaminate treatment effects. Therefore, to test a person's tasting ability (i.e., the true ability to discriminate two methods of tea preparation), a researcher would control factors that could influence the results, such as the temperature of tea, the strength of the tea, the use of sugar and the amount of milk added, in addition to the myriad potential differences that might occur among the cups of tea used in an experiment. As Maxwell and Delaney (1990) pointed out,

> The logic of experimentation up until the time of Fisher dictated that to have a valid experiment here all the cups to be used "must be exactly alike," except for the independent variable being manipulated. Fisher rejected this dictum on two grounds. First, he argued that it was logically impossible to achieve, both in the example and in experimentation in general. . . . Second, Fisher argued that, even if it were conceivable to achieve "exact likeness," or more realistically, "imperceptible difference" on various dimensions of the stimuli, it would in practice be too expensive to attempt. (p. 40)

Instead of controlling for every potential confounding factor, Fisher proposed to *control for nothing*, namely, to employ a method of randomization. Fisher (1935/1971) described his design as follows:

> Our experiment consists in mixing eight cups of tea, four in one way and four in the other, and presenting them to the subject for judgment in a random order. The subject has been told in advance of what the test will consist, namely that she will be asked to taste eight cups, that these shall be four of each kind, and that they shall be presented to her in a random order, that is

in an order not determined arbitrarily by human choice, but by the actual manipulation of the physical apparatus used in games of chance, cards, dice, roulettes, etc., or more expeditiously, from a published collection of random sampling numbers purporting to give the actual results of such a manipulation. Her task is to divide the 8 cups into two sets of 4, agreeing, if possible, with the treatments received. (p. 12)

Before going further, it is crucial to note several important points regarding Fisher's design. First, in this example, the unit of analysis is not individual ($N \neq 1$), but rather the presentation of the tea cups to the tea taster (i.e., $N = 8$, in which a total of 8 cases comprise the sample). Second, there is a treatment assignment process in this example, namely, the order of presentation of tea cups. Using Rosenbaum's (2002b) notation, this is a random variable Z, and any specific presentation of the tea cups to the taster is a realization of Z, or Z = z. For instance, if a specific presentation of the tea cups consists of four cups with milk added first followed by four cups with tea added first, then we may write z = (11110000), where z is just one of many possible assignments. In Rosenbaum's notation, these possible treatment assignments form a set of Ω, and $z \in \Omega$. Determining the total number of elements in Ω (which Rosenbaum denoted as K) is an important task for experimental design and a task that can be accomplished using probability theory. This point will be discussed in more detail elsewhere. Third, there is an actual outcome r, which is the result of the tasting of the eight cups of tea. If the taster gives exactly the same order of tea cups as in the treatment assignment (i.e., she correctly identifies the first four cups as having the milk added first and the next four cups as having the tea added first), then the outcome would be recorded as r = (11110000). Last, the test essentially aims to determine whether the tea taster had the true ability to discriminate the two kinds of tea or whether she made her correct judgment accidentally by guessing. Thus, the null hypothesis (H_0) under testing would be "She has no ability to discriminate," and the test involves finding statistical evidence to reject the null hypothesis at a given significance level.

Building on these explanations, we continue with the tea-tasting test and describe how Fisher implemented his randomized experiment. One important feature of randomized experiments is that, in advance of implementation, the researcher must calculate probable outcomes for each study unit. Fisher (1935/1971) emphasized "forecasting all possible outcomes," even at a design stage when outcome data are completely absent: "In considering the appropriateness of any proposed experimental design, it is always needful to forecast all possible results of the experiment, and to have decided without ambiguity what interpretation shall be placed upon each one of them" (p. 12).

The key of such calculation is to know the total number of elements in the set of Ω (i.e., the value of K). In the above example, we simply made an arbitrary example of treatment assignment 11110000, although many other treatment assignments can be easily figured out, such as alternating cups of tea with the milk added first with the cups prepared by adding the tea infusion first (i.e., 10101010), or presenting four cups with tea infusion added first and then four cups with milk added first (i.e., 00001111). In statistics, the counting rules (i.e., permutations and combinations) inform us that the number of total possible ways to present the eight cups can be solved by finding out the number of combinations of eight things taken four at a time, or $_8C_4$, as

$$_nC_r = \frac{n(n-1)(n-2)\ldots(n-r+1)}{r(r-1)(r-2)\ldots 1} = \frac{n!}{r!(n-r)!}.$$

The solution to our problem is

$$K = {_8C_4} = \frac{8!}{4!4!} = 70.^1$$

Therefore, there are 70 possible ways to present the tea taster with four cups with milk added first and four cups with tea added first. We can keep writing 11110000, 10101010, 00001111, ... until we exhaust all 70 ways. Here, 70 is the number of total elements in the set of Ω or all possibilities for a treatment assignment.

To perform a statistical test of "H_0: No ability," Fisher turned to the task of looking into the possible outcomes r. Furthermore, if we define the taster's true ability to taste discriminately as requiring all eight cups she identified to match *exactly* what we presented to her, we can then calculate the probability of having the true outcome. The significance test performed here involves rejecting the null hypothesis, and the null hypothesis is expressed as "no ability." Fisher used the logic of "guessing the outcome right," that is, the taster has no ability to discriminate but makes her outcome correct by guessing. Thus, what is the probability of having the outcome r that is identical to the treatment assignment z? The outcome r should have one set of values from the 70 possible treatment assignments; that is, the taster could guess any outcome from 70 possible outcomes of 11110000, 10101010, 00001111, ... Therefore, the probability of guessing the right outcome is $1/70 = .0124$, which is a very low probability. Now, we can reject the null hypothesis under a small probability of making a Type I error (i.e., the tea taster did have the ability, but we erroneously rejected the "no ability" hypothesis), and the chance is indeed very low (.0124). In other words, based on statistical evidence (i.e., an examination of all possible outcomes), we can reject the "no ability" hypothesis at a statistical significance level of .05. Thus, we may conclude that under such a design, the taster may have true tasting ability ($p < .05$).

Rosenbaum (2002b) used $t(\mathbf{Z}, \mathbf{r})$ to denote the test statistic. In the above test scenario, we required a perfect match—a total of eight agreements—between the treatment (i.e., the order of tea cups presented to the tea-tasting expert) and the outcome (i.e., the actual outcome identified by the taster); therefore, the problem is to find out the probability $\{t(\mathbf{Z}, \mathbf{r}) > 8\}$. This probability can be more formally expressed in Rosenbaum's notation as follows:

$$\text{prob}\{t(\mathbf{Z}, \mathbf{r}) \geq T\} = \frac{|\{z \in \Omega : t(\mathbf{Z}, \mathbf{r}) \geq T\}|}{K}, \text{ or prob}\{t(\mathbf{Z}, \mathbf{r}) \geq 8\} = \frac{1}{70} = .0124.$$

However, if the definition of "true ability" is relaxed to allow for six exact agreements rather than eight agreements (i.e., six cups in the order of outcome match to the order of presentation), we can still calculate the probability or significance in testing the null hypothesis of "no ability." As in the earlier computation, this calculation involves the comparison of actual outcome $\mathbf{r}$ to the treatment assignment $\mathbf{z}$, and the tea taster's outcome could be any one of 70 possible outcomes. Let us assume that the taster gives her outcome as $\mathbf{r} = (11110000)$. We now need to examine how many treatment assignments (i.e., number of $\mathbf{z}$) match this outcome under the relaxed definition of "true ability." The answer to this question is one perfect match (i.e., the match with eight agreements) plus 16 matches with six agreements (Rosenbaum, 2002b, p. 30), for a total of 17 treatment assignments. To illustrate, we provide all 17 treatment assignments that match to the taster's outcome 11110000:

perfect match **1111**0000, and the following assignments with six exact agreements:

01111000, 01110001, 01110010, 01110100, 10110100, 10110010, 10110001, 10111000, 11010010, 11010001, 11010100, 11011000, 11100001, 11100010, 11100100, 11101000,

where bold numbers indicate agreements.[2]

Thus, the probability of having six exact agreements is 17/70 = .243. In Rosenbaum's notation, the calculation is

$$\text{prob}\{t(\mathbf{Z}, \mathbf{r}) \geq T\} = \frac{|\{z \in \Omega : t(\mathbf{Z}, \mathbf{r}) \geq T\}|}{K}, \text{ or prob}\{t(\mathbf{Z}, \mathbf{r}) \geq 6\} = \frac{17}{70} = .243.$$

That is, if we define "true ability" as correctly identifying six out of eight cups of tea, the probability of having a correct outcome increases to .243. The null hypothesis cannot be rejected at a .05 level. In other words, under this relaxed definition, we should be more conservative, or ought to be more reluctant, to declare that the tea taster has true ability. With a sample of eight cups in total and a relaxed definition of "ability," the statistical evidence is simply insufficient for us to reject the null hypothesis, and, therefore, the experimental design is less significant in testing true tasting ability.

We have described Fisher's famous example of randomized experiment in great detail. Our purpose of doing so is twofold. The first is to illustrate the importance of understanding two processes in generating intervention data: (1) the treatment assignment process (i.e., there is a random variable Z, and the total number of possible ways K is inevitably large) makes it possible to know in advance the probability of receiving treatment in a uniform randomized experiment and (2) the process of generating outcome data (i.e., there is an outcome variable r). This topic is revisited both in Chapters 2 and 3, in the discussion of the so-called ignorable treatment assignment, and in Chapter 8, in the discussion of selection bias and sensitivity analysis. The second purpose in providing a detailed description of Fisher's experiment was to call attention to the core elements of randomized experiments. According to Rosenbaum (2002b),

> First, experiments do not require, indeed cannot reasonably require, that experimental units be homogeneous, without variability in their responses.... Second, experiments do not require, indeed, cannot reasonably require, that experimental units be a random sample from a population of units.... Third, for valid inference about the effects of a treatment on the units included in an experiment, it is sufficient to require that treatments be allocated at random to experimental units—these units may be both heterogeneous in their responses and not a sample from a population. Fourth, probability enters the experiment only through the random assignment of treatments, a process controlled by the experimenter. (p. 23)

1.3.2 TYPES OF RANDOMIZED EXPERIMENTS AND STATISTICAL TESTS

Fisher's framework laid the foundation for randomized experimental design. The method has become a gold standard in program evaluation and continues to be an effective and robust means for assessing treatment effects in nearly every field of interest from agriculture, to computer science, to manufacturing, to medicine and social welfare. Furthermore, many sophisticated randomized designs have been developed to estimate various kinds of treatment effects under various settings of data generation. For example, within the category of uniform randomized experiment,[3] in addition to the traditional method of *completely randomized experiment* where stratification is absent (i.e., $S = 1$ and S stands for number of strata), researchers have developed *randomized block experiments* where two or more strata are permissible (i.e., $S \geq 2$), and *paired randomized experiments* in which $n_S = 2$ (i.e., the number of study participants within stratum S is fixed at 2), $m_S = 1$ (i.e., the number of participants receiving treatment within stratum S is fixed at 1), and S could be reasonably large (Rosenbaum, 2002b).

A more important reason for studying randomized experiments is that statistical tests developed through randomized experiments may be performed virtually without assumptions, which is not the case for nonrandomized experiments. The class of randomization tests, as reviewed and summarized by Rosenbaum (2002b), includes

1. Tests for binary outcomes: the *Fisher's* (1935/1971) *exact test*, the *Mantel-Haenszel* (1959) *statistic*, and the *McNemar's* (1947) *test*

2. Tests for an outcome variable that is confined to a small number of values representing a numerical scoring of several ordered categories (i.e., an ordinal variable): the *Mantel's* (1963) *extension of the Mantel-Haenszel test*

3. Tests for a single stratum $S = 1$ where the outcome variable may take many numerical values (i.e., an interval or ratio variable): the *Wilcoxon's* (1945) *rank sum test*

4. Tests for an outcome variable that is ordinal and the number of strata S is large compared with sample size N: the *Hodges and Lehmann* (1962) *test using the signed-rank statistic*

As opposed to drawing inferences using these tests in randomized designs, drawing inferences using these tests in nonrandomized experiments "requires assumptions that are not at all innocuous" (Rosenbaum, 2002b, p. 27).

1.3.3 CRITIQUES OF SOCIAL EXPERIMENTATION

Although the randomized experiment has proven useful in many applications since Fisher's seminal work, the past three decades have witnessed a chorus of challenges to the fundamental assumptions embedded in the experimental approach. In particular, critics have been quick to note the complexities of applying randomized trials in studies conducted with humans rather than mechanical components, agricultural fields, or cups of tea. The dilemma presented in social behavioral studies with human participants is that assigning participants to a control condition means potentially denying treatment or services to those participants; in many settings, such denial of services would be unethical or illegal. Although the original rationale for using a randomized experiment was the infeasibility of controlling covariates, our evaluation needs have returned to the point where covariant control or its variants (e.g., matching) becomes attractive. This is particularly true in social behavioral evaluations.

In a series of publications, Heckman and his colleagues (e.g., Heckman, 1979; Heckman & Smith, 1995) discussed the importance of directly modeling the process of assigning study participants to treatment conditions by using factors that influence participants' decisions regarding program participation. Heckman and his associates challenged the assumption that we can depend on randomization to create groups in which the treated and nontreated participants share the same characteristics under the condition of nontreatment. They questioned the fundamental assumption embedded in the classical experiment: that randomization removes selection bias.

Heckman and Smith (1995) in particular held that social behavioral evaluations need to explicitly address four questions, none of which can be handled suitably by the randomized experiment: (1) What are the effects of factors such as subsidies, advertising, local labor markets, family income, race, and gender on program application decisions? (2) What are the effects of bureaucratic performance standards, local labor markets, and individual characteristics on administrative decisions to accept applicants and place them in specific programs? (3) What are the effects of family background, subsidies, and local market conditions on decisions to drop out of a program and, alternatively, on the length of time required to complete a program? (4) What are the costs of various alternative treatments?

1.4 Why and When a Propensity Score Analysis Is Needed

Drawing causal inferences in observational studies or studies without randomization is challenging, and it is this task that has motivated statisticians and econometricians to explore new analytic methods. The four analytic models that we discuss in this book derive from this work. Although the models differ on the specific means employed, all four models aim to accomplish data balancing when treatment assignment is nonignorable, to evaluate treatment effects using nonrandomized or nonexperimental approaches, and/or to reduce multidimensional covariates to a one-dimensional score called a *propensity score*. To provide a sense of why and when propensity score methods are needed, we use examples drawn from the literature across various social behavioral disciplines. Propensity score analysis is suitable to data analysis and to causal inferences for all these studies. Most of these examples will be revisited throughout this book.

Example 1: Assessing the Impact of Catholic Versus Public Schools on Learning. A long-standing debate in education is whether Catholic schools (or private schools in general) are more effective than public schools in promoting learning.

Obviously, a variety of selections are involved in the formation of "treatment" (i.e., entrance into Catholic schools). To name a few, *self-selection* is a process that lets those who choose to study in Catholic schools receive the treatment; *school selection* is a process that permits schools to select only those students who meet certain requirements, particularly minimum academic standards, to enter into the treatment; *financial selection* is a process that excludes from the treatment those students whose families cannot afford tuition; and *geographic selection* is a process that selects out (i.e., excludes) students who live in areas where no Catholic school exists. Ultimately, the debate on Catholic schools centers on the question of whether differences observed in outcome data (i.e., academic achievement or graduation rates) between Catholic and public schools are attributable to the intervention or attributable to the fact that the Catholic schools serve a different population. In other words, if the differences are attributable to the intervention, findings suggest that Catholic schools promote learning more effectively than public schools, whereas if the differences are attributable to the population served by Catholic schools, findings would show that students currently enrolled in Catholic schools would always demonstrate better academic outcomes regardless of whether they attended private or public schools. It is infeasible to conduct a randomized experiment to answer these questions; however, observational data such as the National Educational Longitudinal Survey (NELS) data are available to researchers interested in this question.

Because observational data lack randomized assignment of participants into treatment conditions, researchers must employ statistical procedures to balance the data before assessing treatment effects. Indeed, numerous published studies have used the NELS data to address the question of Catholic school effectiveness; however, the findings have been contradictory. For instance, using propensity score matching and the NELS data, Morgan (2001) found that the Catholic school effect is the strongest only among those Catholic school students who, according to their observed characteristics, are least likely to attend Catholic schools. However, in a study that used the same NELS data but employed a new method that directly assessed selectivity bias, Altonji, Elder, and Taber (2005) found that attending a Catholic high school substantially increased a student's probability of graduating from high school and, more tentatively, attending college.

Example 2: Assessing the Impact of Poverty on Academic Achievement. Prior research has shown that exposure to poverty and participation in welfare programs have strong impacts on child development. In general, growing up in poverty adversely affects a child's life prospects, and the consequences become more severe with greater exposure to poverty (Duncan, Brooks-Gunn, Yeung, & Smith, 1998; Foster & Furstenberg, 1998, 1999; Smith & Yeung, 1998). Most prior inquiries in this field have applied a multivariate analysis (e.g., multiple regression or regression-type models) to samples of nationally representative data such as the Panel Study of Income Dynamics (PSID) or administrative data, although a few studies employed a correction method such as propensity score analysis (e.g., Yoshikawa, Maguson, Bos, & Hsueh, 2003). Using a multivariate approach with this type of data poses two fundamental problems. First, the bulk of the literature regarding the impact of poverty on children's academic achievement assumes a causal perspective (i.e., poverty is the cause of poor academic achievement); whereas the analysis using a regression model is, at best, correlational. In addition, a regression model or covariance control approach is less robust in handling endogeneity bias. Second, PSID is an observational survey without randomization and, therefore, researchers must take selection bias into consideration when employing PSID data to assess causal effects.

Guo and Lee (2008) have made several efforts to examine the impacts of poverty. First, using PSID data, propensity score models—including optimal propensity score matching, the treatment effects model, and the matching estimator—were used to estimate the impact of poverty. Second, Guo and Lee conducted a more thorough investigation of poverty. That is, in addition to conventional measures of poverty such as the ratio of income to poverty threshold, they examined 30 years of PSID data to create two new variables: (1) the number of years during a caregiver's childhood (i.e., ages 6 to 12 years) that a caregiver used Aid to Families With Dependent Children (AFDC) and (2) the percentage of time a child used AFDC between birth and 1997 (i.e., the time point when academic achievement data were compared). Last, Guo and Lee conducted both efficacy subset analysis and intent-to-treat analysis, and compared findings. Results using these approaches were more revealing than previous studies.

Example 3: Assessing the Impact of a Waiver Demonstration Program. In 1996, the U.S. Congress approved the Personal Responsibility and Work Opportunity Reconciliation Act (PRWORA). Known as welfare reform, PRWORA ended entitlements to cash assistance that were available under the prior welfare policy, AFDC. As part of this initiative, the federal government launched the Waiver Demonstration program, which allowed participating states and counties to use discretionary funding for county-specific demonstration projects of welfare reform—as long as these demonstrations facilitated "cost neutrality." A key feature of the Waiver Demonstration program, as well as several other programs implemented under welfare reform, is that the county has the option of whether to participate in the Waiver Demonstration. Therefore, by definition, the intervention counties and comparison counties at the state level cannot be formed randomly. Counties that chose to participate differed from counties choosing not to participate. Evaluating such a nonrandomized program is daunting. Using a Monte Carlo study, Guo and Wildfire (2005) demonstrated that propensity score matching is a useful analytic approach for such data and an approach that provides less biased findings than an analysis using the state-level population data.

Example 4: Assessing the Well-Being of Children Whose Parents Abuse Substances. A strong, positive association between parental substance abuse and involvement with the child welfare system has been established (e.g., English, Marshall, Brummel, & Coghan, 1998; U.S. Department of Health and Human Services, 1999). Substance abuse may lead to child maltreatment through several mechanisms, such as child neglect that occurs when substance-abusing parents give greater priority to their drug use than to caring for their children; or substance abuse can lead to extreme poverty and inability to provide for a child's basic needs (Magura & Laudet, 1996). Policymakers have long been concerned about the safety of children of substance-abusing parents. Drawing on a nationally representative sample from the National Survey of Child and Adolescent Well-Being (NSCAW), Guo et al. (2006) used a propensity score matching approach to address the question of whether children whose caregivers received substance-abuse services were more likely to have re-reports of maltreatment than children whose caregivers did not use substance-abuse services. Using the same NSCAW data, Guo et al. employed propensity score analysis with nonparametric regression to examine the relationship between the participation of a caregiver in substance-abuse services and subsequent child outcomes; that is, they investigated whether children of caregivers who used substance-abuse services exhibited more behavioral problems than children of caregivers who did not use such services.

Example 5: Estimating the Impact of Multisystemic Therapy (MST). MST is a multifaceted, short-term (4 to 6 months), home- and community-based intervention for families with youths who have severe psychosocial and behavioral problems. Funding for MST in the United States rose from US$5 million in 1995, to approximately US$18 million in 2000, and to US$35 million in 2003. Most evaluations of the program used a randomized experiment approach, and most studies generally supported the efficacy of MST. However, a recent study using a systematic review approach (Littell, 2005) found different results. Among the problems observed in previous studies, two major concerns arose: (1) the variation in the implementation of MST and (2) the integrity of conducting randomized experiments. From a program-evaluation perspective, this latter concern is a common problem in social behavioral evaluations: Randomization is often compromised. Statistical approaches, such as propensity score matching, may be helpful when randomization fails or is impossible (Barth et al., 2007).

Example 6: Assessing Program Outcomes in Group-Randomized Trials. The Social and Character Development (SACD) program was jointly sponsored by the U.S. Department of Education (DOE) and the Centers for Disease Control and Prevention. The SACD intervention project was designed to assess the impact of schoolwide social and character development education in elementary schools. Using a scientific peer review process, seven proposals to implement SACD were chosen by the Institute of Education Sciences in the U.S. DOE, and the research groups associated with each of the seven proposals implemented different SACD programs in primary schools across the country. At each of the seven sites, schools were randomly assigned to receive either the intervention program or a control curricula, and one cohort of students was followed from third grade (beginning in fall 2004) through fifth grade (ending in spring 2007). A total of 84 elementary schools were randomized to intervention and control at seven sites: Illinois (Chicago); New Jersey; New York (Buffalo, New York City, and Rochester); North Carolina; and Tennessee.

Evaluating programs generated by a group randomization design is often challenging, because the unit of analysis is a cluster—such as a school—and sample sizes are so small as to compromise randomization. At one of the seven sites, the investigators of SACD designed the Competency Support Program to use a group randomization design. The total number of schools participating in the study within a school district was determined in advance, and then schools were randomly assigned to treatment conditions within school districts; for each treated school, a school that best matched the treated school on academic yearly progress, percentage of minority students, and percentage of students receiving free or reduced-price lunch was selected as a control school (i.e., data collection only without receiving intervention). In North Carolina, over a 2-year period, this group randomization procedure resulted in a total of 14 schools (Cohort 1, 10 schools; Cohort 2, 4 schools) for the study: seven received the Competency Support Program intervention, and seven received routine curriculum. As it turned out—as is often the case when implementing randomized experiments in social behavioral sciences—the group randomization did not work out as planned. In some school districts, as few as four schools met the study criteria and were eligible for participation. Just by the luck of the draw (i.e., by random assignment), the two intervention schools differed systematically on covariates from the two control schools. Thus, when comparing data from the 10 schools, the investigators found the intervention schools differed from the control schools in significant ways: The intervention schools had lower academic achievement scores on statewide tests (Adequate

Yearly Progress [AYP]), a higher percentage of students of color, a higher percentage of students receiving free or reduced-price lunches, and lower mean scores on behavioral composite scales at baseline. These differences were statistically significant at the .05 level using bivariate tests and logistic regression models. The researchers were confronted with the failure of randomization. Were these selection effects ignored, the evaluation findings would be biased. It is just at this intersection of design (i.e., failure of randomization) and data analysis that propensity score approaches become very helpful.

1.5 Computing Software Packages

At the time of this book going to the press, few software packages offer comprehensive procedures to handle the statistical analyses described in subsequent chapters. Our review of software packages indicated that *Stata* (StataCorp, 2008) and *R* (The R Foundation for Statistical Computing, 2008) offer the most comprehensive computational facilities. Other packages, such as *SAS*, may offer user-developed macros or procedures targeting specific problems (e.g., *SAS Proc Assign* may be used to implement optimal matching), but they do not offer the variety of analysis options that is available in *Stata* and *R*.

Table 1.1 lists the *Stata* and *R* procedures available for implementing the analyses described in this book. We have chosen to use *Stata* to illustrate most approaches. We chose *Stata* based on our experience with the software and our conclusion that it is the most convenient software package. Specifically, *Stata's* programs **heckman** and **treatreg** can be used to solve problems described in Chapter 4; **psmatch2, boost, imbalance,** and **hodgesl** can be used to solve problems described in Chapter 5; **nnmatch** can be used to solve problems described in Chapter 6; **psmatch2** can be used to solve problems described in Chapter 7; and **rbounds** can be used to perform Rosenbaum's (2002b) sensitivity analysis described in Chapter 8. In each of these chapters, we will provide examples and an overview of *Stata* syntax. We provide illustrative examples for one *R* procedure (i.e., **optmatch**), because this is the only procedure available for conducting optimal matching within *R* and *Stata*. All syntax files and illustrative data can be downloaded from this book's companion Web page http://ssw.unc.edu/psa.

1.6 Plan of the Book

Chapter 2 offers a conceptual framework for the development of scientific approaches to causal analysis, namely, the Neyman-Rubin counterfactual framework. In addition, the chapter reviews a closely related, and recently

Table 1.1   *Stata* and *R* Procedures by Analytic Methods

	Procedure Name and Useful References	
Chapter and Methods	*Stata*	*R*
Chapter 4		
Heckman (1978, 1979) sample selection model	*heckman* (StataCorp, 2003)	*sampleSelection* (Toomet & Henningsen, 2008)
Maddala (1983) treatment effect model	*treatreg* (StataCorp, 2003)	
Chapter 5		
Rosenbaum and Rubin's (1983) propensity score matching	*psmatch2* (Leuven & Sianesi, 2003)	*cem* (Dehejia & Wahba, 1999; Iacus, King, & Porro, 2008)
		Matching (Sekhon, 2007)
		MatchIt (Ho, Imai, King, & Stuart, 2004)
		PSAgraphics (Helmreich & Pruzek, 2008)
		WhatIf (King & Zeng, 2006, 2007)
		USPS (Obenchain, 2007)
Generalized boosted regression	*boost* (Schonlau, 2007)	*gbm* (McCaffrey, Ridgeway, & Morral, 2004)
Optimal matching (Rosenbaum, 2002b)		*optmatch* (Hansen, 2007)
Postmatching covariance imbalance check (Haviland, Nagin, & Rosenbaum, 2007)	*imbalance* (Guo, 2008b)	
Hodges-Lehmann aligned-rank test after optimal matching (Haviland, Nagin, & Rosenbaum, 2007; Lehmann, 2006)	*hodgesl* (Guo, 2008a)	
Chapter 6		
Matching estimators (Abadie & Imbens, 2002, 2006)	*nnmatch* (Abadie, Drukker, Herr, & Imbens, 2004)	*Matching* (Sekhon, 2007)

Table 1.1 (Continued)

Chapter 7		
Kernel-based matching (Heckman, Ichimura, & Todd, 1997, 1998)	*psmatch2* (Leuven & Sianesi, 2003)	
Chapter 8		
Rosenbaum's (2002b) sensitivity analysis	*rbounds* (Gangl, 2007)	*rbounds* (Keele, 2008)

developed, framework that aims to guide scientific inquiry of causal inferences: the econometric model of causality (Heckman, 2005). Next, the chapter includes a discussion of two fundamental assumptions embedded in nearly all outcome-oriented program evaluations: the ignorable treatment assignment assumption and the stable unit treatment value assumption (SUTVA). Violations of these assumptions pose challenges to estimation of counterfactuals.

Chapter 3 focuses on the issue of ignorable treatment assignment from the other side of the coin: strategies for data balancing when treatment effects can only be assessed in a nonexperimental design. This chapter aims to answer the key question of what kind of statistical methods should be considered to remedy the estimation of counterfactuals, when treatment assignment is not ignorable. Moreover, the chapter describes three closely related but methodologically distinctive approaches: *ordinary least square* (OLS) regression, matching, and stratification. The discussion includes a comparison of estimated treatment effects of the three methods under five scenarios. These methods involve making simple corrections when assignment is not ignorable, and they serve as a starting point for discussing the data issues and features of more sophisticated approaches, such as the four advanced models described later in the book. The chapter serves as a review of preliminary concepts that are a necessary foundation for learning more advanced approaches.

Chapters 4 through 7 present statistical theories using examples to illustrate each of the four advanced models covered in this book. Chapter 4 describes and illustrates Heckman's sample selection model in its original version (i.e., the model aims to correct for sample selection) and the revised Heckman model developed to evaluate treatment effects. Chapter 5 describes propensity score matching, specifically the creation of matched samples using caliper (or Mahalanobis metric) matching, and recently developed methods including *optimal matching*, fine balance, propensity score weighting, and modeling doses of treatment. Chapter 6 describes a collection of matching estimators developed by Abadie and Imbens (2002), who provide an extension of Mahalanobis metric matching. Among the attractive features of this procedure is its provision of standard errors for various treatment effects. Chapter 7 describes propensity score analysis with nonparametric regression. Specifically, it describes the two-time-period difference-in-differences approach developed by Heckman and his colleagues (1997, 1998).

Chapter 8 reviews selection bias, which is the core problem all statistical methods described in this book aim to resolve. This chapter gives the selection-bias problem a more rigorous treatment: We simulate two settings of data generation (i.e., selection on observables and selection on unobservables) and compare the performance of four models under these settings using Monte Carlo studies. Hidden selection bias is a problem that fundamentally distinguishes observational studies from randomized experiments. When key variables are missing, researchers inevitably stand on thin ice when drawing inferences about causal effects in observational studies. However, Rosenbaum's (2002b) sensitivity analysis, which is illustrated in Chapter 8, is a useful tool for testing the sensitivity of study findings to hidden selection. This chapter reviews assumptions for all four models and demonstrates practical strategies for model comparison.

Finally, Chapter 9 focuses on continuing issues and challenges in the field. It reviews debates on whether propensity score analysis can be employed as a replacement for randomized experiments. It comments on recent advances. And it suggests directions for the development of new approaches to observational studies.

Notes

1. Excel can be used to calculate the number of combinations of 8 things taken 4 at a time by typing the following in a cell: =COMBIN(8,4), and Excel returns the number 70.

2. If the tea taster gives an outcome other than 11110000, then the number of assignments having six exact agreements remains 17. However, there will be a different set of 17 assignments than those presented here.

3. Uniform here refers to equal probability for elements in the study population to receive treatment.

2

Counterfactual Framework and Assumptions

This chapter examines conceptual frameworks that guide the estimation of treatment effects as well as important assumptions that are embedded in observational studies. Section 2.1 defines causality and describes threats to internal validity. In addition, it reviews concepts that are generally discussed in the evaluation literature, emphasizing their links to statistical analysis. Section 2.2 summarizes the key features of the Neyman-Rubin counterfactual framework. Section 2.3 discusses the *ignorable treatment assignment* assumption. Section 2.4 describes the *stable unit treatment value assumption* (SUTVA). Section 2.5 provides an overview of statistical approaches developed to handle selection bias. With the aim of showing the larger context in which new evaluation methods are developed, this includes a variety of models, including the four models covered in this book. Section 2.6 reviews the underlying logic of statistical inference for both randomized experiments and observational studies. Section 2.7 summarizes a range of treatment effects and extends the discussion of the SUTVA. We examine treatment effects by underscoring the maxim that different research questions imply different treatment effects and different analytic models must be matched to the kinds of effects expected. Section 2.8 reviews Heckman's econometric model of causality, which is a comprehensive framework for drawing causal inference. Section 2.9 concludes the chapter with a summary of key points.

2.1 Causality, Internal Validity, and Threats

Program evaluation is essentially the study of cause-and-effect relationships. It aims to answer this key question: To what extent can the *net difference* observed in outcomes between treated and nontreated groups be attributed to the intervention, given that all other things are held constant (or ceteris paribus)? Causality in this context simply refers to the net gain or loss observed in the outcome of the treatment group that can be attributed to manipulable variables in the intervention. Treatment in this setting ranges from receipt of a well-specified program to falling into a general state such as "being a welfare recipient," as long as such a state can be defined as a result of manipulations of the intervention (e.g., a mother of young children who receives cash assistance under the program of AFDC or Temporary Aid to Needy Families [TANF]). Rubin (1986) argued that there can be no causation without manipulation. According to Rubin, thinking about actual manipulations forces an initial definition of units and treatments, which is essential in determining whether a program truly produces an observed outcome.

Students from any social behavioral discipline may have learned from their earliest research course that association should not be interpreted as the equivalent of causation. The fact that two variables, such as A and B, are highly correlated does not necessarily mean that one is a cause and the other is an effect. The existence of a high correlation between A and B may be the result of the following conditions: (a) Both A and B are determined by a third variable, C, and by controlling for C, the high correlation between A and B disappears. If that's the case, we say that the correlation is spurious. (b) A causes B. In this case, even though we control for another set of variables, we still observe a high association between A and B. (c) In addition, it is possible that B causes A, in which case the correlation itself does not inform us about the direction of causality.

A widely accepted definition of causation was given by Lazarsfeld (1959), who described three criteria for a causal relationship. (1) A causal relationship between two variables must have temporal order, in which the cause must precede the effect in time (i.e., if A is a cause and B an effect, then A must occur before B). (2) The two variables should be empirically correlated with one another. And (3), most important, the observed empirical correlation between two variables cannot be explained away as the result of a third variable that causes both A and B. In other words, the relationship is not spurious.

According to Pearl (2000), the notion that regularity of succession or correlation is not sufficient for causation dates back to the 18th century when Hume (1748/1959) argued,

> We may define a cause to be an object followed by another, and where all the objects, similar to the first, are followed by object similar to the second. Or, in other words, where, if the first object had not been, the second never had existed. (sec. VII)

Based on the three criteria for causation, Campbell (1957) and his colleagues developed the concept of *internal validity*, which serves a paramount role in program evaluation. Conceptually, internal validity shares common features with causation.

> We use the term internal validity to refer to inferences about whether observed covariation between A and B reflects a causal relationship from A to B in the form in which the variables were manipulated or measured. To support such an inference, the researcher must show that A preceded B in time, that A covaries with B . . . and that no other explanations for the relationship are plausible. (Shadish et al., 2002, p. 53)

In program evaluation and observational studies in general, researchers are concerned about threats to internal validity. These threats are factors affecting outcomes other than intervention or the focal stimuli. In other words, threats to internal validity are other possible reasons to think that the relationship between A and B is not causal, that the relationship could have occurred in the absence of the treatment, and that the relationship between A and B could have led to the same outcomes that were observed for the treatment. Nine well-known threats to internal validity are ambiguous temporal precedence, selection, history, maturation, regression, attrition, testing, instrumentation, and additive and interactive effects of threats to internal validity (Shadish et al., 2002, pp. 54–55).

It is noteworthy that many of these threats have been carefully examined in the statistical literature, although statisticians and econometricians have used different terms to describe them. For instance, Heckman, LaLonde, and Smith (1999) referred to the testing threat as the *Hawthorne effect*, meaning that an agent's behavior is affected by the act of participating in an experiment. Rosenbaum (2002b) distinguished between two types of bias that are frequently found in observational studies: *overt bias* and *hidden bias*. Overt bias can be seen in the data at hand, whereas the hidden bias cannot be seen because the required information was not observed or recorded. Although different in their potential for detection, both types of bias are induced by the fact that "the treated and control groups differ prior to treatment in ways that matter for the outcomes under study" (Rosenbaum, 2002b, p. 71). Suffice it to say that Rosenbaum's "ways that matter for the outcomes under study" encompass one or more of the nine threats to internal validity.

This book adopts a convention of the field that defines *selection threat* broadly. That is, when we refer to *selection bias*, we mean a process that involves one or more of the nine threats listed above and not necessarily the more limited definition of selection threat alone. In this sense, then, selection bias may take one or more of the following forms: self-selection, bureaucratic selection, geographic selection, attrition selection, instrumental selection, or measurement selection.

2.2 Counterfactuals and the Neyman-Rubin Counterfactual Framework

Having defined causality, we now present a key conceptual framework developed to investigate causality: the counterfactual framework. Counterfactuals are at the heart of any scientific inquiry. Galileo was perhaps the first scientist who used the thought experiment and the idealized method of controlled variation to define causal effects (Heckman, 1996). In philosophy, the practice of inquiring about causality through counterfactuals stems from the early Greek philosophers such as Aristotle (Holland, 1986). Hume (1748/1959) also was discontent with the regularity of factual account and thought that the counterfactual criterion was less problematic and more illuminating. According to Pearl (2000), Hume's idea of basing causality on counterfactuals was adopted by John Stuart Mill (1843), and it was embellished in the works of David Lewis (1973, 1986). Lewis (1986) called for abandoning the regularity account altogether, and for interpreting "*A* has caused *B*" as "*B* would not have occurred if it were not for *A*."

In statistics, researchers generally credit the development of the counterfactual framework to Neyman (1923) and Rubin (1974, 1978, 1980b, 1986) and call it the *Neyman-Rubin counterfactual framework of causality*. Other scholars who made independent contributions to the development of this framework come from a variety of disciplines, including Fisher (1935/1971) and Cox (1958) from statistics, Thurstone (1930) from psychometrics, and Haavelmo (1943), Roy (1951), and Quandt (1958, 1972) from economics. Holland (1986), Sobel (1996), and Winship and Morgan (1999) have provided detailed reviews of the history and development of the counterfactual framework.

So what is a counterfactual? A counterfactual is a *potential* outcome, or the state of affairs that would have happened in the absence of the cause (Shadish et al., 2002). Thus, for a participant in the treatment condition, a counterfactual is the potential outcome under the condition of control; for a participant in the control condition, a counterfactual is the potential outcome under the condition of treatment. Note that the definition uses the subjunctive mood (i.e., contingent on "would have happened . . ."), which means that the counterfactual is not observed in real data. Indeed, it is a missing value. Therefore, the fundamental task of any evaluation is to use known information to impute a missing value for a hypothetical and not observed outcome.

Neyman-Rubin's framework emphasizes that individuals selected into either treatment or nontreatment groups have potential outcomes in both states: that is, the one in which they are observed and the one in which they are not observed. More formally, assume that each person i under evaluation would have two potential outcomes (Y_{0i}, Y_{1i}) that correspond, respectively, to the potential outcomes in the untreated and treated states. Let $W_i = 1$ denote the receipt of treatment, $W_i = 0$ denote nonreceipt, and Y_i indicate the measured outcome variable. The Neyman-Rubin counterfactual framework can then be expressed as the following model:[1]

$$Y_i = W_i Y_{1i} + (1 - W_i) Y_{0i}. \tag{2.1}$$

In the above equation, W_i is a dichotomous variable; therefore, both the terms W_i and $(1 - W_i)$ serve as a *switcher*. Basically, the equation indicates which of the two outcomes would be observed in the real data, depending on the treatment condition or the "on/off" status of the switch. The key message conveyed in this equation is that to infer a causal relationship between W_i (the cause) and Y_i (the outcome) the analyst cannot directly link Y_{1i} to W_i under the condition $W_i = 1$; instead, the analyst must check the outcome of Y_{0i} under the condition of $W_i = 0$, and compare Y_{0i} with Y_{1i}. For example, we might hypothesize that a child i who comes from a low-income family has low academic achievement. Here, the treatment variable is $W_i = 1$ if the child lives in poverty; the academic achievement $Y_{1i} < p$ if the child has a low academic achievement, where p is a cutoff value defining a low test score, and $Y_{1i} > p$ otherwise. To make a causal statement that being poor ($W_i = 1$) causes low academic achievement $Y_{1i} < p$, the researcher must examine the outcome under the status of not being poor. That is, the task is to determine the child's academic outcome Y_{0i} under the condition of $W_i = 0$, and ask the question, "What would have happened had the child not lived in a poor family?" If the answer to the question is $Y_{0i} > p$, then the researcher can have confidence that $W_i = 1$ causes $Y_{1i} < p$.

The above argument gives rise to many issues that we will examine in detail. The most critical issue is that Y_{0i} is not observed. Holland (1986, p. 947) called this issue the *fundamental problem of causal inference*. How could a researcher possibly know $Y_{0i} > p$? The Neyman-Rubin counterfactual framework holds that a researcher can estimate the counterfactual by examining the average outcome of the treatment participants and the average outcome of the nontreatment participants in the population. That is, the researcher can assess the counterfactual by evaluating the difference in mean outcomes between the two groups or "averaging out" the outcome values of all individuals in the same condition. Specifically, let $E(Y_0|W=0)$ denote the mean outcome of the individuals who comprise the nontreatment group, and $E(Y_1|W = 1)$ denote the mean outcome of the individuals who comprise the treatment group. Because both outcomes in the above formulation (i.e., $E(Y_0|W = 0)$ and $E(Y_1|W = 1)$) are observable, we can then define the treatment effect as a mean difference:

$$\tau = E(Y_1|W = 1) - E(Y_0|W = 0), \tag{2.2}$$

where τ denotes treatment effect. This formula is called the *standard estimator for the average treatment effect*. It is worth noting that under this framework, the evaluation of $E(Y_1|W = 1) - E(Y_0|W = 0)$ can be understood as an effort that uses $E(Y_0|W = 0)$ to estimate the counterfactual $E(Y_0|W = 1)$. The central interest of the evaluation is not in $E(Y_0|W = 0)$ but in $E(Y_0|W = 1)$.

Returning to our example with the hypothetical child, the solution to the dilemma of not observing the academic achievement for child i in the condition of not being poor is resolved by examining the average academic achievement for all poor children in addition to the average academic achievement of all nonpoor children in a well-defined population. If the comparison of two mean outcomes leads to $\tau = E(Y_1|W = 1) - E(Y_0|W = 0) < 0$, or the mean outcome of all poor children is a low academic achievement, then the researcher can infer that poverty causes low academic achievement and also can provide support for hypotheses advanced under resources theories (e.g., Wolock & Horowitz, 1981).

In summary, the Neyman-Rubin framework offers a practical way to evaluate the counterfactuals. Working with data from a sample that represents the population of interest (i.e., using y_1 and y_0 as sample variables denoting, respectively, the population variables Y_1 and Y_0, and w as a sample variable denoting W), we can further define the standard estimator for the average treatment effect as the difference between two estimated means from sample data:

$$\hat{\tau} = E(\hat{y}_1|w = 1) - E(\hat{y}_0|w = 0). \tag{2.3}$$

The Neyman-Rubin counterfactual framework provides a useful tool not only for the development of various approaches to estimating potential outcomes but also for a discussion of whether assumptions embedded in randomized experiments are plausible when applied to social behavioral studies. In this regard, at least eight issues emerge.

1. In the above exposition, we expressed the evaluation of causal effects in an overly simplified fashion that did not take into consideration any covariates or threats to internal validity. In our hypothetical example where poor economic condition causes low academic achievement, many confounding factors might influence achievement. For instance, parental education could covary with income status, and it could affect academic achievement. When

covariates are entered into an equation, evaluators must impose additional assumptions. These include the *ignorable treatment assignment assumption* and the SUTVA, which we clarify in the next two sections. Without assumptions, the counterfactual framework leads us nowhere. Indeed, it is violations of these assumptions that have motivated statisticians and econometricians to develop new approaches.

2. In the standard estimator $E(Y_1|W=1) - E(Y_0|W=0)$, the primary interest of researchers is focused on the average outcome of treatment participants *if* they had not participated (i.e., $E(Y_0|W=1)$). Because this term is unobservable, evaluators use $E(Y_0|W=0)$ as a proxy. It is important to understand when the standard estimator consistently estimates the true average treatment effect for the population. Winship and Morgan (1999) decomposed the average treatment effect in the population into a weighted average of the average treatment effect for those in the treatment group and the average treatment effect for those in the control group as[2]

$$\begin{aligned}
\bar{\tau} &= \pi(\bar{\tau}|W=1) + (1-\pi)(\bar{\tau}|W=0) \\
&= \pi[E(Y_1|W=1) - E(Y_0|W=1)] + (1-\pi) \\
&\quad [E(Y_1|W=0) - E(Y_0|W=0)] \\
&= [\pi E(Y_1|W=1) + (1-\pi)E(Y_1|W=0)] \\
&\quad - [\pi E(Y_0|W=1) + (1-\pi)E(Y_0|W=0)] \\
&= E(Y_1) - E(Y_0),
\end{aligned} \tag{2.4}$$

where π is equal to the proportion of the population that would be assigned to the treatment group, and by the definition of the counterfactual model, let $E(Y_1|W=0)$ and $E(Y_0|W=1)$ be defined analogously to $E(Y_1|W=1)$ and $E(Y_0|W=0)$. The quantities $E(Y_1|W=0)$ and $E(Y_0|W=1)$ that appear in the second and third lines of Equation 2.4 cannot be directly calculated because they are unobservable values of Y. Furthermore, and again on the basis of the definition of the counterfactual model, if we assume that $E(Y_1|W=1) = E(Y_1|W=0)$ and $E(Y_0|W=0) = E(Y_0|W=1)$, then through substitution starting in the third line of Equation 2.4, we have[3]

$$\begin{aligned}
\bar{\tau} &= [\pi E(Y_1|W=1) + (1-\pi)E(Y_1|W=0)] - [\pi E(Y_0|W=1) \\
&\quad + (1-\pi)E(Y_0|W=0)] \\
&= [\pi E(Y_1|W=1) + (1-\pi)E(Y_1|W=1)] - [\pi E(Y_0|W=0) \\
&\quad + (1-\pi)E(Y_0|W=0)] \\
&= E(Y_1|W=1) - E(Y_0|W=0).
\end{aligned} \tag{2.5}$$

Thus, a sufficient condition for the standard estimator to consistently estimate the true average treatment effect in the population is that $E(Y_1|W=1) = E(Y_1|W=0)$ and $E(Y_0|W=0) = E(Y_0|W=1)$. This condition, as shown by numerous statisticians such as Fisher (1925), Kempthorne (1952), and Cox (1958), is met in the classical randomized experiment.[4] Randomization works in a way that makes the assumption about $E(Y_1|W=1) = E(Y_1|W=0)$ and $E(Y_0|W=0) = E(Y_0|W=1)$ plausible. When study participants are randomly assigned either to the treatment condition or to the nontreatment condition, certain physical randomization processes are carried out so that the determination of the condition to which participant i is exposed is regarded as statistically independent of all other variables, including the outcomes Y_1 and Y_0.

3. The real debate in statistical methods applied to observational studies centers on the validity of extending the assumption about randomization (i.e., that the process yields results independent of all other variables) to analysis of social behavioral evaluations. Or, to put it differently, whether the researcher engaged in social behavioral evaluations can continue to assume that $E(Y_0|W=0) = E(Y_0|W=1)$ and $E(Y_1|W=1) = E(Y_1|W=0)$. Not surprisingly, supporters of randomization as the central method for evaluating social behavioral programs answer "yes," whereas proponents of the nonexperimental approach answer "no" to this question. The classical experimental approach assumes no selection bias, and therefore, $E(Y_0|W=1) = E(Y_0|W=0)$. The assumption of no selection bias is indeed true because of the mechanism and logic behind randomization. However, many authors challenge the plausibility of this assumption in social behavioral evaluations. Heckman and Smith (1995) showed that the average outcome for the treated group under the condition of nontreatment is not the same as the average outcome of the nontreated group, precisely $E(Y_0|W=1) \neq E(Y_0|W=0)$, because of selection bias.

4. Rubin extended the counterfactual framework to a more general case—that is, allowing the framework to be applicable to observational studies. Unlike a randomized experiment, an observational study involves complicated situations that require a more rigorous approach to data analysis. Less rigorous approaches are open to criticism; for instance, Sobel (1996) criticized the common practice in sociology that uses a dummy variable (i.e., treatment vs. nontreatment) to evaluate the treatment effect in a regression model (or a regression-type model such as a path analysis or structural equation model) using survey data. As shown in the next section, the primary problem of such an approach is that the dummy treatment variable is specified by these models as exogenous, but in fact it is not. According to Sobel (2005),

> The incorporation of Neyman's notation into the modern literature on causal inference is due to Rubin (1974, 1977, 1978, 1980b), who, using this notation, saw the applicability of the work from the statistical literature on experimental

design to observational studies and gave explicit consideration to the key role of the treatment assignment mechanism in causal inference, thereby extending this work to observational studies. To be sure, previous workers in statistics and economics (and elsewhere) understood well in a less formal way the problems of making causal inferences in observational studies where respondents selected themselves into treatment groups, as evidenced, for example, by Cochran's work on matching and Heckman's work on sample selection bias. But Rubin's work was a critical breakthrough. (p. 100)

5. In the above exposition, we used the most common and convenient statistic (i.e., the mean) to express various counterfactuals and the ways in which counterfactuals are approximated. The average causal effect τ is an average, and as such, according to Holland (1986, p. 949), "enjoys all of the advantages and disadvantages of averages." One such disadvantage is the insensitivity of an average to the variability of the causal effect. If the variability in individuals' causal effects $(Y_i|W_i = 1) - (Y_i|W_i = 0)$ is large over all units, then $\tau = E(Y_1|W = 1) - E(Y_0|W = 0)$ may not well represent the causal effect of a specific unit (say, u_0). "If u_0 is the unit of interest, then τ may be irrelevant, no matter how carefully we estimate it!" (Holland, 1986, p. 949). This important point is expanded in Section 2.7, but we want to emphasize that the variability of the treatment effect at the individual level, or violation of an assumption about a constant treatment effect across individuals, can make the estimation of average treatment effects biased; therefore, it is important to distinguish among various types of treatment effects. In short, different statistical approaches employ counterfactuals of different groups to estimate different types of treatment effects.

6. Another limitation of using an average lies in the statistic mean itself. Although means are conventional, distributions of treatment parameters are also of considerable interest (Heckman, 2005, p. 20). In several articles, Heckman and his colleagues (Heckman, 2005; Heckman, Ichimura, & Todd, 1997; Heckman et al., 1999; Heckman, Smith, & Clements, 1997) have discussed the limitation of reliance on means (e.g., disruption bias leading to changed outcomes or the *Hawthorne* effect) and have suggested using other summary measures of the distribution of counterfactuals such as (a) the proportion of participants in Program A who benefit from the program relative to some alternative B, (b) the proportion of the total population that benefits from Program B compared with Program A, (c) selected quantiles of the impact distribution, and (d) the distribution of gains at selected base state values.

7. The Neyman-Rubin framework expressed in Equation 2.1 is the basic model. However, there are variants that can accommodate more complicated situations. For instance, Rosenbaum (2002b) developed a counterfactual model in which stratification is present and where s stands for the number of strata:

$$Y_{si} = W_{si}Y_{s1i} + (1 - W_{si})Y_s0_i. \tag{2.6}$$

Under this formulation, Equation 2.1 is the simplest case where s equals 1, or stratification is absent.[5]

8. The Neyman-Rubin counterfactual framework is mainly a useful tool for the statistical exploration of causal effects. However, by no means does this framework exclude the importance of using substantive theories to guide causal inferences. Identifying an appropriate set of covariates and choosing an appropriate model for data analysis are primarily tasks of developing theories based on prior studies in the substantive area. As Cochran (1965) argued,

> When summarizing the results of a study that shows an association consistent with a causal hypothesis, the investigator should always list and discuss all alternative explanations of his results (including different hypotheses and biases in the results) that occur to him. (sec. 5)

Dating from Fisher's work, statisticians have long acknowledged the importance of having a good theory of the treatment assignment mechanism (Sobel, 2005). Rosenbaum (2005) emphasized the importance of using theory in observational studies and encouraged evaluators to "be specific" on which variables to match and which variables to control using substantive theories. Thus, similar to all scientific inquiries, the counterfactual framework is reliable only under the guidance of appropriate theories and substantive knowledge.

2.3 The Ignorable Treatment Assignment Assumption

By thinking of the central challenge of all evaluations as estimating the missing outcomes for participants—each of whom is missing an observed outcome for either the treatment or nontreatment condition—the evaluation problem becomes a missing data issue. Consider the standard estimator of the average treatment effect: $\tau = E(Y_1|W = 1) - E(Y_0|W = 0)$. Many sources of error contribute to the bias of τ. It is for this reason that the researcher has to make a few fundamental assumptions to apply the Neyman-Rubin counterfactual model to actual evaluations. One such assumption is the *ignorable treatment assignment* assumption (Rosenbaum & Rubin, 1983). In the literature, this assumption is sometimes presented as part of the SUTVA (e.g., Rubin, 1986); however, we explicitly treat it as a separate assumption because of its importance. Moreover, the ignorable treatment assignment has been explored in detail as an assumption fundamental to the evaluation of treatment effects, particularly in the econometric literature. Our discussion follows this tradition.

The assumption can be expressed as

$$(Y_0, Y_1) \perp W|\mathbf{X}. \tag{2.7}$$

The assumption says that conditional on covariates X, the assignment of study participants to binary treatment conditions (i.e., treatment vs. nontreatment) is independent of the outcome of nontreatment (Y_0) and the outcome of treatment (Y_1).

A variety of terms have emerged to describe this assumption: *unconfoundedness* (Rosenbaum & Rubin, 1983), *selection on observables* (Barnow, Cain, & Goldberger, 1980), *conditional independence* (Lechner, 1999), and *exogeneity* (Imbens, 2004). These terms can be used interchangeably to denote the key idea that assignment to one condition or another is independent of the potential outcomes if observable covariates are held constant.

The researcher conducting a randomized experiment can be reasonably confident that the ignorable treatment assignment assumption holds because randomization typically balances the data between the treated and control groups and makes the treatment assignment independent of the outcomes under the two conditions (Rosenbaum, 2002b; Rosenbaum & Rubin, 1983). However, the ignorable treatment assignment assumption is often violated in quasi-experimental designs and in observational studies because the creation of a comparison group follows a natural process that confounds group assignment with outcomes. Thus, the researcher's first task in any evaluation is to check the tenability of the independence between the treatment assignment and outcomes under different conditions. A widely employed approach to this problem is to conduct bivariate analysis using the dichotomous treatment variable (W) as one and each independent variable available to the analyst (i.e., each variable in the vector X, one at a time) as another. Chi-square tests may be applied to the case where X is a categorical variable, and an independent-sample t test or Wilcoxon rank-sum (Mann-Whitney) test may be applied to the case where X is a continuous variable. Whenever the null hypothesis is rejected as showing the existence of significant difference between the treated and nontreated groups on the variable under examination, the researcher may conclude that there is a correlation between treatment assignment and outcome that is conditional on an observed covariate; therefore, the treatment assignment is not ignorable and taking remedial measures to correct the violation is warranted. Although this method is popular, it is worth noting that Rosenbaum (2002b) cautioned that no statistical evidence exists that supports the validity of this convention.

To demonstrate that the ignorable treatment assignment is nothing more than the same assumption of ordinary least squares (OLS) regression about the independence of the error term from an independent variable, we present evidence of the associative relation between the two assumptions. In the OLS context, the assumption is also known as *contemporaneous independence* of the error term from the independent variable or more generally *exogeneity*.

To analyze observational data, an OLS regression model using a dichotomous indicator is not the best choice. To understand this problem, consider the following

OLS regression model: $Y_i = \alpha + \tau W_i + X_i'\beta + e_i$, where W_i is a dichotomous variable indicating treatment, and X_i is the vector of independent variables for case i. In observational data, because researchers have no control over the assignment of treatment conditions, W is often highly correlated with Y. A statistical control is a modeling process that attempts to extract the independent contribution of explanatory variables (i.e., the vector X) to the outcome Y to determine the net effect of τ. When the ignorable treatment assignment assumption is violated and the correlation between W and e is not equal to 0, the OLS estimator of treatment effect τ is biased and inconsistent. More formally, under this condition there are three problems associated with the OLS estimator.

First, when the treatment assignment is not ignorable, the use of the dummy variable W leads to endogeneity bias. In the above regression equation, the dummy variable W is conceptualized as an exogenous variable. In fact, it is a dummy endogenous variable. The nonignorable treatment assignment implies a mechanism of selection; that is, there are other factors determining W. W is merely an observed variable that is determined by a latent variable W^* in such a way that $W = 1$, if $W^* > C$, and $W = 0$, otherwise, where C is a constant reflecting a cutoff value of *utility function*. Factors determining W^* should be explicitly taken into consideration in the modeling process. Conceptualizing W as a dummy endogenous variable motivated Heckman (1978, 1979) to develop the sample selection model and Maddala (1983) to develop the treatment effect model. Both models attempt to correct for the endogeneity bias. See Chapter 4 for a discussion of these models.

Second, the presence of the endogeneity problem (i.e., the independent variable is not exogenous and is correlated with the error term of the regression) leads to a biased and inconsistent estimation of the regression coefficient. Our demonstration of the adverse consequence follows Berk (2004). For ease of exposition, assume all variables are mean centered, and there is one predictor in the model:

$$y|x = \beta_1 x + e. \tag{2.8}$$

The least squares estimate of $\hat{\beta}_1$ is

$$\hat{\beta}_1 = \frac{\sum_{i=1}^{n} x_i y_i}{\sum_{i=1}^{n} x_i^2}. \tag{2.9}$$

Substituting Equation 2.8 into Equation 2.9 and simplifying, the result is

$$\hat{\beta}_1 = \beta_1 + \frac{\sum_{i=1}^{n} x_i e_i}{\sum_{i=1}^{n} x_i^2}. \tag{2.10}$$

If x and e are correlated, the expected value for the far right-hand term will be nonzero, and the numerator will not go to zero as the sample size increases without limit. The least squares estimate then will be biased and inconsistent. The presence of a nonzero correlation between x and e may be due to one or more of the following reasons: (a) the result of random measurement error in x, (b) one or more omitted variables correlated with x and y, (c) the incorrect functional form, and (d) a number of other problems (Berk, 2004).

This problem is also known as *asymptotical bias*, which is a term that is analogous to inconsistency. Kennedy (2003) explained that when contemporaneous correlation is present, "the OLS procedure, in assigning 'credit' to regressors for explaining variation in the dependent variable, assigns, in error, some of the regressors with which that disturbance is contemporaneously correlated" (p. 158). Suppose that the correlation between the independent variable and the error term is positive. When the error is higher, the dependent variable is also higher, and owing to the correlation between the error and the independent variable, the independent variable is likely to be higher, which implies that too much credit for making the dependent variable higher is likely to be assigned to the independent variable. Figure 2.1 illustrates this scenario. If the error term and the independent variable are positively correlated, negative values of the error will tend to correspond to low values of the independent variable and positive values of the error will tend to correspond to high values of the independent variable, which will create data patterns similar to that shown in the figure. The OLS estimating line clearly overestimates the slope of the true relationship. Obviously, the estimating line in this hypothetical example provides a much better fit to the sample data than does the true relationship, which causes the variance of the error term to be underestimated.

Finally, in observational studies, because researchers have no control over the assignment of treatment conditions, W is often correlated with Y. A statistical control is a modeling process that attempts to extract the independent contribution of explanatory variables to the outcome to determine the net effect of τ. Although the researcher aims to control for all important variables by using a well-specified vector X, the omission of important controls often occurs and results in a specification error. The consequence of omitting relevant variables is a biased estimation of the regression coefficient. We follow

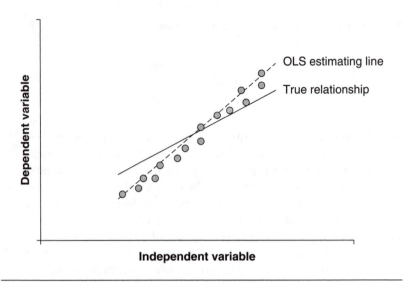

Figure 2.1   Positive Contemporaneous Correlation

SOURCE: Kennedy, P., *A Guide to Econometrics,* Fifth Edition, Figure 9.1, page 158. Copyright © 2003 Massachusetts Institute of Technology. Reprinted by permission of The MIT Press.

Greene (2003, pp. 148–149) to show why this is the case. Suppose that a correctly specified regression model would be

$$y = X_1\beta_1 + X_2\beta_2 + \varepsilon, \tag{2.11}$$

where the two parts of X have K_1 and K_2 columns, respectively. If we regress y on X_1 without including X_2, then the estimator is

$$b_1 = (X_1'X_1)^{-1}X_1'y = \beta_1 + (X_1'X_1)^{-1}X_1'X_2\beta_2 + (X_1'X_1)^{-1}X_1'\varepsilon. \tag{2.12}$$

Taking the expectation, we see that unless $X_1'X_2 = 0$ or $\beta_2 = 0$, b_1 is biased. The well-known result is the omitted variable formula

$$E[b_1|X] = \beta_1 + P_{1.2}\beta_2, \tag{2.13}$$

where

$$P_{1.2} = (X_1'X_1)^{-1}X_1'X_2. \tag{2.14}$$

Each column of the $K_1 \times K_2$ matrix $P_{1.2}$ is the column of slopes in the least squares regression of the corresponding column of X_2 on the column of X_1.

When the ignorable treatment assignment assumption is violated, remedial action is needed and the use of statistical controls with OLS regression is not the best choice. In Section 2.5, we review approaches that have been developed to correct for biases under this condition (e.g., the Heckman sample selection model directly modeling the endogenous dummy treatment condition) or approaches that relax the fundamental assumption to focus on a special type of treatment effect (e.g., average treatment effect for the treated [TT] rather than sample average treatment effect).

2.4 The Stable Unit Treatment Value Assumption

The *stable unit treatment value assumption* (SUTVA) was labeled and formally presented by Rubin in 1980. Rubin (1986) later extended this assumption, arguing that it plays a key role in deciding which questions are adequately formulated to have causal answers. Only under SUTVA is the representation of outcomes by the Neyman-Rubin counterfactual model adequate.

Formally, consider the situation with N units indexed by $i = 1, \ldots, N$; T treatments indexed by $w = 1, \ldots, T$; and outcome variable Y, whose possible values are represented by $Y_{iw}(w = 1, \ldots, T; i = 1, \ldots, N)$.[6] SUTVA is simply the a priori assumption that the value of Y for unit i when exposed to treatment w will be the same no matter what mechanism is used to assign treatment w to unit i and no matter what treatments the other units receive, and this holds for all $i = 1, \ldots, N$ and all $w = 1, \ldots, T$.

As it turns out, SUTVA basically imposes *exclusive restrictions*. Heckman (2005, p. 11) interprets these exclusive restrictions as the following two circumstances: (1) SUTVA rules out social interactions and general equilibrium effects and (2) SUTVA rules out any effect of the assignment mechanism on potential outcomes.

We previously examined the importance of the second restriction (ignorable treatment assignment) in Section 2.3. The following section explains the importance of the first restriction and describes the conditions under which the assumption is violated.

According to Rubin (1986), SUTVA is violated when unrepresented versions of treatment exist (i.e., Y_{iw} depends on which version of treatment w is received) or when there is interference between units (i.e., Y_{iw} depends on whether i' received treatment w or w', where $i \neq i'$ and $w \neq w'$). The classic example of violation of SUTVA is the analysis of treatment effects in agricultural research, such as rainfall that surreptitiously carries fertilizer from a treated plot to an adjacent untreated plot. In social behavioral evaluations, SUTVA is violated when a treatment alters social or environmental conditions that, in turn, alter potential outcomes. Winship and Morgan (1999) illustrated this idea by describing the impact of a large job training program on local labor markets:

> Consider the case where a large job training program is offered in a metropolitan area with a competitive labor market. As the supply of graduates from the program increases, the wage that employers will be willing to pay graduates of the program will decrease. When such complex effects are present, the powerful simplicity of the counterfactual framework vanishes. (p. 663)

SUTVA is both an assumption that facilitates investigation or estimation of counterfactuals and a conceptual perspective that underscores the importance of analyzing differential treatment effects with appropriate estimators. We return to SUTVA as a conceptual perspective in Section 2.7.

It is noteworthy that Heckman and his colleagues (Heckman et al., 1999) treated SUTVA as a strong assumption and presented evidence against the assumption. The limitations imposed by the strong assumption may be overcome by relaxed assumptions (Heckman & Vytlacil, 2005).

2.5 Methods to Estimate Treatment Effects

As previously discussed in Section 2.3, violating the ignorable treatment assignment assumption has adverse consequences. Indeed, when treatment assignment is not ignorable, the OLS regression estimate of treatment effect is biased and inconsistent. Furthermore, the consequences are worse when important predictors are omitted, and in an observational study when hidden selection bias is present (Rosenbaum, 2002b). What can be done? This question served as the original motivation for statisticians and econometricians to develop new methods for program evaluation. As a part of this work, new analytic models have been designed for observational studies and, more generally, for nonexperimental approaches that may be used in place of randomized experiments. When treatment assignment is nonignorable, the growing consensus among statisticians and econometricians is that OLS regression or simple covariance control is no longer the method of choice, though this statement runs the risk of oversimplification.

2.5.1 THE FOUR MODELS

The four models presented in this book are methods that aim (a) to relax the nonignorable treatment assignment assumption by considering estimation and inference that does not rely on strong assumptions requiring distributional and functional forms, (b) to rebalance assigned conditions so that they become more akin to data generated by randomization, and (c) to estimate counterfactuals that represent different treatment effects of interest by using selected statistics (i.e., means or proportions).

In estimating counterfactuals, the four models have the following core features.

1. *Heckman's sample selection model* (1978, 1979) and its revised version *estimating treatment effects* (Maddala, 1983). The crucial features of these models are (a) an explicit modeling of the structure of selection, (b) a switching regression that seeks exogenous factors determining the switch of study participants between two regimes (i.e., the treated and nontreated regimes), and (c) the use of the conditional probability of receiving treatment in the estimation of treatment effects.

2. *The propensity score matching model* (Rosenbaum & Rubin, 1983). The fundamental feature of the propensity score matching model is that it balances data through resampling or matching nontreated participants to treated ones on probabilities of receiving treatment (i.e., the propensity scores) and permits follow-up bivariate or multivariate analysis (e.g., stratified analysis of outcomes within quintiles of propensity scores, OLS regression, survival modeling, structural equation modeling, hierarchical linear modeling) as would be performed on a sample generated by a randomized experiment. Reducing the dimensionality of covariates to a one-dimensional score—the propensity—is a substantial contribution that leverages matching. This is a process described by Rubin (2008) as the design of observational studies to approximate randomized trials: "A crucial idea when trying to estimate causal effects from an observational data set is to conceptualize the observational data set as having arisen from a complex randomized experiment, where the rules used to assign the treatment conditions have been lost and must be reconstructed" (p. 815). From this perspective, the estimation of propensity scores and use of propensity score matching is the "most basic ingredient of an unconfounded assignment mechanism" (Rubin, 2008, p. 813).

3. *Matching estimators* (Abadie & Imbens, 2002, 2006). The key feature of this method is to directly impute counterfactuals for both treated and nontreated participants by using a vector norm with a positive definite matrix (i.e., the Mahalanobis metric or the inverse of sample variance matrix). Various types of treatment effects may be estimated: (a) the sample average treatment effect (SATE), (b) the sample average treatment effect for the treated (SATT), (c) the sample average treatment effect for the controls (SATC), and (d) the equivalent effects for the population (i.e., population average treatment effect [PATE], population average treatment effect for the treated [PATT], and population average treatment effect for the controls [PATC]). Standard errors corresponding to these sample average treatment effects are developed and used in significance tests.

4. *Propensity score analysis with nonparametric regression* (Heckman et al., 1997, 1998). The critical feature of this method is the comparison of each treated

participant to all nontreated participants based on distances between propensity scores. A nonparametric regression such as local linear matching is used to produce an estimate of the average treatment effect for the treatment group. By applying the method to data at two time points, this approach estimates the average treatment effect for the treated in a dynamic fashion, known as *difference-in-differences*.

It is worth noting that the four methods were not originally developed to correct for nonignorable treatment assignment. Quite the contrary, some of these models still assume that treatment assignment is strongly ignorable. According to Rosenbaum and Rubin (1983), showing "strong ignorability" allows analysts to evaluate a nonrandomized experiment as if it had come from a randomized experiment. However, in many evaluations, this assumption cannot be justified. Notwithstanding, in most studies, we wish to conduct analyses under the assumption of ignorability (Abadie, Drukker, Herr, & Imbens, 2004, p. 292).

Instead of correcting for the violation of the assumption about strongly ignorable treatment assignment, the corrective approaches (i.e., the four methods covered in this book) take various measures to control selection bias, including (a) relaxation of the assumption (e.g., instead of assuming *conditional independence* or *full independence,* Heckman et al., 1997, 1998, assumed *mean independence* by only requiring that conditional on covariates, the mean outcome under control condition for the treated cases be equal to the mean outcome under the treated condition for the controls); (b) modeling the treatment assignment process directly by treating the dummy treatment condition as an endogenous variable and using a two-step estimating procedure (i.e., the Heckman sample selection model); (c) developing a one-dimensional propensity score so that biases due to observable covariates can be removed by conditioning solely on the propensity score (i.e., the Rosenbaum and Rubin's propensity score matching model and the Heckman and colleagues' propensity score analysis with nonparametric regression); and (d) employing bias-corrected matching with a robust variance estimator to balance covariates between treatment conditions (i.e., the matching estimators). Because of these features, the four models offer advantages over OLS regression, regression-type models, and other simple corrective methods such as those discussed in Chapter 3. Rapidly being developed and refined, they are showing usefulness when compared with traditional approaches. Parenthetically, all these methods correct for overt selection bias only and do nothing to correct for hidden selection bias. It is for this reason that the randomized experiment remains a gold standard. When properly implemented, it corrects for both types of selection bias.

2.5.2 OTHER BALANCING METHODS

We chose to include these four models in this text because they are robust, efficient, and effective in addressing questions that arise commonly in social behavioral evaluations. Although the selection is based on our own experience, many applications can be found in biostatistics, economics, education, epidemiology, medicine, nursing, psychology, public health, social work, and sociology.

There are certainly other models that accomplish the same goal of balancing data. To offer a larger perspective, we provide a brief review of additional models.

Imbens (2004) summarized five groups of models that serve the common goal of estimating average treatment effects: (1) regression estimators that rely on consistent estimation of key regression functions; (2) matching estimators that compare outcomes across pairs of matched treated and control units, with each unit matched to a fixed number of observations in the opposite treatment; (3) estimators characterized by a central role of the propensity score (i.e., there are four leading approaches in this category: weighting by the reciprocal of the propensity score, blocking on the propensity score, regression on the propensity score, and matching on the propensity score); (4) estimators that rely on a combination of these methods, typically combining regression with one of its alternatives; and (5) Bayesian approaches to inference for average treatment effects. In addition, Winship and Morgan (1999) reviewed five methods, including research designs that are intended to improve causal interpretation in the context of nonignorable treatment assignment. These include (1) regression discontinuity designs, (2) instrumental variables (IV) approaches, (3) interrupted time series designs, (4) differential rate of growth models, and (5) analysis of covariance models.

Of these models, the IV approach shares common features with some models discussed in this book, particularly, the switching regression model. We discuss IV and its connection to the switching regression model in Chapter 4. The IV approach is among the earliest attempts in statistics and econometrics to address the endogeneity bias problem, and it has been shown to be useful in estimating local average treatment effects (Angrist, Imbens, & Rubin, 1996).

2.6 The Underlying Logic of Statistical Inference

When a treatment is found to be effective (or not effective), evaluators often want to generalize the finding to the population represented by the sample. They ask whether or not the treatment effect is zero (i.e., perform a nondirectional test) or is greater (less) than some cutoff value (i.e., perform a directional test) in the population. This is commonly known as *statistical inference*, a process of estimating unknown population parameters from known sample statistics. Typically, such an inference involves the calculation of a standard error to conduct a hypothesis test or to estimate a confidence interval.

The statistical inference of treatment effects stems from the tradition of randomized experimentation developed by Sir Ronald Fisher (1935/1971). The procedure is called a *permutation test* (also known as a *randomization test*, a *rerandomization test*, or an *exact test*) in that it makes a series of assumptions about the sample. When generalizing, researchers often find that one or more of these assumptions are violated, and thus, they have to develop strategies for statistical inference that deal with estimation when assumptions are differentially tenable. In this section, we review the underlying logic of statistical inference for both randomized experiments and observational studies. We argue that much of the statistical inference in observational studies follows the logic of statistical inference for randomized experiments and that checking the tenability of assumptions embedded in permutation tests is crucial in drawing statistical inferences for observational studies.

Statistical inference always involves a comparison of sample statistics to statistics from a *reference distribution*. Although in testing treatment effects from a randomized experiment, researchers often employ a parametric distribution (such as the normal distribution, the t distribution, and the F distribution) to perform a so-called parametric test, such a parametric distribution is not the reference distribution per se; rather it is an approximation of a randomization distribution. Researchers use parametric distributions in significance testing because these distributions "are approximations to randomization distributions—they are good approximations to the extent that they reproduce randomization inferences with reduced computational effort" (Rosenbaum, 2002a, p. 289). Strictly speaking, all statistical tests performed in randomized experiments are nonparametric tests using randomization distributions as a reference. Permutation tests are based on reference distributions developed by calculating all possible values of a test statistic under rearrangements of the "labels" on the observed data points. In other words, the method by which treatments are allocated to participants in an experimental design is mirrored in the analysis of that design. If the labels are exchangeable under the null hypothesis, then the resulting tests yield exact significance levels. Confidence intervals can then be derived from the tests.

Recall the permutation test of a British woman's tea-tasting ability (see Section 1.3.1). To reject the null hypothesis that the taster has no ability in discriminating two kinds of tea (or equivalently, testing the hypothesis that she makes a correct judgment by accidentally guessing it right), the evaluator lists all 70 possible ways $\left(\text{i.e., } {}_nC_r = {}_8C_4 = \dfrac{n!}{r!(n-r)!} = \dfrac{8!}{4!4!} = 70\right)$ of presenting eight cups of tea with four cups adding the milk first and four cups adding the tea infusion first. That is, the evaluator builds up a reference distribution of "11110000, 10101010, 00001111, . . ." that contains 70 elements in the series. The inference is drawn on a basis of the following logic: The taster could guess (choose) any one outcome out of the 70 possible ones; the probability of guessing the right outcome is $1/70 = .0124$, which is a low probability; thus, the null hypothesis of "no ability" can be rejected at a statistical significance level of $p < .05$. If the definition of "true ability" is relaxed to allow for six exact agreements rather than eight exact agreements (i.e., six cups are selected in an order that matches the order of actual presentation), then there are a total of 17 possible ways to have six agreements, and the probability of falsely rejecting the null hypothesis increases to $17/70 = .243$. The null hypothesis cannot be rejected at a .05 level. Under this relaxed definition, we should be more conservative, or ought to be more reluctant, in declaring that the tea taster has true ability.

All randomization tests listed in Section 1.3.2 (i.e., the Fisher's exact test, the Mantel-Haenszel test, the McNemar's test, the Mantel's extension of the Mantel-Haenszel test, the Wilcoxon's rank-sum test, and the Hodges and Lehmann signed rank test) are permutation tests that use randomization distributions as references and calculate all possible values of the test statistic to draw an inference. For this reason, this type of test is called *nonparametric*—it relies on distributions of all possible outcomes. In contrast, *parametric tests* employ parametric distributions as references. To illustrate, we now follow Lehmann to show the underlying logic of statistical inference employed in the Wilcoxon's rank-sum test (Lehmann, 2006).

The Wilcoxon's rank-sum test may be used to evaluate an outcome variable that takes many numerical values (i.e., an interval or ratio variable). To evaluate treatment effects, N participants (patients, students, etc.) are divided at random into a group of size n that receives a treatment and a control group of size m that does not receive treatment. At the termination of the study, the participants are ranked according to some response that measures treatment effectiveness. The null hypothesis of no treatment effect is rejected, and the superiority of the treatment is acknowledged, if in this ranking the n treated participants rank sufficiently high. The significance test calculates the statistical significance or probability of falsely rejecting the null hypothesis based on the following equation: $P_H(k = c) = \dfrac{w}{{}_N C_n} = \dfrac{w}{N!/n!(N-n)!}$, where k is the sum of treated participants' ranks under the null hypothesis of no treatment effect, c is a prespecified value at which one wants to evaluate its probability, and w is the frequency (i.e., number of times) of having value k under the null hypothesis. Precisely, if there were no treatment effect, then we could think of each participant's rank as attached before assignments to treatment and control are made. Suppose we have a total of $N = 5$ participants, $n = 3$ are assigned to treatment, and $m = 2$ are assigned to control. Under the null hypothesis of no treatment effect, the five participants may be ranked as 1, 2, 3, 4, and 5. With five participants taken three at a time to form the treatment group, there are a total of 10 possible groupings $\left(\text{i.e., } {_N}C_n = \dfrac{N!}{n!(N-n)!} = \dfrac{5!}{3!2!} = 10\right)$ of outcome ranks under the null hypothesis:

Treated	(3, 4, 5)	(2, 4, 5)	(1, 4, 5)	(2, 3, 5)	(1, 3, 5)
Control	(1, 2)	(1, 3)	(2, 3)	(1, 4)	(2, 4)
Treated	(2, 3, 4)	(1, 3, 4)	(1, 2, 4)	(1, 2, 3)	(1, 2, 5)
Control	(1, 5)	(2, 5)	(3, 5)	(4, 5)	(3, 4)

The rank sum of treated participants corresponding to each of the above groups may look like the following:

Treatment ranks	3, 4, 5	2, 4, 5	1, 4, 5	2, 3, 5	1, 3, 5	2, 3, 4	1, 3, 4	1, 2, 4	1, 2, 3	1, 2, 5
Rank sum k	12	11	10	10	9	9	8	7	6	8

The probabilities of taking various rank sum values under the null hypothesis of no treatment effect $\left(\text{i.e., } P_H(k=c) = \dfrac{w}{{_N}C_n} = \dfrac{w}{N!/n!(N-n)!}\right)$ are displayed below:

K	6	7	8	9	10	11	12
$P_H(k=c)$	.1	.1	.2	.2	.2	.1	.1

For instance, under the null hypothesis of no treatment effect, there are two possible ways to have a rank sum $k = 10$ (i.e., $w = 2$, when the treatment group is composed of treated participants whose ranks are (1, 4, 5) or is composed of treated participants whose ranks are (2, 3, 5)). Because there are a total of 10 possible ways to form the treatment and control groups, the probability of having a rank sum $k = 10$ is $2/10 = .2$. The above probabilities constitute the randomization distribution (i.e., the reference distribution) for this permutation test. From any real sample, one will observe a realized outcome that takes any one of the seven k values (i.e., 6, 7, . . . , 12). Thus, a significance test of no treatment effect is to compare the observed rank sum from the sample data with the above distribution and check the probability of having such a rank sum from the reference. If the probability is small, then one can reject the null hypothesis and conclude that in the population the treatment effect is not equal to zero.

Suppose that the intervention being evaluated is an educational program that aims to promote academic achievement. After implementing the intervention, the program officer observes that the three treated participants have academic test scores of 90, 95, 99, and the two control participants have test scores of 87, 89, respectively. Converting these outcome values to ranks, the three treated participants have ranks of 3, 4, 5 and the two control participants have ranks of 1, 2, respectively. Thus, the rank sum of the treated group observed from the sample is $3 + 4 + 5 = 12$. This observed statistic is then compared with the reference distribution, and the probability of having a rank sum of 12 under the null hypothesis of no treatment effect is $P_H(k = 12) = .1$. Because this probability is small, we can reject the null hypothesis of no treatment effect at a significance level of .1 and conclude that the intervention may be effective in the population. Note that in the above illustration, we used very small numbers of N, n, and m, and thus the statistical significance for this example cannot reach the conventional level of .05—the smallest probability in the distribution of this illustrating example is .1. In typical evaluations, N, n, and m tend to be larger, and a significance level of .05 can be attained.

The Wilcoxon rank-sum test, as described above, employs a randomization distribution based on the null hypothesis of no treatment effect. The exact probability of having a rank sum equal to a specific value is calculated, and such a calculation is based on all possible arrangements of N participants into n and m. The probabilities of having all possible values of rank sum based on all possible arrangements of N participants into n and m are then calculated, and it is these probabilities that constitute the reference for significance testing. Comparing the observed rank sum of treated participants from a real sample with the reference, evaluators draw a conclusion about whether or not they can reject the null hypothesis of no treatment effect at a statistically significant level.

The illustrations above show a primary feature of statistical inference involving permutation tests: These tests build up a distribution that exhausts all possible arrangements of study participants under a given N, n, and m, and calculate all possible probabilities of having a particular outcome (e.g., the specific rank sum of treated participants) under the null hypothesis of no treatment effect. This provides a significance test for treatment in a realized sample. To make the statistical inference valid, we must ensure that the sample being evaluated meets certain assumptions. At the minimum, these assumptions include the following: (a) The sample is a real random sample from a well-defined population, (b) each participant has a known probability of receiving treatment, (c) treatment assignment is strongly ignorable, (d) the individual-level treatment effect (i.e., the difference between observed and potential outcomes $\tau_i = Y_{1i} - Y_{0i}$) is constant, (e) there is a stable unit treatment value, and (f) probabilities of receiving treatment overlap between treated and control groups.

When a randomized experiment in the strict form of Fisher's definition is implemented, all the above assumptions are met, and therefore, statistical inference using permutation tests is valid. Challenges arise when evaluators move from randomized experiments to observational studies, because in the latter case one or more of the above assumptions are not tenable.

So what is the underlying logic of statistical inference for observational studies? To answer this question, we draw on perspectives from Rosenbaum (2002a, 2002b) and from Imbens (2004).

Rosenbaum's framework follows the logic used in the randomized experiments and is an extension of permutation tests to observational studies. To begin with, Rosenbaum examines covariance adjustment in completely randomized experiments. In the above examples, for simplicity of exposition, we did not use any covariates. In real evaluations of randomized experiments, evaluators typically would have covariates and want to control them in the analysis. Rosenbaum shows that testing the null hypothesis of no treatment effect in studies with covariates follows the permutation approach, with the added task of fitting a linear or generalized linear model. After fitting a linear model that controls for covariates, the residuals for both conditions (treatment and control groups) are fixed and known; and therefore, one can apply the Wilcoxon's rank-sum test or similar permutation tests (e.g., the Hodges-Lehmann aligned rank test) to model-fitted residuals.

A propensity score adjustment can be combined with the permutation approach in observational studies with overt bias. "Overt bias . . . can be seen in the data at hand—for instance, prior to treatment, treated participants are observed to have lower incomes than controls" (Rosenbaum, 2002b, p. 71). In this context, one can balance groups by estimating a propensity score, which is a conditional probability of receiving treatment given observed covariates, and then perform conditional permutation tests using a matched sample. Once again, the statistical inference employs the same logic applied to randomized experiments. We describe in detail three such permutation tests after an optimal matching on propensity scores (see Chapter 5): regression adjustment of difference scores based on a sample created by optimal pair matching, outcome analysis using the Hodges-Lehmann aligned rank test based on a sample created by optimal matching, and regression adjustment using the Hodges-Lehmann aligned rank test based on a sample created by optimal matching.

Finally, Rosenbaum considers statistical inference in observational studies with hidden bias. Hidden bias is similar to overt bias but it cannot be seen in the data at hand, because measures that might have revealed a selection effect were omitted from data collection. When bias exists but it is not observable, one can still perform propensity score matching and conduct statistical tests by comparing treatment and control participants matched on propensity scores. But caution is warranted and sensitivity analyses should be undertaken before generalizing findings to a population. Surprisingly and importantly, the core component of Rosenbaum's sensitivity analysis involves permutation tests, which include the McNemar's test, the Wilcoxon's signed-rank test, and the Hodges-Lehmann point and interval estimates for matched pairs, sign-score methods for matching with multiple controls, sensitivity analysis for matching with multiple controls when responses are continuous variables, and sensitivity analysis for comparing two unmatched groups. We illustrate these methods in Chapter 8.

In 2004, Imbens reviewed inference approaches using nonparametric methods to estimate average treatment effects under the unconfoundedness assumption (i.e., the ignorable treatment assignment assumption). He discusses advances in generating sampling distributions by bootstrapping (a method for estimating the sampling distribution of an estimator by sampling with replacement from the original sample) and observes,

> There is little formal evidence specific for these estimators, but, given that the estimators are asymptotically linear, it is likely that bootstrapping will lead to valid standard errors and confidence intervals at least for the regression propensity score methods. Bootstrapping may be more complicated for matching estimators, as the process introduces discreteness in the distribution that will lead to ties in the matching algorithm. (p. 21)

Furthermore, Imbens, Abadie, and others show that the variance estimation employed in the matching estimators (Abadie & Imbens, 2002, 2006) requires no additional nonparametric estimation and may be a good alternative to estimators using bootstrapping. Finally, in the absence of consensus on the best estimation methods, Imbens challenges the field to provide implementable versions of the various estimators that do not require choosing bandwidths (i.e., a user-specified parameter in implementing kernel-based matching, see Chapter 7) or other smoothing parameters and to improve estimation methods so that they can be applied with a large number of covariates and varying degrees of smoothness in the conditional means of the potential outcomes and the propensity scores.

In summary, understanding the logic of statistical inference underscores in turn the importance of checking the tenability of statistical assumptions. In general, current estimation methods rely on permutation tests, which have roots in randomized experimentation. We know too little about estimation when a reference distribution is generated by bootstrapping, but this seems promising. Inference becomes especially challenging when nonparametric estimation requires making subjective decisions, such as specifications of bandwidth size, when data contain a large number of covariates and when sample sizes are small. Caution seems particularly warranted in observational studies. Omission of important variables and measurement error in the covariates—both of which are difficult to detect—justify use of sensitivity analysis.

2.7 Types of Treatment Effects

Unlike many texts that address treatment effects as the net difference between the mean scores of participants in treatment and control conditions, we introduce and discuss a variety of treatment effects. This may seem pedantic, but there are at least three reasons why distinguishing, both conceptually and methodologically, among types of treatment effects is important. First, distinguishing among types of treatment effects is important because of the limitation in solving the fundamental problem of causal inference (see Section 2.2). Recall that at the individual level the researcher cannot observe both potential outcomes (i.e., outcomes under the condition of treatment and outcomes under the condition of nontreatment) and thus has to rely on group averages to evaluate counterfactuals. The estimation of treatment effects so derived at the population level uses averages or $\tau = E(Y_1|W = 1) - E(Y_0|W = 0)$. As such, the variability in individuals' causal effects $(Y_i|W_i = 1) - (Y_i|W_i = 0)$ would affect the accuracy of an estimated treatment effect. If the variability is large over all units, then $\tau = E(Y_1|W = 1) - E(Y_0|W = 0)$ may not represent the causal effect of a specific unit very well, and under many evaluation circumstances, treatment effects of certain units (groups) serve a central interest. Therefore, it is critical to ask which effect is represented by the standard estimator. It is clear that the effect represented by the standard estimator may not be the same as those arising from the researcher's interest. Second, there are inevitably different ways to define groups and to use different averages to represent counterfactuals. Treatment effects and their surrogate counterfactuals are then multifaceted. Last, SUTVA is both an assumption and a perspective for the evaluation of treatment effects. As such, when social interaction is absent, SUTVA implies that different versions of treatment (or different dosages of the same treatment) should result in different outcomes. This is the rationale that leads evaluators to distinguish two different effects: *program efficacy* versus *program effectiveness*.

Based on our review of the literature, the following seven treatment effects are most frequently discussed by researchers in the field. Although some are related, the key notion is that researchers should distinguish between different effects. That is, we should recognize that different effects require different estimation methods, and by the same token, different estimation methods estimate different effects.

1. *Average treatment effect (ATE) or average causal effect:* This is the core effect estimated by the standard estimator

$$ATE = \tau = E(Y_1|W = 1) - E(Y_0|W = 0).$$

Under certain assumptions, one can also write it as

$$ATE = E[(Y_1|W = 1) - (Y_0|W = 0)|X].$$

2. In most fields, evaluators are interested in evaluating program effectiveness, which indicates how well an intervention works when implemented under conditions of actual application (Shadish et al., 2002, p. 507). Program effectiveness can be measured by the *intent-to-treat* (ITT) effect. ITT is generally analogous to ATE: "Statisticians have long known that when data are collected using randomized experiments, the difference between the treatment group mean and the control group mean on the outcome is an unbiased estimate of the ITT" (Sobel, 2005, p. 114). In other words, the standard estimator employs counterfactuals (either estimation of the missing-value outcome at the individual level or mean difference between the treated and nontreated groups) to evaluate the overall effectiveness of an intervention as implemented.

3. Over the past 25 years, evaluators have also become sensitive to the differences between effectiveness and efficacy. The treatment assigned to a study participant may not be implemented in the way it was designed. The term *efficacy* is used to indicate how well an intervention works when it is implemented under conditions of ideal application (Shadish et al., 2002, p. 507). Measuring *efficacy effect* (EE) requires a careful monitoring of program implementation and taking measures to warrant intervention fidelity. EE plays a central role in the so-called *efficacy subset analysis* (ESA) that deliberately measures impact on the basis of treatment exposure or dose.

4. Average treatment effect for the treated (TT) can be expressed as

$$E[(Y_1 - Y_0)|X, W = 1].$$

Heckman (1992, 1996, 1997, 2005) argued that in a variety of policy contexts, it is the TT that is of substantive interest. The essence of this argument is that in deciding whether a policy is beneficial, our interest is not whether on average the program is beneficial for all individuals but whether it is beneficial for those individuals who are assigned or who would assign themselves to the treatment (Winship & Morgan, 1999, p. 666). The key notion here is TT $\neq$ ATE.

5. *Average treatment effect for the untreated (TUT) is an effect parallel to TT for the untreated:*

$$E[(Y_1 - Y_0)|X, W = 0].$$

Although estimating TUT is not as important as TT, noting the existence of such an effect is a direct application of the Neyman-Rubin model. In policy research, the estimation of TUT addresses (conditionally and unconditionally) the question of how extension of a program to nonparticipants as a group might affect their outcomes (Heckman, 2005, p. 19). The matching estimators described in Chapter 6 offer a direct estimate of TUT.

6. *Marginal treatment effect (MTE) or its special case of the treatment effect for people at the margin of indifference (EOTM):* In some policy and practice situations, it is important to distinguish between marginal and average returns (Heckman, 2005). For instance, the average student going to college may do better (i.e., have higher grades) than the marginal student who is indifferent about going to school or not. In some circumstances, we wish to evaluate the

impact of a program at the margins. Heckman and Vytlacil (1999, 2005) have shown that MTE plays a central role in organizing and interpreting a wide variety of evaluation estimators.

7. *Local average treatment effect (LATE):* Angrist et al. (1996) outlined a framework for causal inference where assignment to binary treatment is ignorable, but compliance with the assignment is not perfect so that the receipt of treatment is nonignorable. LATE is defined as *the average causal effect for compliers.* It is not the average treatment effect either for the entire population or for a subpopulation identifiable from observed values. Using the instrumental variables approach, Angrist et al. demonstrated how to estimate LATE.

To illustrate the importance of distinguishing different treatment effects, we invoke an example originally developed by Rosenbaum (2002b, pp. 181–183). Using hypothetical data in which responses under the treatment and control conditions are known, it demonstrates the inequality of four effects:

$$EE \neq ITT \ (ATE) \neq TT \neq \text{Naive ATE}.$$

Consider a randomized trial in which patients with chronic obstructive pulmonary disease are encouraged to exercise. Table 2.1 presents an artificial data set of 10 patients (i.e., $N = 10$ and $i = 1, \ldots, 10$). The treatment, W_i, is encouragement to exercise: $W_i = 1$ signifying encouragement and $W_i = 0$ signifying no encouragement. The assignment of treatment conditions to patients is randomized. The pair (d_{1i}, d_{0i}) indicates whether patient i would exercise, with or without encouragement, where 1 signifies exercise and 0 indicates no exercise. For example, $i = 1$ would exercise whether encouraged or not, $(d_{1i}, d_{0i}) = (1, 1)$, whereas $i = 10$ would not exercise in either case, $(d_{1i}, d_{0i}) = (0, 0)$, but $i = 3$ exercises only if encouraged, $(d_{1i}, d_{0i}) = (1, 0)$.

The response, (Y_{1i}, Y_{0i}), is a measure of lung function, or forced expiratory volume on a conventional scale, with higher numbers signifying better lung function. By design, the efficacy effect is known in advance ($EE = 5$); that is, switching from no exercise to exercise raises lung function by 5. Note that counterfactuals in this example are hypothesized to be known. For $i = 3$, $W_i = 1$ or exercise is encouraged, $Y_{1i} = 64$ is the outcome under the condition of exercise, and $Y_{0i} = 59$ is the counterfactual (i.e., if the patient did not exercise, the outcome would have been 59), and for this case the observed outcome $R_i = 64$. In contrast, for $i = 4$, $W_i = 0$ or exercise is not encouraged, $Y_{1i} = 62$ is the counterfactual and $Y_{0i} = 57$ is the outcome under the condition of no exercise, and for this case the observed outcome $R_i = 57$. D_i is a measure of compliance with the treatment, $D_i = 0$ signifying exercise actually not performed, and $D_i = 1$ signifying exercise was performed. So for $i = 2$, even though $W_i = 0$ (no treatment, or exercise is not encouraged), the patient exercised anyway. Likewise, for $i = 10$, even though exercise is encouraged and

Table 2.1   An Artificial Example of Noncompliance With Encouragement (W_i) to Exercise (D_i)

i	d_{1i}	d_{0i}	Y_{1i}	Y_{0i}	W_i	D_i	R_i
1	1	1	71	71	1	1	71
2	1	1	68	68	0	1	68
3	1	0	64	59	1	1	64
4	1	0	62	57	0	0	57
5	1	0	59	54	0	0	54
6	1	0	58	53	1	1	58
7	1	0	56	51	1	1	56
8	1	0	56	51	0	0	51
9	0	0	42	42	0	0	42
10	0	0	39	39	1	0	39

SOURCE: Rosenbaum (2002b, p. 182). Reprinted with kind permission of Springer Science + Business Media.

$W_i = 1$, the patient did not exercise, $D_i = 0$. Comparing the difference between W_i and D_i for each i gives a sense of intervention fidelity. In addition, on the basis of the existence of discrepancies in fidelity, program evaluators claim that treatment effectiveness is not equal to treatment efficacy.

Rosenbaum goes further to examine which patients responded to encouragement. Patients $i = 1$ and $i = 2$ would have the best lung function without encouragement, and they will exercise with or without encouragement. Patients $i = 9$ and $i = 10$ would have the poorest lung function without encouragement, and they will not exercise even when encouraged. Patients $i = 3, 4, \ldots, 8$ have intermediate lung function without exercise, and they exercise only when encouraged. The key point noted by Rosenbaum is that although treatment assignment or encouragement, W_i, is randomized, compliance with assigned treatment, (d_{1i}, d_{0i}), is strongly confounded by the health of the patient. Therefore, in this context, how can we estimate the efficacy?

To estimate a naive ATE, we might ignore the treatment state (i.e., ignoring W_i) and (naively) take the difference between the mean response of patients who exercised and those who did not exercise (i.e., using D_i as a grouping variable). In this context and using the standard estimator, we would estimate the naive ATE as

$$\frac{71 + 68 + 64 + 58 + 56}{5} - \frac{57 + 54 + 51 + 42 + 39}{5} = \frac{317 - 243}{5} = \frac{74}{5} = 14.8,$$

which is nearly three times the true effect of 5. The problem with this estimate is that the people who exercised were in better health than the people who did not exercise.

Alternatively, a researcher might ignore the level of compliance with the treatment and use the treatment state W_i to obtain ATE (i.e., taking the mean difference between those who were encouraged and those who were not). In this context and using the standard estimator, we find that the estimated ATE is nothing more than the intent-to-treat (ITT) effect:

$$\frac{71 + 64 + 58 + 56 + 39}{5} - \frac{68 + 57 + 54 + 51 + 42}{5}$$

$$= \frac{288 - 272}{5} = \frac{16}{5} = 3.2,$$

which is much less than the true effect of 5. This calculation demonstrates that ITT is an estimate of program effectiveness but not of program efficacy.

Finally, a researcher might ignore the level of compliance and estimate the average treatment effect for the treated (TT) by taking the average differences between Y_{1i} and Y_{0i} for the five treated patients:

$$\frac{(71 - 71) + (64 - 59) + (58 - 53) + (56 - 51) + (39 - 39)}{5}$$

$$= \frac{0 + 5 + 5 + 5 + 0}{5} = \frac{15}{5} = 3.$$

Although TT is substantially lower than efficacy, this is an effect that serves a central substantive interest in many policy and practice evaluations.

In sum, this example illustrates the fundamental differences among four treatment effects, EE ≠ ITT (ATE) ≠ TT ≠ Naive ATE, and one similarity, ITT = ATE. Our purpose for showing this example is not to argue which estimate is the best but to show the importance of estimating appropriate treatment effects using appropriate methods suitable for research questions.

2.8 Heckman's Econometric Model of Causality

In Chapter 1, we described two traditions in drawing causal inferences: the econometric tradition that relies on structural equation modeling and the statistical tradition that relies on randomized experiment. The economist Heckman (2005) developed a conceptual framework for causal inference that he called the *scientific model of causality*. In this work, Heckman sharply contrasted his model with the statistical approach—primarily the Neyman-Rubin counterfactual model—and advocated for an econometric approach that directly models the selection process. Heckman argued that the statistical literature on causal inferences was incomplete because it had not attempted to model the structure or process by which participants are selected into treatments. Heckman further argued that the statistical literature confused the task of identifying causal models from population distributions (where the sampling variability of empirical distributions is irrelevant) with the task of identifying causal models from actual data (where sampling variability is an issue). Because this model is relatively new, and debate about it continues, we will highlight its main features in this section. The brevity of our presentation is necessitated by the fact that the model is a comprehensive framework and includes forecasting the impact of interventions in new environments, a topic that exceeds the scope of this book. We concentrate on Heckman's critique of the Neyman-Rubin model, which is a focal point of this chapter.

First, Heckman (2005, pp. 9–21) developed a notation system for his scientific model that explicitly encompassed variables and functions that were not defined or treated comprehensively in prior literature. In this system, Heckman defined outcomes for persons in a universe of individuals and corresponding to possible treatments within a set of treatments where assignment is subject to certain rules, the valuation associated with each possible treatment outcome, including both private evaluations based on personal utility and evaluations by others (e.g., the "social planner"), and the selection mechanism appropriate under alternative policy conditions. Using this notation system and assumptions, Heckman further defined both individual-level treatment (causal) effects and population-level treatment effects.

Second, Heckman (2005, p. 3) specified three distinct tasks in the analysis of causal models: (1) defining the set of hypotheticals or counterfactuals, which requires a scientific theory; (2) identifying parameters (causal or otherwise) from hypothetical population data, which requires mathematical analysis of point or set identification; and (3) identifying parameters from real data, which requires estimation and testing theory.

Third, Heckman (2005, pp. 7–9) distinguished three broad classes of policy evaluation questions: (1) evaluating the impact of previous interventions on outcomes, including their impact in terms of welfare (i.e., a problem of internal validity); (2) forecasting the impacts (constructing counterfactual states) of interventions implemented in one environment on other environments, including their impacts in terms of welfare (i.e., a problem of external validity); and (3) forecasting the impacts of interventions (constructing counterfactual states associated with interventions) never historically experienced for other environments, including impacts in terms of welfare (i.e., using history to forecast the consequences of new policies).

Fourth, Heckman (2005, pp. 35–38) contrasted his scientific model (hereafter denoted as H) with the Neyman-Rubin model (hereafter denoted as NR) in terms of six basic assumptions. Specifically, NR assumes (1) a set of counterfactuals defined for *ex post* outcomes (no evaluations of outcomes or specification of treatment selection rules); (2) no social interactions; (3) invariance of counterfactual to assignment of treatment; (4) evaluating the impact of historical interventions on outcomes, including their impact in terms of welfare is the only problem of interest; (5) mean causal effects are the only objects of interest; and (6) there is no simultaneity in causal effects, that is, outcomes cannot cause each other reciprocally. In contrast, H (1) decomposes outcomes under competing states (policies or treatments) into their determinants; (2) considers valuation of outcomes as an essential ingredient of any study of causal inference; (3) models the choice of treatment and uses choice data to infer subjective valuations of treatment; (4) uses the relationship between outcomes and treatment choice equations to motivate, justify, and interpret alternative identifying strategies; (5) explicitly accounts for the arrival of information through *ex ante* and *ex post* analyses; (6) considers distributional causal parameters as well as mean effects; (7) addresses all three policy evaluation problems; and (8) allows for nonrecursive (simultaneous) causal models. The comparison of the NR and H models is summarized and extended in Table 2.2.

Table 2.2   Econometric Versus Statistical Causal Models

	Statistical Causal Model	*Econometric Models*
Sources of randomness	Implicit	Explicit
Models of conditional counterfactuals	Implicit	Explicit
Mechanism of intervention for determining counterfactuals	Hypothetical randomization	Many mechanisms of hypothetical interventions, including a randomization mechanism that is explicitly modeled
Treatment of interdependence	Recursive	Recursive or simultaneous systems
Social/market interactions	Ignored	Modeled in general equilibrium frameworks
Projections to different populations?	Does not project	Projects
Parametric?	Nonparametric	Becoming nonparametric
Range of questions answered	One focused treatment effect	In principle, answers many possible questions

SOURCE: Heckman (2005, p. 87). Reprinted with permission of Wiley-Blackwell.

Finally, Heckman (2005, pp. 50–85) discussed the identification problem and various estimators to evaluate different types of treatment effects. In Section 2.7, we have highlighted the main effects of interest that are commonly found in the literature (i.e., ATE, TT, TUT, MTE, and LATE). Heckman carefully weighed the implicit assumptions underlying four widely used methods of causal inference applied to data in the evaluation of these effects: matching, control functions, the instrumental variable method, and the method of directed acyclic graphs (i.e., Pearl, 2000).

The scientific model of causality may be a tipping-point development in the field of program evaluation. Perhaps the most important contribution of the model is its comprehensive investigation of the estimation problem, effects of interest, and estimation methods under a general framework. This is pioneering. Although it is too early to make judgments about the model's strengths and limitations, it is stimulating widespread discussion, debate, and methodological innovation. To conclude, we cite Sobel's (2005) comment that, to a great extent, coincides with our opinion:

> Heckman argues for the use of an approach to causal inference in which structural models play a central role. It is worth remembering that these models are often powerful in part because they make strong assumptions. . . . But I do not want to argue that structural modeling is not useful, nor do I want to suggest that methodologists should bear complete responsibilities for the use of the tools they have fashioned. To my mind, both structural modeling and approaches that feature weaker assumptions have their place, and in some circumstances, one will be more appropriate than the other. Which approach is more reasonable in a particular case will often depend on the feasibility of conducting a randomized study, what we can actually say about the reasonableness of invoking various assumptions, as well as the question facing the investigator (which might be dictated by a third party, such as a policy maker). An investigator's tastes and preferences may also come into play. A cautious and risk-averse investigator may care primarily about being right, even if this limits the conclusions he or she draws, whereas another investigator who wants (or is required) to address a bigger question may have (or need to have) a greater tolerance for uncertainty about the validity of his or her conclusions. (pp. 127–128)

2.9 Conclusions

This chapter examined the Neyman-Rubin counterfactual framework, the ignorable treatment assignment assumption, the SUTVA assumption, the underlying logic of statistical inference, and the econometric model of causality. We began with an overview of the counterfactual perspective that serves as a useful tool for the evaluation of treatment effects, and we ended with a brief review of Heckman's comprehensive and controversial scientific model of causal inference. It is obvious that there are disagreements among research scholars. In particular, debate between the econometric and statistical traditions continues to play a central role in the development of estimation methods. Specifically, we have emphasized the importance of disentangling treatment effects from treatment assignment and evaluating different treatment effects suitable to evaluation objectives under competing assumptions. The debate between the econometric and statistical schools of thought often centers on whether assumptions are too strong and to what extent we may relax assumptions. The disagreements reflect the complexities and challenges of program evaluation. We will revisit these challenges throughout this book.

Notes

1. In the literature, there are notation differences in expressing this and other models. To avoid confusion, we will use consistent notation in the text and present the original notation in footnotes. Equation 2.1 was expressed by Heckman and Vytlacil (1999, p. 4730) as

$$Y_i = D_i Y_{1i} + (1 - D_i)Y_{0i}.$$

2. In Winship and Morgan's (1999, p. 665) notation, Equation 2.4 is expressed as

$$
\begin{aligned}
\bar{\delta} &= \pi \bar{\delta}_{i \in T} + (1 - \pi)\bar{\delta}_{i \in C} \\
&= \pi(\bar{Y}^t_{i \in T} - \bar{Y}^c_{i \in T}) + (1 - \pi)(\bar{Y}^t_{i \in C} - \bar{Y}^c_{i \in C}) \\
&= [\pi \bar{Y}^t_{i \in T} + (1 - \pi)\bar{Y}^t_{i \in C}] - [\pi \bar{Y}^c_{i \in T} + (1 - \pi)(\bar{Y}^c_{i \in C})] \\
&= \bar{Y}^t - \bar{Y}^c.
\end{aligned}
$$

3. In Winship and Morgan's (1999) notation, Equation 2.5 is expressed as

$$
\begin{aligned}
\bar{\delta} &= [\pi \bar{Y}^t_{i \in T} + (1 - \pi)\bar{Y}^t_{i \in C}] - [\pi \bar{Y}^c_{i \in T} + (1 - \pi)(\bar{Y}^c_{i \in C})] \\
&= [\pi \bar{Y}^t_{i \in T} + (1 - \pi)\bar{Y}^t_{i \in T}] - [\pi \bar{Y}^c_{i \in C} + (1 - \pi)(\bar{Y}^c_{i \in C})] \\
&= \bar{Y}^t_{i \in T} - \bar{Y}^c_{i \in C}.
\end{aligned}
$$

4. Holland (1986) provides a thorough review of these statisticians' work under the context of randomized experiment.

5. In Rosenbaum's (2002b, p. 41) notation, Equation 2.6 is expressed as $R_{si} = Z_{si} r_{Tsi} - (1 - Z_{si}) r_{Csi}$.

6. We have changed notation to make the presentation of SUTVA consistent with the notation system adopted in this chapter. In Rubin's original presentation, he used u in place of i and t in place of w.

3

Conventional
Methods for Data Balancing

A s preparation for understanding advances in data balancing, this chapter
reviews conventional approaches to the analysis of observational data. In
Chapter 2, we examined the Neyman-Rubin counterfactual framework and
its associated assumptions. Although the assumption of "no social interactions"
is often deemed too strong, researchers generally agree that the *ignorable
treatment assignment assumption* is necessary for program evaluations,
including observational studies. Given the importance of this assumption in all
emerging approaches, this chapter scrutinizes the mechanisms that produce
violations of ignorable treatment assignment as well as conventional corrective
approaches. We do so by creating data under five scenarios, and we show how
to balance data by using three common methods, namely, ordinary least
squares (OLS) regression, matching, and stratification. All three methods provide for control of covariates and may lead to unbiased estimation of treatment
effect. All three methods are seemingly different but essentially aim to
accomplish common objectives. All three also have limitations in handling
selection bias. Understanding these methods is helpful in developing an understanding of advanced models. The key message this chapter conveys is that
covariance control does not necessarily correct for nonignorable treatment
assignment, and it is for this reason that advanced methods for estimating
treatment effects should be considered in practice.

Section 3.1 presents a heuristic example to address the question of why
data balancing is necessary. Section 3.2 presents the three correction models
(i.e., OLS regression, matching, and stratification). Section 3.3 describes the
procedure for data simulation used in this chapter, particularly the five scenarios
under which the ignorable treatment assignment assumption is violated to
varying degrees. Section 3.4 shows the biases of each method in estimating treatment effect under each of the five scenarios (i.e., the results of data simulation).

Section 3.5 summarizes the implications of data simulation. Section 3.6 is a succinct review of important aspects of running OLS regression, including important assumptions embedded in the OLS model, and a review of Berk's (2004) work on the pitfalls in running regression. Section 3.7 concludes with a summary of key points.

3.1 Why Is Data Balancing Necessary? A Heuristic Example

To illustrate the importance of controlling for covariates in observational studies—which is analogous to balancing data—we repeat the famous example first published by Cochran (1968) and repeatedly cited by others. Rubin (1997) also used this example to show the value of data balancing and the utility of stratification. The importance of controlling for covariates is underscored by Table 3.1, which presents a comparison of mortality rates for two groups of smokers (i.e., cigarette smokers, and cigar and pipe smokers) and one nonsmoking group that were measured in three countries: Canada, the United Kingdom, and the United States. Rubin (1997) argued that analyses of these data failed to control for age—a crucial covariate of mortality rate—and that, as a result, the observed data appear to show that cigarette smoking is *good* for health, especially relative to cigar and pipe smoking. For example, the Canadian data showed that the cigar and pipe smokers had a 35.5% mortality rate, whereas the mortality rates for the cigarette smokers and nonsmokers were not only similar but also much lower than those of the cigar and pipe smokers (20.5% and 20.2%, respectively). The pattern of mortality was consistent in the data from the other two countries. However, this finding is contradictory to our knowledge about the adverse consequences of smoking. Why is this the case? The primary reason is that age is an important confounding variable that affects the outcome of mortality rate. Note that the cigar and pipe smokers in Canada had the oldest average age (i.e., 65.9 years), while the average age of the nonsmokers was in the middle range for the three groups (i.e., 54.9 years), and the cigarette smokers had the youngest average age (i.e., 50.5 years). As presented, the unadjusted mortality rates are known as *crude death rates* because they do not take into account the age distribution of the three groups.

To conduct a meaningful evaluation, we must *balance the data* to control for covariance. Balancing implies data manipulation that, in this case, would render trivial the confounding effect of age. Balancing raises the question of what the mortality rates will look like if we force all three groups to follow the same age distribution. To address the balancing problem, both Cochran (1968) and Rubin (1997) used a method called *stratification*, which is synonymous with

Table 3.1 Comparison of Mortality Rates for Three Smoking Groups in Three Databases

	Canadian Study			United Kingdom Study			United States Study		
Variable	*Nonsmokers*	*Cigarette Smokers*	*Cigar and Pipe Smokers*	*Nonsmokers*	*Cigarette Smokers*	*Cigar and Pipe Smokers*	*Nonsmokers*	*Cigarette Smokers*	*Cigar and Pipe Smokers*
Mortality rates per 1,000 person-years, %	20.2	20.5	35.5	11.3	14.1	20.7	13.5	13.5	17.4
Average age (years)	54.9	50.5	65.9	49.1	49.8	55.7	57.0	53.2	59.7
Adjusted mortality rates using subclasses, %									
2 Subclasses	20.2	26.4	24.0	11.3	12.7	13.6	13.5	16.4	14.9
3 Subclasses	20.2	28.3	21.2	11.3	12.8	12.0	13.5	17.7	14.2
9–11 Subclasses	20.2	29.5	19.8	11.3	14.8	11.0	13.5	21.2	13.7

SOURCES: Cochran (1968, tables 1–3) and Rubin (1997, p. 758).

subclassification. The lower panel of Table 3.1 shows the adjusted mortality rates for all three groups after applying three schemes of stratification. Cochran tested three schemes by stratifying the sample into 2 subclasses, 3 subclasses, and 9 to 11 subclasses. All three subclassifications successfully removed estimation bias, under which the apparent advantages of cigarette smoking disappear, and nonsmokers are shown as the healthiest group.

The stratification method used by Cochran is one of the three methods that are the focus of this chapter, and we elaborate on stratification in subsequent sections. The key message conveyed by this example is that estimation bias can be substantial if the researcher fails to control for key covariates in data from observational studies.

There are a number of other methods available to accomplish the same objective of controlling for covariates. One popular method in demography is *age standardization.* The core idea of age standardization is simple: choose one age distribution from the three groups, and, for each age-group, multiply the proportion of persons in the group from the standard population to each age-specific death rate, then sum up all products. The resulting number is the adjusted mortality rate or standardized mortality rate for which age is no longer a confounding variable. To illustrate this concept, we created an artificial data set (Table 3.2) that simulates the same problem as found in Table 3.1. Note that in Table 3.2, the unadjusted mortality rate is the highest for cigar and pipe smokers, while the groups of cigarette smokers and nonsmokers appear to have the same rate. In addition, note that the simulated data follow the same pattern of age distribution as in the Table 3.1 data, where the cigar and pipe smokers are the oldest (i.e., 25% of this group were 61 years or older), while the cigarette smokers are the youngest (i.e., 8.33% of this group were 61 years or older).

To conduct age standardization, we can choose the age distribution from any one of the three groups as a standard. Even when a different age is chosen— that is, a different standard is selected—the results will be the same in terms of the impact of smoking on mortality rate. In our illustration, we selected the age distribution of the cigarette smokers as the standard. Table 3.3 shows the results of the standardization.

Using the age distribution of the cigarette smokers as the standard, we obtained three adjusted mortality rates (Table 3.3). The adjusted mortality rate of the cigarette smokers (i.e., 27.17 per 1,000) is the same as the unadjusted rate because both rates were based on the same age distribution. The adjusted mortality rate of the cigar and pipe smokers (or more formally "the adjusted mortality rate of cigar and pipe smokers with the cigarette smokers' age distribution as standard") is 26.06 per 1,000, which is much lower than the unadjusted rate of 38.5 per 1,000; and the adjusted mortality rate of the nonsmokers is 22.59, which is also lower than the unadjusted rate of 27.5 per 1,000.

Table 3.2   Artificial Data of Mortality Rates for Three Smoking Groups

Age	Nonsmokers		Cigarette Smokers		Cigar and Pipe Smokers		Nonsmokers		Cigarette Smokers		Cigar and Pipe Smokers	
	Number of Persons	Age Distribution	Number of Persons	Age Distribution	Number of Persons	Age Distribution	Number of Deaths	Age-Specific Death Rate (ASDR)	Number of Deaths	Age-Specific Death Rate (ASDR)	Number of Deaths	Age-Specific Death Rate (ASDR)
18–40	500	0.0833	1,500	0.2500	500	0.0833	15	0.0300	50	0.0333	16	0.0320
41–60	4,500	0.7500	4,000	0.6667	4,000	0.6667	40	0.0089	45	0.0113	55	0.0138
61+	1,000	0.1667	500	0.0833	1,500	0.2500	110	0.1100	68	0.1360	160	0.1067
Total	6,000		6,000		6,000		165		163		231	
Unadjusted mortality rate (per 1,000)								27.50		27.17		38.50

Table 3.3 Adjusted Mortality Rates Using the Age Standardization Method (i.e., Adjustment Based on the Cigarette Smokers' Age Distribution)

Age	Age Distribution of Cigarette Smokers	Age-Specific Death Rate			Standardization Using Cigarette Smokers' Age Distribution		
		Nonsmokers	Cigarette Smokers	Cigar and Pipe Smokers	Nonsmokers	Cigarette Smokers	Cigar and Pipe Smokers
	(a)	(b)	(c)	(d)	$(a \star b)$	$(a \star c)$	$(a \star d)$
18–40	0.2500	0.0300	0.0333	0.0320	0.0075	0.0083	0.0080
41–60	0.6667	0.0089	0.0113	0.0138	0.0059	0.0075	0.0092
61+	0.0833	0.1100	0.1360	0.1067	0.0092	0.0113	0.0089
Adjusted mortality rate (per 1,000)					22.59	27.17	26.06

Simple corrective methods such as covariate standardization often work well, if the researcher is confident that the sources of confounding are known, and that all confounding variables are observed (i.e., measured). When confounding is overtly understood and measured, Rosenbaum (2002b) refers to it as *overt bias*. Indeed, overt bias may be well removed by simple approaches. The following section describes three methods that are equally straightforward and work well when sources of bias are understood and observed in measurement.

3.2 Three Methods of Data Balancing

This section formally describes three conventional methods that help balance data—that is, OLS regression, matching, and stratification. Our interest centers on the key questions of how each method operates to balance data and to what extent each method accomplishes this goal.

3.2.1 THE ORDINARY LEAST SQUARES REGRESSION

The material presented in this section is not new and can be found in most textbooks on regression. Let $Y = X\beta + e$ represent a population regression model, where Y is an $(n \times 1)$ vector of the dependent variable for the n participants, X is an $(n \times p)$ matrix containing a unit column (i.e., all elements in the column take value 1) and $p - 1$ independent variables, e is an $(n \times 1)$ vector of the error term, and β is a $(p \times 1)$ vector of regression parameters containing one intercept and $p - 1$ slopes. Assuming repeated sampling and fixed X, and $e \sim iid$, $N(0, \sigma^2 I_n)$, where I_n is an $(n \times n)$ identity matrix and σ^2 is a scalar, so that $\sigma^2 I_n = E(ee')$ is the variance-covariance matrix of the error term. With the observed data of Y and X, we can use the least squares criterion to choose the estimate of the parameter vector β that makes the sum of the squared errors of the error vector e a minimum, that is, we minimize the quadratic form of the error vector:

$$\ell = e'e = (Y - X\beta)'(Y - X\beta)$$
$$= Y'Y - 2\beta'X'Y + \beta'X'X\beta.$$

Taking the partial derivative of ℓ with respect to β,

$$\frac{1}{2}\frac{\partial \ell}{\partial \beta} = X'X\beta - X'Y,$$

and letting the partial derivative be zero,

$$\frac{1}{2}\frac{\partial \ell}{\partial \beta} = 0,$$

we obtain the optimizing vector β, that is, $\beta = (X'X)^{-1}X'Y$. If we have sample data and use lowercase letters to represent sample variables and statistics, we have the sample estimated vector of regression coefficients as

$$b = (x'x)^{-1}x'y.$$

The variance-covariance matrix of b, $s^2\{b\}$ of an order $(p \times p)$, can be obtained by the following matrix:

$$s^2\{b\} = MSE(x'x)^{-1} = \begin{bmatrix} s^2\{b_0\} & & \\ s\{b_1, b_0\} & s^2\{b_1\} & \\ \cdots & \cdots & \cdots \\ s\{b_{p-1}, b_0\} & \cdots & s\{b_{p-1}, b_{p-2}\} & s^2\{b_{p-1}\} \end{bmatrix},$$

where $MSE = SSE/(n - p)$, and $SSE = y'[I - H]y$, in which I is an identity matrix with the order of $(n \times n)$, and $H_{(n \times n)} = x(x'x)^{-1}x'$. Taking the square root of each estimated variance, the researcher obtains standard errors (se) of the estimated regression coefficients, as $se(b_0) = \sqrt{s^2\{b_0\}}$, $se(b_1) = \sqrt{s^2\{b_1\}}$, $se(b_2) = \sqrt{s^2\{b_2\}}$, $se(b_{p-1}) = \sqrt{s^2\{b_{p-1}\}}$.

With estimated regression coefficients and corresponding standard errors, we can perform statistical significance tests or estimate the confidence intervals of the coefficients as follows:

1. *Two-tailed test (when hypothetical direction of β is unknown):*

   H_0: $\beta_1 = 0$ (meaning: X_1 has no impact on Y),

   H_a: $\beta_1 \neq 0$ (meaning: X_1 has an impact on Y).

   Calculate $t^* = b_1/se\{b_1\}$. If $|t^*| \leq t(1 - \alpha/2; n - p)$, conclude H_0; if $|t^*| > t(1 - \alpha/2; n - p)$, conclude H_a.

2. *One-tailed test (when one can assume a direction for β based on theory):*

   H_0: $\beta_1 \leq 0$ (meaning: X_1 has either a negative or no impact on Y),

   H_a: $\beta_1 > 0$ (meaning: X_1 has a positive impact on Y).

   Calculate $t^* = b_1/se\{b_1\}$. If $|t^*| \leq t(1 - \alpha; n - p)$, conclude H_0; if $|t^*| > t(1 - \alpha; n - p)$, conclude H_a.

3. *Confidence interval of b:*

   $(1 - \alpha) \times 100\%$ confidence interval can be computed as follows:

$$b_1 \pm t\left(1 - \frac{\alpha}{2}; n - p\right)[se(b_1)]$$

$$b_2 \pm t\left(1 - \frac{\alpha}{2}; n - p\right)[se(b_2)]$$

$$\vdots$$

$$b_{p-1} \pm t\left(1 - \frac{\alpha}{2}; n - p\right)[se(b_{p-1})].$$

The Gauss-Markov theorem reveals the *best linear unbiased estimator* or BLUE property of the OLS regression. The theorem states that given the assumptions of the classical linear regression model, the least squares estimators, in the class of unbiased linear estimators, have minimum variance, that is, they are BLUEs.

Using this setup for the OLS regression estimator, the researcher can control for the covariates and balance data. We now rewrite the regression model $Y = X\beta + e$ to $Y = \alpha + \tau W + X_1\beta + e$ by treating W as a separate variable and taking it out from the original matrix X. Here W is a dichotomous variable indicating treatment condition (i.e., $W = 1$, if treated; and $W = 0$ otherwise). Thus, X_1 now contains one variable less than X and contains no unit column. If the model is appropriately specified—that is, if the X_1 contains all variables affecting outcome Y (i.e., one or more of the threats to internal validity have been taken care of by a correct formulation of the X_1 matrix)—and if all variables take a correct functional form, then τ will be an unbiased and consistent estimate of the average treatment effect of the sample, or $\hat{\tau} = E(\hat{y}_1|w = 1) - E(\hat{y}_0|w = 0)$. However, it is important to note that we can draw this conclusion only when all other assumptions of OLS regression are met, which is questionable under many conditions. For now, we will assume that all other assumptions of OLS regression have been met. Recall the interpretation of a regression coefficient: Other things being equal (or ceteris paribus), a one-unit increase in variable x_1 will decrease (or increase, depending on the sign of the coefficient) b_1 units in the dependent variable y. This is an appealing feature of OLS: By using least squares minimization, a single regression coefficient captures the net impact of an independent variable on the dependent variable. This mechanism is exactly how OLS regression controls for covariates and balances data. If the researcher successfully includes all covariates, and if the regression model meets other assumptions, then $\hat{\tau}$ is an unbiased and consistent estimate of the average treatment effect.

In sum, the key feature of OLS regression, or more precisely the mechanism for balancing data through regression, is to include important covariates in the regression equation and to ensure that the key assumptions embedded in the regression model are plausible. By doing so, the regression coefficient of the dichotomous treatment variable indicates the average treatment effect of the sample. Employing the regression approach requires making strong assumptions that are often violated in the real world. Key assumptions and issues related to running regressions are examined in Section 3.6.

3.2.2 MATCHING

Before the development of the new estimation methods for observational data, matching was the most frequently used conventional method for handling observational data. For instance, Rossi and Freeman (1989) described how to balance data by conducting ex post matching based on observed covariates. Matching can be performed with or without stratification (Rosenbaum, 2002b, p. 80). The central idea of the method is to match each treated participant ($x_i | w_i = 1$) to n nontreated participants ($x_j | w_j = 0$) on x, where x is a vector of matching variables (i.e., variables that covary with the treatment), and then compare the average of y of the treated participants with the average of y of the matched nontreated participants. The resultant difference is an estimate of the sample average treatment effect. Under this condition, the standard estimator may be rewritten as

$$\hat{\tau}_{match} = E(\hat{y}_{match,1} | w_{match} = 1) - E(\hat{y}_{match,0} | w_{match} = 0) \, ,$$

where the subscript "match" indicates matched subsample. For $w_{match} = 1$, the group comprises all treated participants whose matches are found (i.e., the group excludes treated participants without matches), and for $w_{match} = 0$, the group is composed of all nontreated participants who were matched to treated participants. Denoting M as the original sample size, M_{match} as the matched sample size, $N_{w=1}$ the number of treated participants before matching, $N_{w=0}$ the number of untreated participants before matching, $N_{match,w=1}$ the number of treated participants after matching, and $N_{match,w=0}$ the number of untreated participants after matching, we have $N_{match,w=1} < N_{w=1}$ (because some treated participants cannot find a match in the nontreated group), and $M_{match} < M$ (because of the loss of both treated and untreated participants due to nonmatching). The following equality is also true: $N_{match,w=0} = n(N_{match,w=1})$ depending on how many matches (i.e., n) the researcher chooses to use. If the researcher chooses to match one participant from the untreated pool to each treated participant (i.e., $n = 1$), then $N_{match,w=0} = N_{match,w=1}$; if one chooses to match four participants from the untreated pool to each treated participant (i.e., $n = 4$), then $N_{match,w=1} = 4(N_{match,w=0})$, or $N_{match,w=0}$ is four times as large as $N_{match,w=1}$.

The choice of n deserves scrutiny. If we choose $n = 1$ to perform a one-to-one match, then we lose variation in multiple matches who all match to a treated participant. If we choose a large n such as $n = 6$, then we may not find six matches for each treated participant. Abadie et al. (2004) developed a procedure that allows the researcher to specify a maximum number of matches. Abadie and Imbens's (2002) data simulation suggested that $n = 4$ (i.e., matching up to 4 untreated participants for each treated participant) usually worked well in terms of mean-squared error. Rosenbaum (2002b) recommended the use of optimal matching to solve the problem. These issues are reviewed and extended in later chapters.

In sum, the key feature of matching, or more precisely the mechanism for balancing data through matching, involves identifying untreated participants who are similar on covariates to treated participants and using the mean outcome of the nontreated group as a proxy to estimate the counterfactual of the treated group.

3.2.3 STRATIFICATION

Stratification is a procedure that groups participants into strata on the basis of a covariate x (Rosenbaum, 2002b). From the M participants, select $N < M$ participants and group them into S nonoverlapping strata with n_s participants in stratum s. In selecting the N units and assigning them to strata, use only the x variable and possibly a table of random numbers. A stratification formed in this way is called *stratification on x*. In addition, *an exact stratification on x* has strata that are homogeneous in x, so two participants are included in the same stratum only when both have the same value of x; that is, $x_{si} = x_{sj}$ for all s, i, and j. Exact stratification on x is practical only when x is of low dimensionality and its coordinates are discrete; otherwise, it will be difficult to locate many participants with the same x. For instance, exact stratification is feasible when we have a discrete age variable with three groups ($x = 1$, if ages 18 to 40 years; $x = 2$, if ages 41 to 60 years; and $x = 3$, if ages 61 years or older).

For the purpose of balancing data, we introduce a stratification procedure based on the percentile of x, called *quartile* or *quintile stratification*, that is designed to handle situations where x is a continuous variable. The quintile stratification procedure involves the following five steps:

1. Sort data on x so that all participants are in an ascending order of x.

2. Choose participants whose values on x are equal to or less than the first quintile of variable x to form the first stratum (i.e., Stratum 1 contains all participants whose x values are equal to or less than the value of the 20th percentile of x); then choose participants whose values on x fall into the range bounded by the second quintile to form the second stratum (i.e., Stratum 2 contains all participants whose x is in a range between values of the 21st percentile and the 40th percentile of x); keep working in this fashion until the fifth stratum is formed.

3. For each stratum s ($s = 1, 2, \ldots, 5$), perform the standard estimator to calculate the difference of mean outcome y between the treated and nontreated groups, that is, $\hat{\tau}_s = E(\hat{y}_{s1} | w_s = 1) - E(\hat{y}_{s0} | w_s = 0)$, where $s = 1, 2, 3, 4, 5$.

4. Calculate the arithmetic mean of the five means to obtain the average treatment effect of the sample using the equation

$$\hat{\tau} = \sum_{s=1}^{S} \frac{n_s}{M} \left[\overline{Y}_{1s} - \overline{Y}_{0s} \right],$$

and calculate the variance of the estimated average treatment effect using the equation

$$\mathrm{Var}(\hat{\tau}) = \sum_{s=1}^{S} \left(\frac{n_s}{M} \right)^2 \mathrm{Var} \left[\overline{Y}_{1s} - \overline{Y}_{0s} \right].$$

5. Use these statistics to perform a statistical significance test to determine whether the sample average treatment effect is statistically significant.

Quartile stratification is performed in a similar fashion, except the statistic quartile is used and four strata are created.

There are more sophisticated methods involving stratification. For instance, in the example using data from the three smoking groups, Cochran's (1968) method for balancing the mortality rate data (see Table 3.1) was to choose the number of strata so that each stratum contained a reasonable number of participants from each treatment condition. This explains why there were three stratification schemes in Table 3.1 (i.e., 2 subclasses, 3 subclasses, and 9 to 11 subclasses). Cochran offered theoretical results showing that when the treatment and the control groups overlapped in their covariate distributions (i.e., age in his example), comparisons using five or six subclasses typically removed 90% or more of the bias presented in the raw comparisons.

In sum, the key feature of stratification, or more precisely the mechanism for balancing data through stratification, is to make participants within a stratum as homogeneous as possible in terms of an observed covariate; then the mean outcome of the nontreated group is used as a proxy to estimate the counterfactual for the treated group. Exact stratification produces homogeneity only for discrete covariates. For instance, if gender is a covariate, stratification will produce two homogeneous strata: all females in one stratum and all males in another. However, this is not the case for a continuous covariate. With a continuous covariate x and using a quintile stratification, participants within stratum $s = 1$ are not exactly the same in terms of x. In these circumstances, we must assume that within-stratum differences on x are ignorable, and participants falling into the same stratum are "similar enough." The level of exactness on x can be improved by increasing the number of strata S, but the literature has suggested that $S = 5$ typically works well (Rosenbaum & Rubin, 1984, 1985; Rubin, 1997).

3.3 Design of the Data Simulation

To help demonstrate the importance of data balancing, and how conventional methods can be used to correct for bias under different settings of data generation, we designed a data simulation with a threefold purpose. First, we show how conventional correction methods work when selection bias is overt and is controlled properly. Second, we use the simulation to show conditions under which the conventional methods *do not work*. From this, we demonstrate the need for more sophisticated methods for balancing data. Third, we use these simulated data and their conditions as an organizing theme for the book. We illustrate the reasons why new methods were developed, the specific problem each method was designed to resolve, and the contributions made by each method.

Unlike using real data, the simulation approach creates artificial data based on designed scenarios. The advantage of using artificial data is that the true value of the treatment effect is known in advance. This allows us to evaluate bias directly. In addition, this simulation assumes that we are working with population data, so sampling variability is ignored. For now, we do not examine the sensitivity of estimated standard errors. However, we return to this issue in Chapter 8.

The data generation is based on the following regression model:

$$y_i = \alpha + \beta x_i + \tau w_i + e_i,$$

where x is a covariate or control variable, and w is a dummy variable indicating treatment conditions ($w = 1$ treated, and $w = 0$ control). The simulation creates data of y, x, w, and e according to the following known parameters: $\alpha = 10$, $\beta = 1.5$, and $\tau = 2$. That is, we created data using the following equation:

$$y_i = 10 + 1.5x_i + 2w_i + e_i.$$

We describe five scenarios that assume different relationships among y, x, w, and e. In each of the scenarios, the sample size is fixed at 400. Because the parameters are known, we can compare the model estimated treatment effect $\hat{\tau}$ with the true value $\tau = 2$ to evaluate bias for each scenario. The five scenarios are described below.

Scenario 1: $w \perp e|x$, $x \perp e$, $x \perp w$, *and* $e \sim$ *iid,* $N(0, 1)$. This scenario assumes the following four conditions: (1) conditional on covariate x, the treatment variable w is independent of the error term e, (2) there is no correlation between covariate x and the error term e, (3) there is no correlation between covariate x and the treatment variable w, and (4) the error term is identically and independently distributed and follows a normal distribution with mean of zero and variance of one. This is an ideal condition. The crucial assumption is $w \perp e|x$, which simulates the ignorable treatment assignment. In addition, this models the data generated from a randomized study, because these conditions are likely to be met only in data obtained from a randomized experiment. To ensure that the data generation strictly meets the assumption of $w \perp e|x$, we force x to be an ordinal variable taking values 1, 2, 3, 4. The remaining scenarios simulate conditions that are not ideal and involve relaxing one or more of the assumptions included in Scenario 1.

Scenario 2: $\rho_{we} \neq 0|x$, $x \perp e$, $x \perp w$, *and* $e \sim$ *iid, N(0, 1).* In this scenario, the ignorable treatment assignment assumption (i.e., $w \perp y|x$, which is reflected by $w \perp e|x$) is violated. The scenario indicates that conditional on covariate x, the correlation between the treatment variable w and the error term e is not equal to zero ($\rho_{we} \neq 0|x$). All other conditions of Scenario 2 are identical to those of Scenario 1. Thus, Scenario 2 simulates the contemporaneous correlation between the error term and the treatment variable. We expect that the estimated treatment effect will be biased and inconsistent.[1]

Scenario 3: $\rho_{wx} \neq 0$, $x \perp e$, $w \perp e$, *and* $e \sim$ *iid, N(0, 1).* This scenario relaxes the Scenario 1 condition of independence between the covariate x and the treatment variable w. To simulate real situations, we also allow the covariate x to be a continuous variable to take values on more than four ordinal levels. All other conditions in Scenario 3 remain the same as those in Scenario 1. Scenario 3 simulates the condition of multicollinearity, which is a typical occurrence when using real data. In this scenario, the ignorable treatment assignment assumption still holds because x is the only source of the correlation with w, and x is used as a control variable. Thus, we can expect that the results will be unbiased, although it is interesting to see how different methods react to multicollinearity.

Scenario 4: $\rho_{we} \neq 0|x$, $\rho_{wx} \neq 0$, $x \perp e$, *and* $e \sim$ *iid, N(0, 1).* This scenario differs from Scenario 1 by relaxing two of the conditions: (1) the independence between the covariate x and the treatment variable w and (2) the independence between the treatment variable w and the error term e. Similar to Scenario 3, the covariate x is a continuous variable. All other conditions in Scenario 4 remain the same as those in Scenario 1. In addition, Scenario 4 differs from Scenario 3 in that it relaxes the assumption of independence between the treatment variable w and the error term e (i.e., it changes $w \perp e$ to $\rho_{we} \neq 0|x$). Thus, under this scenario ignorable treatment assignment no longer holds, which makes Scenario 4 similar to Scenario 2. In addition, Scenario 4 simulates two data problems: (1) linear correlation between independent variables and (2) the nonignorable treatment assignment. We expect that the estimated treatment effect will be biased and inconsistent.

Scenario 5: $\rho_{xe} \neq 0$, $\rho_{we} \neq 0$, $\rho_{wx} \neq 0$, and e ~ iid, N(0, 1). Scenario 5 relaxes three of the conditions: (1) independence between the covariate x and the treatment variable w, (2) independence between the treatment variable w and the error term e, and (3) independence of the covariate x and the error term e. In addition, we allow the covariate x to be a continuous variable. This scenario is a further relaxation of assumptions embedded in Scenario 4 (i.e., it changes $x \perp e$ to $\rho_{xe} \neq 0$), and this is the worst-case scenario. Because the ignorable treatment assignment assumption is violated, and the problem of multicollinearity is present, we expect the estimated treatment effect will be biased and inconsistent.

The programming syntax for the data simulation is available at the companion Web page for this book. We employed the simple matching estimator (Abadie, et al., 2004) to produce results. Abadie et al.'s algorithm is slightly different from the conventional matching estimator because it matches with replacement (see Chapter 6). Readers may replicate the analysis to verify the findings.

3.4 Results of the Data Simulation

Table 3.4 presents descriptive statistics for the data and estimated treatment effects produced using the three methods under Scenario 1. The descriptive statistics show that conditions described by the design of Scenario 1 are met: that is, w is not correlated with e ($\rho_{we} = .01$); x is not correlated with e ($\rho_{xe} = .03$); x is not correlated with w ($\rho_{xw} = .03$); the mean of e is close to 0; and the standard deviation of e is close to 1. Figure 3.1 is the scatterplot of data x and y under Scenario 1.

The results show that under the ideal conditions of Scenario 1, all three methods accurately estimate the treatment effect, and all biases are close to zero. Among the three methods, regression worked the best with a bias of .011; stratification worked second best with a bias of .016; and matching worked relatively the worst with a bias of .029. Note that although the three methods are technically different, each method estimated a treatment effect very close to the true parameter. The conditions of data generation under Scenario 1 are stringent. In reality, a well-designed and -implemented randomized experiment is likely the only kind of study that could produce data in such an ideal fashion.

Table 3.5 presents descriptive statistics of the data and estimated treatment effects using three methods under Scenario 2. The descriptive statistics show that conditions described by Scenario 2 are met: that is, w is correlated with e ($\rho_{we} = .61$); x is not correlated with e ($\rho_{xe} = -.04$); x is not correlated with w ($\rho_{xw} = .02$); the mean of e is close to 0; and the standard deviation of e is close to 1. Figure 3.2 is the scatterplot of data x and y under Scenario 2.

Table 3.4 Data Description and Estimated Effects by Three Methods: Scenario 1

| | Description of Variables | | | | Estimated Effect | | | | | |
| | Correlation, Mean, & SD | | | | Regression | | Matching | | Stratification | |
	y	x	w	e	Variable	B	n	Tx Effect	Stratum	Tx Effect
y	—				x	1.529	n = 1	2.029	1	2.251
x	.78	—			w(Tx effect)	2.011	n = 4	2.029	2	1.587
w	.47	.03	—		Constant	9.967			3	1.978
e	.46	.03	.01						4	2.246
Mean	14.876	2.500	.540	.046					Sample	2.016
					Bias	0.011	Bias n = 1	0.029	Bias	0.016
SD	2.226	1.119	.499	.963			Bias n = 4	0.029		

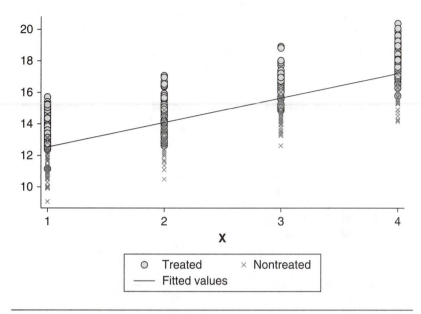

Figure 3.1   Scatterplot of Data Under Scenario 1

Under the conditions of Scenario 2, each of the three methods produced a biased estimate of the treatment effect. Specifically, all three methods estimated a treatment effect that was 1.13 units higher than the true value, or a 57% overestimation. In particular, the OLS result shows that the bias is in an upward direction (i.e., inflated treatment effect), when the treatment variable w is positively correlated with the error (i.e., $\rho_{we} = .61$). The issue of inflated treatment effect was discussed theoretically in Chapter 2, and we showed that when the error term and the independent variable are positively correlated, the OLS estimated slope is higher than the true slope (see Figure 2.1). This upward bias is exactly what happens when conditions are similar to those in Scenario 2.

Imagine a real study such as Scenario 2. In an observational data collection (where an independent variable of interest is not manipulated), suppose w is a state indicating addiction to illegal drugs, y is a measure of mental status (i.e., a high value on y indicates more psychological problems), and the data were obtained by surveying a representative sample of a population. Because the data are observational and exist naturally (i.e., representing a population) without randomization, participants who used illegal drugs were likely to have high values on y. Although x is an important covariate of y (i.e., $\rho_{xy} = .69$), specifying only one of such covariates in the regression model is not sufficient. When we note that there is a high correlation between the treatment and the error term (i.e., $\rho_{we} = .61$), it becomes clear that additional covariates were omitted in the regression model. Under this condition, in addition to nonignorable

Table 3.5 Data Description and Estimated Effects by Three Methods: Scenario 2

Description of Variables					Estimated Effect					
Correlation, Mean, & SD					Regression		Matching		Stratification	
	y	x	w	e	Variable	B	n	Tx Effect	Stratum	Tx Effect
y	—				x	1.455	n = 1	3.129	1	3.084
x	.69	—			w(Tx effect)	3.130	n = 4	3.129	2	3.205
w	.67	.02	—		Constant	9.495			3	3.075
e	.61	−.04	.61	—					4	3.159
Mean	14.777	2.500	.525	−.023					Sample	3.131
SD	2.396	1.119	.500	.932	Bias	1.13	Bias n = 1	1.129	Bias	1.131
							Bias n = 4	1.129		

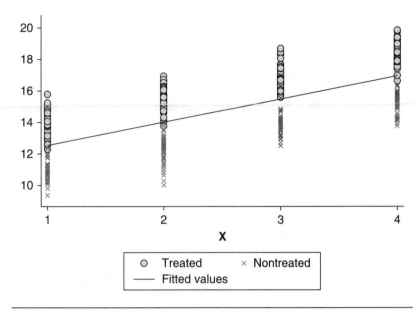

Figure 3.2   Scatterplot of Data Under Scenario 2

treatment assignment, we encounter the problem of *omitted covariates*, which is also known as *selection on unobservables* (see Section 2.3). As a result of the combination of these two problems, the estimated treatment effect (i.e., the net impact of addiction to illegal drugs on psychological problems) is biased upward. Although theory might tell us that abusing drugs causes mental problems, our data would overestimate the impact of drug abuse on mental status. This overestimation is true regardless of which one of the three analytic methods is used. It is for this reason that we can conclude that conventional correction methods are not appropriate (i.e., do not work well) in conditions such as those described for Scenario 2. Under this scenario, the researcher must consider using more advanced analytic approaches, though these too are subject to bias as a result of unobserved heterogeneity, and it is advisable to conduct sensitivity analyses to assess the potential effect of hidden selection. We will discuss sensitivity analyses in Chapter 8.

Table 3.6 presents descriptive statistics of the data and estimated treatment effects produced using three analytic methods under the conditions of Scenario 3. The descriptive statistics show that conditions described by Scenario 3 are met: that is, w is correlated with x ($\rho_{wx} = .56$); x is not correlated with e ($\rho_{xe} = .07$); w is not correlated with e ($\rho_{we} = .03$); the mean of e is close to 0; and the standard deviation of e is close to 1. Figure 3.3 is the scatterplot of data x and y under Scenario 3.

Table 3.6   Data Description and Estimated Effects by Three Methods: Scenario 3

| | Description of Variables | | | | Estimated Effect | | | | | |
| | Correlation, Mean, & SD | | | | Regression | | Matching | | Stratification | |
	y	x	w	e	Variable	B	n	Tx Effect	Stratum	Tx Effect
y	—				x	1.574	$n=1$	2.052	1	2.384
x	.86	—			w(Tx effect)	1.975	$n=4$	2.075	2	2.303
w	.75	.56	—		Constant	9.931			3	1.864
e	.44	.07	.03	—					4	2.518
									Sample	2.267
Mean	11.119	.081	.538	-.077	Bias	-0.025	Bias $n=1$	0.052	Bias	0.267
SD	2.488	1.009	.499	.971			Bias $n=4$	0.057		

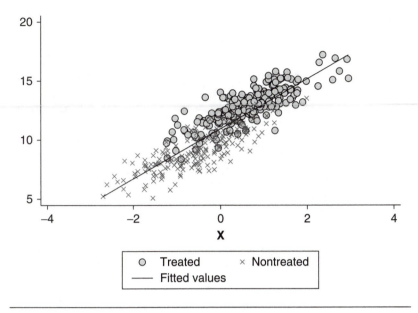

Figure 3.3   Scatterplot of Data Under Scenario 3

Under Scenario 3, the ignorable treatment assignment still holds; therefore, the estimated treatment effect produced using the regression method is unbiased. Furthermore, the presence of multicollinearity does not affect the estimation of treatment effect, even though it affects the significance test; however, the significance test is irrelevant in the current discussion about the population parameters. These results confirm that among the three analytic methods, regression worked the best, with a bias of −.025 (i.e., a bias slightly downward); matching worked reasonably well, with a bias of .052; and stratification worked in a problematic way with a bias of .267 (or 13% larger than the true value). This example suggests that the methods react differently to multicollinearity and are not equally good in terms of bias correction.

Table 3.7 presents descriptive statistics and estimated treatment effects using three methods under Scenario 4. The descriptive statistics show that conditions described by Scenario 4 are met: that is, w is correlated with e ($\rho_{we} = .59$); w is correlated with x ($\rho_{wx} = .55$); x is not correlated with e ($\rho_{xe} = .06$); the mean of e is close to 0; and the standard deviation of e is close to 1. Figure 3.4 is the scatterplot of data x and y under Scenario 4. When this scatterplot is compared with the scatterplot for Scenario 3, we see a more systematic concentration of data points, as all treated cases are clustered in the upper panel, and all untreated cases are clustered in the lower panel. This pattern is produced by the correlation between w and e. This scenario is more likely to occur in real applications when researchers think that they have controlled for covariates, but the available

Table 3.7   Data Description and Estimated Effects by Three Methods: Scenario 4

Description of Variables					Estimated Effect						
					Regression		Matching		Stratification		
Correlation, Mean, & SD					Variable	B	n	Tx Effect	Stratum	Tx Effect	
	y	x	w	e							
y	—				x	1.095	n = 1	3.866	1	4.237	
x	.77	—			w(Tx effect)	3.647	n = 4	3.918	2	3.651	
w	.90	.55	—		Constant	9.178			3	3.794	
e	.63	.06	.59	—					4	4.416	
Mean	11.260	.119	.535	.011	Bias	1.647	Bias n = 1	1.866	Sample	4.024	
SD	2.678	.976	.499	1.023			Bias n = 4	1.918	Bias	2.024	

76

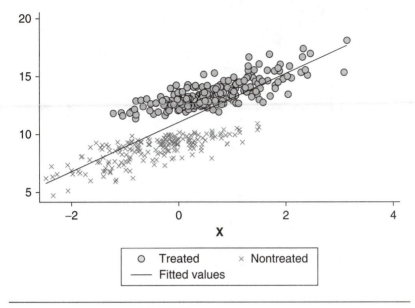

Figure 3.4   Scatterplot of Data Under Scenario 4

controls are not sufficient and/or relevant covariates are omitted. Thus, we encounter the same problem of selection on unobservables as that in Scenario 2.

These results show that all three methods are biased. Under the conditions of Scenario 4, the three methods are ranked for bias as follows: regression produced the lowest bias of 1.647, matching produced higher bias at 1.866, and stratification produced the highest bias at 2.024. The findings are clear. When treatment assignment is not ignorable and when multicollinearity is present, no conventional method produces unbiased estimation of treatment effects.

Table 3.8 presents descriptive statistics and estimated treatment effects using three methods under Scenario 5. The descriptive statistics show that conditions described by the design of Scenario 5 are met: that is, x is correlated with e ($\rho_{xe} = .74$); w is correlated with e ($\rho_{we} = .56$); w is correlated with x ($\rho_{wx} = .57$); the mean of e is close to 0; and the standard deviation of e is close to 1. Figure 3.5 is the scatterplot of data x and y under Scenario 5.

Note that Scenario 5 relaxes one assumption of Scenario 4 (i.e., it changes $x \perp e$ to $\rho_{xe} \neq 0$). Under this scenario, both w and x correlate with the error term, and the two independent variables w and x correlate with one another. Among all five scenarios, Scenario 5 assumes the weakest conditions for data generation, which would lead us to expect that the results would be the worst. Indeed, the estimated treatment effects produced with each method are biased, and the methods are ranked in terms of bias as follows: regression produced the lowest bias (i.e., .416); matching produced the second lowest bias estimates

Table 3.8 Data Description and Estimated Effects by Three Methods: Scenario 5

Description of Variables					Estimated Effect					
					Regression		Matching		Stratification	
Correlation, Mean, & SD	y	x	w	e	Variable	B	n	Tx Effect	Stratum	Tx Effect
y	—				x	2.156	n = 1	2.492	1	3.044
x	.92	—			w(Tx effect)	2.416	n = 4	2.747	2	2.441
w	.79	.57	—		Constant	9.758			3	2.689
e	.88	.74	.56	—					4	3.025
Mean	11.281	.095	.545	.048	Bias	0.416	Bias n = 1	0.492	Sample	2.800
SD	3.064	.991	.499	1.036			Bias n = 4	0.747	Bias	0.800

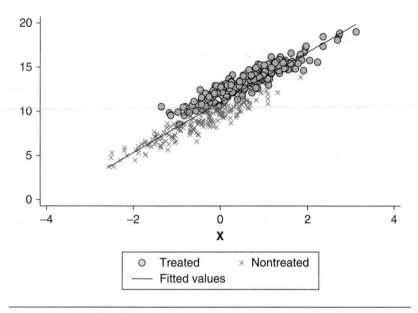

Figure 3.5   Scatterplot of Data Under Scenario 5

(i.e., .492 for $n = 1$ and .747 for $n = 4$, respectively); and stratification produced the highest bias (i.e., .800). However, contrary to our expectation, all biases were lower than those produced by the methods under Scenario 4. Although Scenario 5 is the worst-case scenario, the results are not the most severe. This interesting finding indicates that data conditions often work in complicated ways, and results may not conform to our expectations. Overall, our findings underscore the importance of assumptions, the risk of biased parameter estimation when conventional corrective methods are used, and the need to develop models with greater sophistication.

3.5 Implications of the Data Simulation

Our data simulation, or a three-methods-and-five-scenarios design, illuminates the conditions under which common data balancing strategies correct selection bias. We purposely created challenging data environments, but they are not dissimilar to those encountered in routine program evaluation. Contending with such challenges has motivated statisticians and econometricians to seek new methods. The data simulation has at least three implications in developing these methods.

First, simple methods of data balancing work well only under ideal conditions (i.e., Scenario 1), and under ideal conditions, all three methods work equally well. When the ignorable treatment assignment assumption holds but independent variables are correlated (i.e., Scenario 3), the regression and matching methods provide unbiased estimation of treatment effect. However, estimation produced with the stratification method is problematic. When the ignorable treatment assignment assumption is violated (i.e., Scenarios 2, 4, and 5), none of the three conventional correction methods provides an unbiased estimation of the treatment effect.

Second, covariance control does not automatically correct for nonignorable treatment assignment. In all five scenarios, x is highly correlated with y (i.e., ρ_{xy} ranges from .69 to .92), indicating that x is an important covariate of y in all scenarios. However, there are only two scenarios under which the ignorable treatment assignment assumption holds (i.e., Scenarios 1 and 3). The data simulation clearly shows that only under these two scenarios did three conventional methods produce unbiased estimates. Under all other scenarios, the three methods failed. Thus, common methods controlling for covariance do not necessarily correct for nonignorable treatment assignment.

Finally, the findings suggest that we must understand data generation before running regressions and other statistical models. As Berk (2004) noted, data generation is important for inference drawn from analyses:

> [Both statistical inference and causal inference] depend fundamentally on how the data used in the regression analysis were generated: How did nature and the investigator together produce the data then subjected to a regression analysis? Was there a real intervention? Can the data be properly viewed as a random sample from a real population? How were key variables measured? There is precious little in the data themselves that bear on such matters. One must bring to the regression analysis an enormous amount of information about data, and this information will, in practice, do some very heavy lifting. (pp. 1–2)

In addition, the three-methods-and-five-scenarios design offers a useful tool for thinking about the key features of new approaches for evaluation. We use this design as an organizing theme to assess the methods described in this book.

1. In the data simulation, we only have one control variable x, whereas in practice there are typically many control variables. When the number of control variables increases, the conventional ex post matching often fails because, under such conditions, it is difficult to match treated cases to untreated cases on multiple characteristics: This problem is known as the *dimensionality of matching*. A simple example helps illustrate this problem. If you use only three demographic variables (i.e., age prior to receiving treatment, gender, and race/ethnicity), you may be able to match 90% of the treated participants to the untreated participants, assuming that the sample is of a reasonably large size (e.g., $N = 500$) and the two groups have equal size. Even with three matching variables, chances are that some cases would be lost because they are so different that no match would be available. Depending on the nature of the sample and,

of course, depending on how accurate you want matching on age to be (i.e., Should matches be exact with, say, 25.09-year-olds, or could the age be matched within an age range, such as matching participants who fall within 20 to 30 years of age?), 90% is merely a guess for the percentage of possible matches. Now suppose you add one matching variable, which is a depression scale ranging from 40 to 100. With this addition, it is likely that you will match still fewer treated participants to the untreated participants. With every added dimension, the percentage of successful matches declines.

This dimensionality problem motivated researchers to develop several new approaches. These include (a) a two-step estimator that uses probabilities of receiving treatment (Heckman, 1978, 1979; Maddala, 1983), (b) propensity score matching (Rosenbaum & Rubin, 1983), and (c) the use of vector norms (Abadie & Imbens, 2002, 2006; Rubin, 1980a). These approaches to the problem of dimensionality in matching are examined in Chapters 4, 5, and 6, respectively.

2. Propensity score matching may be thought of as a slightly more complex method that combines the two conventional methods of regression and matching. That is, the analyst first creates propensity scores for all study participants, such that multiple characteristics are reduced to a one-dimensional score. The analyst then matches the scores between treated and nontreated cases to create a new sample. Last, the analyst performs a secondary analysis, such as regression, on the matched sample. Note that in the second stage, many kinds of multivariate analysis may be performed (e.g., regression-type models such as the random coefficients model, multiple-group structural equation modeling, survival analysis, generalized linear models). This method of propensity score matching is described in Chapter 5.

3. Propensity score matching can also be thought of as a combination of matching and stratification. When multiple matching variables are reduced to a one-dimensional propensity score, the analyst can match treated cases to nontreated cases on the one-dimensional score, and then use stratification or subclassification to estimate treatment effect. As such, the analyst does not use regression or other types of second-stage modeling (Rosenbaum & Rubin, 1984). This method of propensity score matching is also described in Chapter 5.

4. The analyst can also use a single method of matching to estimate the treatment effect (i.e., a method without using regression and stratification). Sophisticated approaches have been developed to improve matching as a method for estimation of the treatment effect. Abadie and Imbens (2002, 2006) developed *matching estimators* that use the vector norm (i.e., the Mahalanobis metric distance or the inverse of sample variance matrix) to estimate the average treatment effect for the treated group, the control group, and the sample, and similar effects for the population. The matching estimator method is described in Chapter 6. Heckman, Ichimura, and Todd (1997, 1998)

developed an alternative approach for matching using *nonparametric regression,* or more precisely, *local linear regression,* to match treated participants to untreated participants. This method provides a robust and efficient estimator of the average treatment effect for the treated group. Matching with local linear regression is described in Chapter 7.

5. In the data simulation, *w* was a dichotomous variable indicating treatment conditions: treated or nontreated. The propensity score method can easily be expanded to include multiple treatment conditions or to include the analysis of doses of treatment (Imbens, 2000; Rosenbaum, 2002b). This method of modeling treatment dosage is described in Chapter 5.

6. Rubin (1997) summarized three limitations of propensity score matching, one of which is that propensity score matching cannot control for unobserved selection bias. That is, propensity score matching cannot adjust for hidden selection bias, and therefore, there is no solution to Scenario 2. (We give further attention to this problem in Chapter 8, where we compare sampling properties of four models through a Monte Carlo study.) Though selection bias is hidden, the analyst is able to evaluate the magnitude of bias through *sensitivity analysis,* which is seminal work developed by Rosenbaum (2002b). As mentioned earlier, the sensitivity analysis method is described in Chapter 8.

7. The second limitation of propensity score matching that Rubin (1997) identified is that the method differentially handles covariates based on the relation to outcome. That is, the method does not handle a covariate that is related to treatment assignment but not to outcome. This problem can be viewed as a variant of Scenario 4, for which there is no desirable solution currently available; however, James Robins's marginal structural modeling appears promising. We revisit this issue in Chapter 9.

8. Scenario 5 features a combination of several weak assumptions about data generation. Although the data generation is complicated, this scenario is realistic and likely to occur in practice. We revisit this scenario in Chapter 8, and underscore that—even with advanced methods—unobserved heterogeneity always holds the potential to bias effect estimations.

3.6 Key Issues Regarding the Application of OLS Regression

Among all statistical approaches, the OLS regression model is perhaps the most important because it not only serves as the foundation for advanced models but also is the key to understanding new approaches for the evaluation of treatment effects. We have seen conditions under which the OLS regression provides (or does not provide) unbiased estimation of treatment effects. To conclude the current chapter, we review important assumptions embedded in the OLS regression, and fundamental issues with regard to the application of OLS regression. Our review follows Kennedy (2003) and Berk (2004).

When applying the OLS regression model, five basic assumptions must be made about the data structure:

1. The dependent variable is a linear function of a specific set of independent variables plus a disturbance term.

2. The expected value of the disturbance term is zero.

3. The disturbances have a uniform variance and are uncorrelated with one another.

4. The observations on the independent variables are considered fixed in repeated samples.

5. The number of observations is greater than the number of independent variables, and no exact linear relationships exist between independent variables (Kennedy, 2003).

These assumptions are crucial for consistent and unbiased estimation in regression models. In practice, it is important to understand the conditions under which these assumptions are violated, to be attentive to conducting diagnostic tests to detect violations, and to take remedial actions if violations are observed.

From Berk (2004), regression may be used to draw causal inferences when four conditions are satisfied. First, "clear definitions of the relevant concepts and . . . good information about how the data were generated" are required (Berk, 2004, p. 101). Causal effects involve not only the relationships between inputs and outputs, not only the relationships between observable and unobservable outcomes, but also a number of additional features related to how nature is supposed to operate. Berk recommended using Freedman's response schedules framework to infer causal effects. The available theories, including those used in economics, are mostly silent on how to characterize the role of the errors in the regression formulation. In contrast, Freedman's response schedule framework requires that errors be treated as more than a nuisance to be dispatched by a few convenient assumptions. From this perspective, causal stories should include a plausible explanation of how the errors are generated in the natural world.

Second, the ceteris paribus condition must be assumed. In warning that there are typically a host of ceteris paribus conditions that never exist in real life, Berk (2004) argued, "Covariance adjustments are only arithmetic manipulations of the data. If one decides to interpret those manipulations as if some confounder were actually being fixed, a very cooperative empirical world is required" (p. 115).

Third, causal relationships must be anticipated in data generation and analysis. Berk (2004, p. 196) objected in part to Pearl's (2000) claim that the analyst can routinely carry out causal inference with regression analysis (or more generally, structural equation modeling) of observational data.

Last, Berk (2004) argued that credible causal inferences cannot be drawn from regression findings alone. The output from regression analysis is merely a way to characterize the conditional distribution of the response variable, given a set of predictors. Standardized coefficients do not represent the causal importance of a variable. Contributions to explained variance for different predictors do not represent the causal importance of a variable. A good over-all fit does not demonstrate that a causal model is correct. Moreover, there are no regression diagnostics, he argued, through which causal effects can be demonstrated with confidence. There are no specification tests through which causal effects can be demonstrated. There are no mathematical formalisms through which causal effects can be demonstrated. In short, Berk argues that causal inference rests first and foremost on a credible response schedule. Without a credible response schedule and in the absence of a well-supported model (i.e., hypothesized relationships based on prior research and theory), the potential problems with regression-based causal models outweigh existing remedial strategies (Berk, 2004, p. 224).

3.7 Conclusions

In this chapter, we reviewed conditions under which the fundamental assumption of ignorable treatment assignment is violated, and we discussed three conventional approaches (i.e., regression, matching, and stratification) to balance data in the presence of nonignorable treatment assignment (i.e., selection bias). In a simulation based on five scenarios of data generation, conventional methods worked well only in an ideal scenario in which the ignorable treatment assignment assumption is met. Unfortunately, these ideal conditions are most likely to be satisfied in a randomized experiment. In social behavior evaluations involving the use of observational data, they are unlikely to be satisfied. Thus, the findings suggest that the use of OLS regression to control for covariates and balance data in the absence of a randomized design may be unwarranted in many settings where a causal inference is desired. New methods to estimate treatment effects when treatment assignment is nonignorable are needed.

Note

1. However, in our current data simulation, we cannot see the consequence of inconsistency, because the current simulation does not look into sampling variability and sampling properties when sample size becomes extremely large.

4

Sample Selection and Related Models

This chapter describes three models: the sample selection model, the treatment effect model, and the instrumental variables approach. Heckman's (1974, 1978, 1979) sample selection model was developed using an econometric framework for handling limited dependent variables. It was designed to address the problem of estimating the average wage of women using data collected from a population of women in which housewives were excluded by self-selection. Based on this data set, Heckman's original model focused on the *incidental truncation* of a dependent variable. Maddala (1983) extended the sample selection perspective to the evaluation of treatment effectiveness. We review Heckman's model first because it not only offers a theoretical framework for modeling sample selection but is also based on what was at the time a pioneering approach to correcting selection bias. Equally important, Heckman's model lays the groundwork for understanding the treatment effect model. The sample selection model is among the most important contributions to program evaluation; however, the treatment effect model is the focus of this chapter because this model offers practical solutions to various types of evaluation problems. Although the instrumental variables approach is similar in some ways to the sample selection model, it is often conceptualized as a different method. We included it in this chapter for the convenience of the discussion.

Section 4.1 describes the main features of the Heckman model. Section 4.2 reviews the treatment effect model. Section 4.3 reviews the instrumental variables approach. Section 4.4 provides an overview of the *Stata* programs that are applicable for estimating the models described here. Examples in Section 4.5 illustrate the treatment effect model and show how to use this model to solve typical evaluation problems. Section 4.6 concludes with a review of key points.

4.1 The Sample Selection Model

Undoubtedly, Heckman's sample selection model is among the more significant work in 20th-century program evaluation. The sample selection model triggered both a rich theoretical discussion on modeling selection bias and the development of new statistical procedures that address the problem of selection bias. Heckman's key contributions to program evaluation include the following: (a) he provided a theoretical framework that emphasized the importance of modeling the dummy endogenous variable; (b) his model was the first attempt that estimated the probability (i.e., the propensity score) of a participant being in one of the two conditions indicated by the endogenous dummy variable, and then used the estimated propensity score model to estimate coefficients of the regression model; (c) he treated the unobserved selection factors as a problem of specification error or a problem of omitted variables, and corrected for bias in the estimation of the outcome equation by explicitly using information gained from the model of sample selection; and (d) he developed a creative two-step procedure by using the simple least squares algorithm. To understand Heckman's model, we first review concepts related to the handling of limited dependent variables.

4.1.1 TRUNCATION, CENSORING, AND INCIDENTAL TRUNCATION

Limited dependent variables are common in social and health data. The primary characteristics of such variables are censoring and truncation. *Truncation*, which is an effect of data gathering rather than data generation, occurs when sample data are drawn from a subset of a larger population of interest. Thus, a truncated distribution is the part of a larger, untruncated distribution. For instance, assume that an income survey was administered to a limited subset of the population (e.g., those whose incomes are above poverty threshold). In the data from such a survey, the dependent variable will be observed only for a portion of the whole distribution. The task of modeling is to use that limited information—a *truncated distribution*—to infer the income distribution for the entire population.

Censoring occurs when all values in a certain range of a dependent variable are transformed to a single value. Using the above example of population income, censoring differs from truncation in that the data collection may include the entire population, but below-poverty-threshold incomes are coded as zero. Under this condition, researchers may estimate a regression model for a larger population using both the censored and the uncensored data. Censored data are ubiquitous. They include (1) household purchases of durable goods, in which low expenditures for durable goods are censored to a zero value (the

Tobit model, developed by James Tobin in 1958, is the most widely known model for analyzing this kind of dependent variable); (2) number of extramarital affairs, in which the number of affairs beyond a certain value is collapsed into a maximum count; (3) number of hours worked by women in the labor force, in which women who work outside the home for a low number of hours are censored to a zero value; and (4) number of arrests after release from prison, where arrests beyond a certain value are scored as a maximum (Greene, 2003).

The central task of analyzing limited dependent variables is to use the truncated distribution or censored data to infer the untruncated or uncensored distribution for the entire population. In the context of regression analysis, we typically assume that the dependent variable follows a normal distribution. The challenge then is to develop moments (mean and variance) of the truncated or censored normal distribution. Theorems of such moments have been developed and can be found in textbooks on the analysis of limited dependent variables. In these theorems, moments of truncated or censored normal distributions involve a key factor called the *inverse Mills ratio*, or *hazard function*, which is commonly denoted as λ. Heckman's sample selection model uses the inverse Mills ratio to estimate the outcome regression. In Section 4.1.3, we review moments for sample selection data and the inverse Mills ratio.

A concept closely related to truncation and censoring, or a combination of the two concepts, is *incidental truncation*. Indeed, it is often used interchangeably with the term *sample selection*. From Greene (2003), suppose you are funded to conduct a survey of persons with high incomes and that you define eligible respondents as those with net worth of $500,000 or more. This selection by income is a form of truncation—but it is not quite the same as the general case of truncation. The selection criterion (e.g., at least $500,000 net worth) does not exclude those individuals whose current income might be quite low although they had previously accrued high net worth. Greene (2003) explained by saying,

> Still, one would expect that, on average, individuals with a high net worth would have a high income as well. Thus, the average income in this subpopulation would in all likelihood also be misleading as an indication of the income of the typical American. The data in such a survey would be nonrandomly selected or incidentally truncated. (p. 781)

Thus, sample selection or incidental truncation refers to a sample that is not randomly selected. It is in situations of incidental truncation that we encounter the key challenge to the entire process of evaluation, that is, departure of evaluation data from the classic statistical model that assumes a randomized experiment. This challenge underscores the need to model the sample selection process explicitly. We encounter these problems explicitly and implicitly in many data situations. Consider the following from Maddala (1983).

Example 1: Married women in the labor force. This is the problem Heckman (1974) originally considered under the context of shadow prices (i.e., women's reservation wage or the minimum wage rate at which a woman who is at home might accept marketplace employment), market wages, and labor supply. Let y^* be the reservation wage of a housewife based on her valuation of time in the household. Let y be the market wage based on an employer's valuation of her effort in the labor force. According to Heckman, a woman participates in the labor force if $y > y^*$. Otherwise, a woman is not considered a participant in the labor force. In any given sample, we only have observations on y for those women who participate in the labor force, and we have no observation on y for the women not in the labor force. For women not in the labor force, we only know that $y^* \geq y$. In other words, the sample is not randomly selected, and we need to use the sample data to estimate the coefficients in a regression model explaining both y^* and y. As explained below by Maddala (1983), with regard to women who are not in the labor market and who work at home, the problem is truncation, or more precisely incidental truncation, not censoring, because

> we do not have any observations on either the explained variable y or the explanatory variable x in the case of the truncated regression model if the value of y is above (or below) a threshold. . . . In the case of the censored regression model, we have data on the explanatory variables x for all the observations. As for the explained variable y, we have actual observations for some, but for others we know only whether or not they are above (or below) a certain threshold. (pp. 5–6)

Example 2: Effects of unions on wages. Suppose we have data on wages and personal characteristics of workers that include whether the worker is a union member. A naïve way of estimating the effects of unionization on wages is to estimate a regression of wage on the personal characteristics of the workers (e.g., age, race, sex, education, and experience) plus a dummy variable that is defined as $D = 1$ for unionized workers and $D = 0$ otherwise. The problem with this regression model lies in the nature of D. This specification treats the dummy variable D as exogenous when D is not exogenous. In fact, there are likely many factors affecting a worker's decision whether to join the union. As such, the dummy variable is endogenous and should be modeled directly; otherwise, the wage regression estimating the impact of D will be biased. We have seen the consequences of naïve treatment of D as an exogenous variable in both Chapters 2 and 3.

Example 3: Effects of fair-employment laws on the status of African American workers. Consider a regression model (Landes, 1968) relating to the effects of fair-employment legislation on the status of African American workers $y_i = \alpha X_i + \beta D_i + u_i$, where y_i is the wage of African Americans relative to that for whites in state i, X_i is the vector of exogenous variables for state i, $D_i = 1$ if state i has a fair-employment law ($D_i = 0$ otherwise), and u_i is a residual. Here the same problem of the endogeneity of D is found as in our second example, except that the unit of analysis in the previous example is individual, whereas the unit in the current example is state i. Again D_i is treated as exogenous when in fact it is endogenous. "States in which African Americans would fare well without a fair-employment law may be more likely to pass such a law if legislation depends on the consensus" (Maddala, 1983, p. 8). Heckman (1978) observed,

> An important question for the analysis of policy is to determine whether or not measured effects of legislation are due to genuine consequences of legislation or to the spurious effect that the presence of legislation favorable to blacks merely proxies the presence of the pro-black sentiment that would lead to higher status for blacks in any event. (p. 933)

Example 4: Compulsory school attendance laws and academic or other outcomes. The passage of compulsory school attendance legislation is itself an endogenous variable. Similar to Example 3, it should be modeled first. Otherwise estimation of the impact of such legislation on any outcome variable risks bias and inconsistency (Edwards, 1978).

Example 5: Returns of college education. In this example, we are given income for a sample of individuals, some with a college education and others without. Because the decision whether to attend college is a personal choice determined by many factors, the dummy variable (attending vs. not attending) is endogenous and should be modeled first. Without modeling this dummy variable first, the regression of income showing the impact of college education would be biased, regardless of whether the regression model controlled for covariates such as IQ (intelligence quotient) or parental socioeconomic status.

Today, these illustrations are considered classic examples, and they have been frequently cited and discussed in the literature on sample selection. The first three examples were discussed by Heckman (1978, 1979) and motivated his work on sample selection models. These examples share three features: (1) the sample being inferred was not generated randomly; (2) the binary explanatory variable was endogenous rather than exogenous; and (3) sample selection or incidental truncation must be considered in the evaluation of the impact of such a dummy variable. However, there is an important difference between Example 1 and the other four examples. In Example 1, we observe only the outcome variable (i.e., market wage) for women who participate in the labor force (i.e., only for participants whose $D_i = 1$; we do not observe the outcome variable for women whose $D_i = 0$), whereas, in Example 2 through Example 5, the outcome variables (i.e., wages, the wage status of African American workers relative to that of white workers, academic achievement, and income) for both the participants (or states) whose $D_i = 1$ and $D_i = 0$ are observed. Thus, Example 1 is a sample selection model, and the other four examples illustrate the treatment effect model. The key point is the importance of distinguishing between these two types of models: (1) the sample selection model (i.e., the model analyzing outcome data observed only for $D_i = 1$) and (2) the treatment effect model (i.e., the model analyzing outcome data observed for both $D_i = 1$ and $D_i = 0$). Both models share common characteristics and may be viewed as Heckman-type models. However, the treatment effect model focuses on program evaluation, which is not the intent of the sample selection model. This distinction is important when choosing appropriate software. In the Stata software, for example, the sample selection model is estimated by the program *heckman,* and the treatment effect model is estimated by the program *treatreg;* we elaborate on this point in Section 4.4.

4.1.2 WHY IS IT IMPORTANT TO MODEL SAMPLE SELECTION?

Although the topic of sample selection is ubiquitous in both program evaluation and observational studies, the importance of giving it a formal treatment was largely unrecognized until Heckman's (1974, 1976, 1978, 1979) work and the independent work of Rubin (1974, 1978, 1980b, 1986). Recall that, in terms of causal inference, sample selection was not considered a problem in randomized experiments because randomization renders selection effects irrelevant. In nonrandomized studies, Heckman's work emphasized the importance of modeling sample selection by using a two-step procedure or *switching regression,* whereas Rubin's work drew the same conclusion by applying a generalization of the randomized experiment to observational studies.

Heckman focused on two types of selection bias: self-selection bias and selection bias made by data analysts. Heckman (1979) described self-selection bias as follows:

> One observes market wages for working women whose market wage exceeds their home wage at zero hours of work. Similarly, one observes wages for union members who found their nonunion alternative less desirable. The wages of migrants do not, in general, afford a reliable estimate of what nonmigrants would have earned had they migrated. The earnings of manpower trainees do not estimate the earnings that nontrainees would have earned had they opted to become trainees. In each of these examples, wage or earnings functions estimated on selected samples do not in general, estimate population (i.e., random sample) wage functions. (pp. 153–154)

Heckman argued that the second type of bias, selection bias made by data analysts or data processors, operates in much the same fashion as self-selection bias.

In their later work, Heckman and his colleagues generalized the problem of selectivity to a broad range of social experiments and discussed additional types of selection biases (e.g., see Heckman & Smith, 1995). From Maddala (1983), Figure 4.1 describes three types of decisions that create selectivity (i.e., individual selection, administrator selection, and attrition selection).

In summary, Heckman's approach underscores the importance of modeling selection effects. When selectivity is inevitable, such as in observational studies, the parameter estimates from a naive ordinary least squares (OLS) regression model are inconsistent and biased. Alternative analytic strategies that model selection must be explored.

4.1.3 MOMENTS OF AN INCIDENTALLY TRUNCATED BIVARIATE NORMAL DISTRIBUTION

The theorem for moments of the incidentally truncated distribution defines key functions such as the inverse Mills ratio under the setting of a normally distributed variable. Our discussion follows Greene (2003).

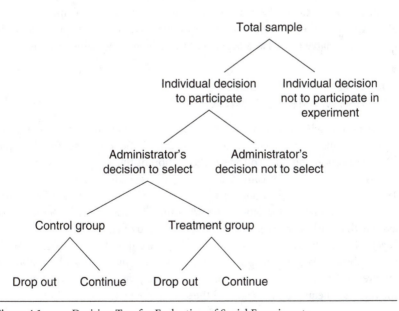

Figure 4.1   Decision Tree for Evaluation of Social Experiments

SOURCE: Maddala (1983, p. 266). Reprinted with the permission of Cambridge University Press.

Suppose that y and z have a bivariate normal distribution with correlation ρ. We are interested in the distribution of y given that z exceeds a particular value a. The truncated joint density of y and z is

$$f(y, z|z > a) = \frac{f(y, z)}{\text{Prob}(z > a)}.$$

Given the truncated joint density of y and z, given that y and z have a bivariate normal distribution with means μ_y and μ_z, standard deviations σ_y and σ_z, and correlation ρ, the moments (mean and variance) of the incidentally truncated variable y are as follows (Greene, 2003, p. 781):

$$
\begin{aligned}
E[y|z > a] &= \mu_y + \rho\sigma_y\lambda(c_z), \\
\text{Var}[y|z > a] &= \sigma_y^2[1 - \rho^2\delta(c_z)],
\end{aligned}
\tag{4.1}
$$

where a is the cutoff threshold, $c_z = (a - \mu_z)/\sigma_z$, $\lambda(c_z) = \phi(c_z)/[1 - \Phi(c_z)]$, $\delta(c_z) = \lambda(c_z)[\lambda(c_z) - c_z]$, $\phi(c_z)$ is the standard normal density function, and $\Phi(c_z)$ is the standard cumulative distribution function.

In the above equations, $\lambda(c_z)$ is called the inverse Mills ratio and is used in Heckman's derivation of his two-step estimator. Note that in this theorem we consider moments of a single variable; in other words, this is a theorem about univariate properties of the incidental truncation of y. Heckman's model applied and expanded the theorem to a multivariate case in which an incidentally truncated variable is used as a dependent variable in a regression analysis.

4.1.4 THE HECKMAN MODEL
AND ITS TWO-STEP ESTIMATOR

A sample selection model always involves two equations: (1) the regression equation considering mechanisms determining the outcome variable and (2) the selection equation considering a portion of the sample whose outcome is observed and mechanisms determining the selection process (Heckman, 1978, 1979). To put this model in context, we revisit the example of the wage earning of women in the labor force (Example 1, Section 4.1.1). Suppose we assume that the hourly *wage* of women is a function of education (*educ*) and age (*age*), whereas the probability of working (equivalent to the probability of wage being observed) is a function of marital status (*married*) and number of children at home (*children*). To express the model, we can write two equations, the regression equation of wage and the selection equation of working:

$$wage = \beta_0 + \beta_1\, educ + \beta_2\, age + u_1 \qquad \text{(regression equation)}.$$

Wage is observed if

$$\gamma_0 + \gamma_1 \ married + \gamma_2 \ children + \gamma_3 \ educ + \gamma_4 \ age + u_2 > 0 \quad \text{(selection equation)}.$$

Note that the selection equation indicates that wage is observed only for those women whose wages were greater than 0 (i.e., women were considered as having participated in the labor force if and only if their wage was above a certain threshold value). Using a zero value in this equation is a normalization convenience and is an alternate way to say that the market wage of women who participated in the labor force was greater than their reservation wage (i.e., $y > y^*$). The fact that the market wage of homemakers (i.e., those not in the paid labor force) was less than their reservation wage (i.e., $y < y^*$) is expressed in the above model through the fact that these women's wage was not observed in the regression equation, that is it was incidentally truncated. The selection model further assumes that u_1 and u_2 are correlated to have a nonzero correlation ρ.

This example can be expanded to a more general case. For the purpose of modeling any sample selection process, two equations are used to express the determinants of outcome y_i:

$$\text{Regression equation: } y_i = x_i \beta + \varepsilon_i, \text{ observed only if } w_i = 1, \qquad (4.2a)$$

$$\text{Selection equation: } w_i^* = z_i \gamma + u_i, \ w_i = 1 \text{ if } w_i^* > 0, \text{ and } w_i = 0 \text{ otherwise} \quad (4.2b)$$

$$\text{Prob}(w_i = 1 | z_i) = \Phi(z_i \gamma)$$

and

$$\text{Prob}(w_i = 0 | z_i) = 1 - \Phi(z_i \gamma),$$

where x_i is a vector of exogenous variables determining outcome y_i, and w_i^* is a latent endogenous variable. If w_i^* is greater than the threshold value (say value 0), then the observed dummy variable $w_i = 1$, and otherwise $w_i = 0$; the regression equation observes value y_i only for $w_i = 1$; z_i is a vector of exogenous variables determining the selection process or the outcome of w_i^*; $\Phi(\cdot)$ is the standard normal cumulative distribution function; and u_j and ε_j are error terms of the two regression equations, and assumed to be bivariate normal, with mean zero and covariance matrix $\begin{bmatrix} \sigma_\varepsilon & \rho \\ \rho & 1 \end{bmatrix}$.

Given incidental truncation and censoring of y, the evaluation task is to use the observed variables (i.e., y, z, x, and probably w) to estimate the regression coefficients β that are applicable to sample participants whose values of w equal both 1 and 0.

The sample selection model can be estimated by either the maximum likelihood method or the least squares method. Heckman's two-step estimator uses the least squares method. We review the two-step estimator first. The maximum likelihood method is reviewed in the next section as a part of a discussion of the treatment effect model.

To facilitate the understanding of Heckman's original contribution, we use his notations that are slightly different from those used in our previous discussion. Heckman first described a general model containing two structural equations. The general model considers continuous latent random variables y_{1i}^* and y_{2i}^*, and may be expressed as follows:

$$
\begin{aligned}
y_{1i}^* &= X_{1i}\alpha_1 + d_i\beta_1 + y_{2i}^*\gamma_1 + U_{1i}, \\
y_{2i}^* &= X_{2i}\alpha_2 + d_i\beta_2 + y_{1i}^*\gamma_2 + U_{2i},
\end{aligned}
\tag{4.3}
$$

where X_{1i} and X_{2i} are row vectors of bounded exogenous variables; d_i is a dummy variable defined by

$$d_i = 1 \text{ if and only if } y_{2i}^* > 0,$$

$$d_i = 0 \text{ otherwise,}$$

and

$$E(U_{ji}) = 0, E(U_{ji}^2) = \sigma_{jj}, E(U_{1i}U_{2i}) = \sigma_{12}, j = 1, 2; i = 1, \ldots, I$$

$$E(U_{ji}U_{j'i'}) = 0, \text{ for } j, j' = 1, 2; i \neq i'.$$

Heckman next discussed six cases where the general model applies. His interest centered on the sample selection model, or Case 6 (Heckman, 1978, p. 934). The primary feature of Case 6 is that structural shifts in the equations are permitted. Furthermore, Heckman allowed that y_{1i}^* was observed, so the variable can be written without an asterisk, as y_{1i}, and y_{2i}^* is not observed. Writing the model in reduced form (i.e., only variables on the right-hand side should be exogenous variables), we have the following equations:

$$
\begin{aligned}
y_{1i} &= X_{1i}\pi_{11} + X_{2i}\pi_{12} + P_i\pi_{13} + V_{1i} + (d_i - P_i)\pi_{13}, \\
y_{2i}^* &= X_{1i}\pi_{21} + X_{2i}\pi_{22} + P_i\pi_{23} + V_{2i} + (d_i - P_i)\pi_{23},
\end{aligned}
\tag{4.4}
$$

where P_i is the conditional probability of $d_i = 1$, and

$$
\pi_{11} = \frac{\alpha_1}{1 - \gamma_1\gamma_2}, \pi_{21} = \frac{\alpha_1\gamma_2}{1 - \gamma_1\gamma_2}, \pi_{12} = \frac{\alpha_2\gamma_1}{1 - \gamma_1\gamma_2}, \pi_{22} = \frac{\alpha_2}{1 - \gamma_1\gamma_2},
$$

$$
\pi_{13} = \frac{\beta_1 + \gamma_1\beta_2}{1 - \gamma_1\gamma_2}, \pi_{23} = \frac{\gamma_2\beta_1 + \beta_2}{1 - \gamma_1\gamma_2}, V_{1i} = \frac{U_{1i} + \gamma_1 U_{2i}}{1 - \gamma_1\gamma_2}, V_{2i} = \frac{\gamma_2 U_{1i} + U_{2i}}{1 - \gamma_1\gamma_2}.
$$

The model assumes that U_{1i} and U_{2i} are bivariate normal random variables. Accordingly, the joint distribution of V_{1i}, V_{2i}, $h(V_{1i}, V_{2i})$, is a bivariate normal density fully characterized by the following assumptions:

$$E(V_{1i}) = 0, \quad E(V_{2i}) = 0, \quad E(V_{1i}^2) = \omega_{11}, \quad E(V_{2i}^2) = \omega_{22}.$$

For the existence of the model, the analyst has to impose restrictions. A necessary and sufficient condition for the model to be defined is that $\pi_{23} = 0 = \gamma_2 \beta_1 + \beta_2$. Heckman called this condition *the principal assumption*. Under this assumption, the model becomes

$$y_{1i} = X_{1i}\pi_{11} + X_{2i}\pi_{12} + P_i\pi_{13} + V_{1i} + (d_i - P_i)\pi_{13}, \tag{4.5a}$$

$$y_{2i}^* = X_{1i}\pi_{21} + X_{2i}\pi_{22} + V_{2i}, \tag{4.5b}$$

where $\pi_{11} \neq 0, \pi_{12} \neq 0, \pi_{21} \neq 0, \pi_{22} \neq 0$.

With the above specifications and assumptions, the model (4.5) can be estimated in two steps:

1. First, estimate Equation 4.5b, which is analogous to solving the problem of a probit model. We estimate the conditional probabilities of the events $d_i = 1$ and $d_i = 0$ by treating y_{2i} as a dummy variable. Doing so, π_{21} and π_{22} are estimated. Subject to the standard requirements for identification and existence of probit estimation, the analyst needs to normalize the equation by $\sqrt{\omega_{22}}$ and estimate:

$$\pi_{21}^* = \frac{\pi_{21}}{\sqrt{\omega_{22}}}, \pi_{22}^* = \frac{\pi_{22}}{\sqrt{\omega_{22}}}.$$

2. Second, estimate Equation 4.5a. Rewrite Equation 4.5a as the conditional expectation of y_{1i} given d_i, X_{1i}, and X_{2i}:

$$E(y_{1i}|X_{1i}, X_{2i}, d_i) = X_{1i}\pi_{11} + X_{2i}\pi_{12} + d_i\pi_{13} + E(V_{1i}|d_i, X_{1i}, X_{2i}). \tag{4.6}$$

Using a result of biserial correlation, $E(V_{1i}|d_i, X_{1i}, X_{2i})$ is estimated:

$$E(V_{1i}|d_i, X_{1i}, X_{2i}) = \frac{\omega_{12}}{\sqrt{\omega_{22}}} (\lambda_i d_i + \tilde{\lambda}_i(1 - d_i)), \tag{4.7}$$

where $\lambda_i = \phi(c_i)/(1 - \Phi(c_i))$ with $c_i = -(X_{1i}\pi_{21}^* + X_{2i}\pi_{22}^*)$, ϕ and Φ are the density and distribution function of a standard normal random variable, respectively, and $\tilde{\lambda}_i = -\lambda_i[\Phi(-c_i) / \Phi(c_i)]$. Because $E(V_{1i}|d_i, X_{1i}, X_{2i})$ can now be estimated, Equation 4.6 can be solved by the standard least squares method. Note that

$\lambda_i = \phi(c_i)/(1 - \Phi(c_i))$ refers to a truncation of y whose truncated z exceeds a particular value a (see Equation 4.1). Under this condition, Equation 4.7 becomes $E(V_{1i}|d_i, X_{1i}, X_{2i}) = (\omega_{12}/\sqrt{\omega_{12}}) \lambda_i d_i$. Using estimated π_{21}^* and π_{22}^* from Step 1, $\lambda_i = \phi(c_i)/(1 - \Phi(c_i))$ is calculated using $c_i = -(X_1\pi_{21}^* + X_2\pi_{22}^*)$. Now in the equation of $E(V_{1i}|d_i, X_{1i}, X_{2i}) = (\omega_{12}/\sqrt{\omega_{12}}) \lambda_i d_i$, because λ_i, d_i, and $\sqrt{\omega_{22}}$ are known, the only coefficient to be determined is ω_{12}; thus solving Equation 4.6 is a matter of estimating the following regression:

$$E(y_{1i}|X_{1i}, X_{2i}, d_i) = X_{1i}\pi_{11} + X_{2i}\pi_{12} + d_i\pi_{13} + \frac{\lambda_i d_i}{\sqrt{\omega_{22}}}\omega_{12}.$$

Therefore, the parameters π_{11}, π_{12}, π_{13}, and ω_{12} can be estimated by using the standard OLS estimator.

A few points are particularly worth noting. First, in Equation 4.5b, V_{2i} is an *error term* or *residuals of the variation* in the latent variable y_{2i}^*, after the variation is explained away by X_{1i} and X_{2i}. This is a specification error or, more precisely, a case of unobserved heterogeneity determining selection bias. This specification error is treated as a true omitted-variable problem and is creatively taken into consideration when estimating the parameters of Equation 4.5a. In other words, *the impact of selection bias is neither thrown away nor assumed to be random but is explicitly used and modeled in the equation estimating the outcome regression.* This treatment for selection bias connotes Heckman's contribution and distinguishes the econometric solution to the selection bias problem from that of the statistical tradition. Important implications of this modeling feature were summarized by Heckman (1979, p. 155). In addition, there are different formulations for estimating the model parameters that were developed after Heckman's original model. For instance, Greene (1981, 2003) constructed consistent estimators of the individual parameter ρ (i.e., the correlation of the two error terms) and σ_ε (i.e., the variance of the error term of the regression equation). However, Heckman's model has become standard in the literature. Last, the same sample selection model can also be estimated by the maximum likelihood estimator (Greene, 1995), which yields results remarkably similar to those produced using the least squares estimator. Given that the maximum likelihood estimator requires more computing time, and computing speed three decades ago was considerably slower than today, Heckman's least squares solution is a remarkable contribution. More important, Heckman's solution was devised within a framework of structural equation modeling that is simple and succinct and that can be used in conjunction with the standard framework of OLS regression.

4.2 Treatment Effect Model

Since the development of the sample selection model, statisticians and econometricians have formulated many new models and estimators. In mimicry of the Tobit or logit models, Greene (2003) suggested that these Heckman-type models might be called "*Heckit*" models. One of the more important of these developments was the direct application of the sample selection model to estimation of treatment effects in observational studies.

The treatment effect model differs from the sample selection model—that is, in the form of Equation 4.2—in two aspects: (1) a dummy variable indicating the treatment condition w_i (i.e., $w_i = 1$ if participant i is in the treatment condition, and $w_i = 0$ otherwise) is directly entered into the regression equation and (2) the outcome variable y_i of the regression equation is observed for both $w_i = 1$ and $w_i = 0$. Specifically, the treatment effect model is expressed in two equations:

$$\text{Regression equation: } y_i = x_i\beta + w_i\delta + \varepsilon_i, \tag{4.8a}$$

$$\text{Selection equation: } w_i^* = z_i\gamma + u_i, \ w_i = 1 \text{ if } w_i^* > 0, \text{ and } w_i = 0 \text{ otherwise} \tag{4.8b}$$

$$\text{Prob}(w_i = 1|z_i) = \Phi(z_i\gamma)$$

and

$$\text{Prob}(w_i = 0|z_i) = 1 - \Phi(z_i\gamma),$$

where ε_i and u_j are bivariate normal with mean zero and covariance matrix $\begin{bmatrix} \sigma_\varepsilon & \rho \\ \rho & 1 \end{bmatrix}$. Given incidental truncation (or sample selection) and that w is an endogenous dummy variable, the evaluation task is to use the observed variables to estimate the regression coefficients β, while controlling for selection bias induced by nonignorable treatment assignment.

Note that the model expressed by Equations 4.8a and 4.8b is a *switching regression*. By substituting w_i in Equation 4.8a with Equation 4.8b, we obtained two different equations of the outcome regression:

$$\text{when } w_i^* > 0, \ w_i = 1: \ y_i = x_i\beta + (z_i\gamma + u_i)\delta + \varepsilon, \tag{4.9a}$$

and

$$\text{when } w_i^* \leq 0, \ w_i = 0: \ y_i = x_i\beta + \varepsilon_i. \tag{4.9b}$$

This is Quandt's (1958, 1972) form of the switching regression model that explicitly states that there are two regimes: treatment and nontreatment. Accordingly, there are separate models for the outcome under each regime: For treated participants, the outcome model is $y_i = x_i\beta + (z_i\gamma + u_i)\delta + \varepsilon_i$; whereas, for nontreated participants, the outcome model is $y_i = x_i\beta + \varepsilon_i$.

The treatment effect model illustrated above can be estimated in a two-step procedure similar to that described for the sample selection model. To increase the efficiency of our exposition of models, we move on to the maximum likelihood estimator. Readers who are interested in the two-step estimator may consult Maddala (1983).

Let $f(\varepsilon, u)$ be the joint density function of ε and u defined by Equations 4.8a and 4.8b. According to Maddala (1983, p. 129), the joint density function of y and w is given by the following:

$$g(y, w=1) = \int_{-\infty}^{z\gamma} f(y - \delta - x\beta, u)\,du,$$

and

$$g(y, w=0) = \int_{z\gamma}^{\infty} f(y - x\beta, u)\,du.$$

Thus, the log likelihood functions for participant i (StataCorp, 2003) are as follows: for $w_i = 1$,

$$= \ln \Phi \left\{ \frac{-z_i\gamma + (y_i - x_i\beta - \delta)\rho/\sigma}{\sqrt{1 - \rho^2}} \right\} - \frac{1}{2}\left(\frac{y_i - x_i\beta - \delta}{\sigma} \right)^2 - \ln(\sqrt{2\pi}\sigma) \quad (4.10a)$$

for $w_i = 0$,

$$l_i = \ln \Phi \left\{ \frac{-z_i\gamma(y_i - x_i\beta)\rho/\sigma}{\sqrt{1 - \rho^2}} \right\} - \frac{1}{2}\left(\frac{y_i - x_i\beta - \delta}{\sigma} \right)^2 - \ln(\sqrt{2\pi}\sigma) \quad (4.10b)$$

The treatment effect model has many applications in program evaluation. In particular, it is useful when evaluators have data that were generated by a nonrandomized experiment and, thus, are faced with the challenge of nonignorable treatment assignment or selection bias. We illustrate the application of the treatment effect model in Section 4.5. However, before that, we briefly review a similar estimator, the instrumental variables approach, which shares common features with the sample selection and treatment effect models.

4.3 Instrumental Variables Estimator

Recall Equation 4.8a or the regression equation of the treatment effect model $y_i = x_i\beta + w_i\delta + \varepsilon_i$. In this model, w is correlated with ε. As discussed in Chapter 2, the consequence of contemporaneous correlation of the independent variable and the error term is biased and inconsistent estimation of β. This problem is the same as that shown in Chapter 3 by three of the scenarios in which treatment assignment was nonignorable. Under Heckit modeling, the solution to this problem is to use vector z to model the latent variable w_i^*. In the Heckit models, z is a vector or a set of variables predicting selection. An alternative approach to the problem is to find a single variable z_1 that is not correlated with ε but, at the same time, is highly predictive of w. If z_1 meets these conditions, then it is called an *instrumental variable* (IV), and Equation 4.8a can be solved by the least squares estimator. We follow Wooldridge (2002) to describe the IV approach.

Formally, consider a linear population model:

$$y = \beta_0 + \beta_1 x_1 + \beta_2 x_2 + \ldots + \beta_K x_K + \varepsilon. \tag{4.11}$$

$$E(\varepsilon) = 0, \operatorname{Cov}(x_j, \varepsilon) = 0, \operatorname{Cov}(x_K, \varepsilon) \neq 0, j = 1, \ldots, K-1.$$

Note that in this model, x_K is correlated with ε (i.e., $\operatorname{Cov}(x_K, \varepsilon) \neq 0$), and x_K is potentially endogenous. To facilitate the discussion, we think of ε as containing one omitted variable that is uncorrelated with all explanatory variables except x_K.[1]

To solve the problem of endogeneity bias, the analyst needs to find an observed variable, z_1, that satisfies the following two conditions: (1) z_1 is uncorrelated with ε, or $\operatorname{Cov}(z_1, \varepsilon) = 0$ and (2) z_1 is correlated with x_K, meaning that the linear projection of x_K onto all exogenous variables exists. Otherwise stated as

$$x_K = \delta_0 + \delta_1 x_1 + \delta_2 x_2 + \cdots + \delta_{K-1} x_{K-1} + \theta_1 z_1 + r_K,$$

where by definition, $E(r_K) = 0$ and r_K is uncorrelated with $x_1, x_2, \ldots$ and x_{K-1}, z_1; the key assumption here is that the coefficient on z_1 is nonzero, or $\theta_1 \neq 0$.

Next, consider the model (i.e., Equation 4.11)

$$y = x\beta + \varepsilon, \tag{4.12}$$

where the constant is absorbed into x so that $x = (1, x_2, \ldots, x_K)$ and z is $1 \times K$ vector of all exogenous variables, or $z = (1, x_2, \ldots, x_{K-1}, z_1)$. The above two conditions about z_1 imply the K population orthogonality conditions, or

$$E(z'\varepsilon) = 0. \tag{4.13}$$

Multiplying Equation 4.12 through by z', taking expectations, and using Equation 4.13, we have

$$[E(z'x)]\beta = E(z'y), \tag{4.14}$$

where $E(z'x)$ is $K \times K$ and $E(z'y)$ is $K \times 1$. Equation 4.14 represents a system of K linear equations in the K unknowns $\beta_1, \ldots, \beta_K$. This system has a unique solution if and only if the $K \times K$ matrix $E(z'x)$ has full rank, or the rank of $E(z'x)$ is K. Under this condition, the solution to β is

$$\beta = [E(z'x)]^{-1} E(z'y).$$

Thus, given a random sample $\{(x_i, y_i, z_i): i = 1, 2, \ldots, N\}$ from the population, the analyst can obtain the instrumental variables estimator of β as

$$\hat{\beta} = \left(N^{-1} \sum_{i=1}^{N} z_i' x_i \right)^{-1} \left(N^{-1} \sum_{i=1}^{N} z_i' y_i \right) = (Z'X)^{-1} Z'Y. \qquad (4.15)$$

The challenge to the application of the IV approach is to find such an instrumental variable, z_1, that is omitted but meets the two conditions listed. It is for this reason that we often consider using a treatment effect model that directly estimates the selection process. Heckman (1997) examined the use of the IV approach to estimate the mean effect of treatment on the treated, the mean effect of treatment on randomly selected persons, and the local average treatment effect. He paid special attention to the economic questions that were addressed by these parameters and concluded that when responses to treatment vary, the standard argument justifying the use of instrumental variables fails unless person-specific responses to treatment do not influence the decision to participate in the program being evaluated. This condition requires that participant gains from a program—which cannot be predicted from variables in outcome equations—have no influence on the participation decisions of program participants.

4.4 Overview of the Stata Programs and Main Features of *treatreg*

Most models described in this chapter can be estimated by the Stata and R packages. Many helpful user-developed programs are also available from the Internet. Within Stata, *heckman* can be used to estimate the sample selection model, and *treatreg* can be used to estimate the treatment effect model.

In Stata, *heckman* was developed to estimate the original Heckman model; that is, it is a model that focuses on incidentally truncated dependent variables. Using wage data collected from a population of employed women in which homemakers were self-selected out, Heckman wanted to estimate determinants of the average wage of the entire female population. Two characteristics distinguish this kind of problem from the treatment effect model: the dependent variable is observed only for a subset of sample participants (e.g., only observed for women in the paid labor force); and the group membership variable is not entered into the regression equation (see Equations 4.2a and 4.2b). Thus, the task fulfilled by *heckman* is different from the task most program evaluators or observational researchers aim to fulfill. Typically, for study samples such as the group of women in the paid labor force, program evaluators or researchers will have observed outcomes for participants in both conditions. Therefore, the treatment membership variable is entered into the regression equation to discern treatment effects. We emphasize these differences because it is *treatreg*, rather than *heckman*, that offers practical solutions to various types of evaluation problems.

Within Stata, *ivreg* and *ivprobit* are used to estimate instrumental variables models using two-stage least squares or conditional maximum likelihood estimators. In this chapter, we have been interested in an IV model that considers one instrument z_1 and treats all x variables as exogenous (see Equation 4.11). However, *ivreg* and *ivprobit* treat z_1 and all x variables as instruments. By doing so, both programs estimate a nonrecursive model that depicts a reciprocal relationship between two endogenous variables. As such, both programs are estimation tools for solving a simultaneous equation problem, or a problem known to most social behavioral scientists as *structural equation modeling*. In essence, *ivreg* and *ivprobit* serve the same function as specialized software packages, such as LISREL, Mplus, EQS, and AMOS. As mentioned earlier, although the IV approach sounds attractive, it is often confounded by a fundamental problem: in practice, it is difficult to find an instrument that is both highly correlated with the treatment condition and independent of the error term of the outcome regression. On balance, we recommend that whenever users find a problem for which the IV approach appears appealing, they can use the Heckit treatment effect model (i.e., *treatreg*) or other models we describe in later chapters. To employ the IV approach describe in Section 4.3 to estimate treatment effects, you must develop programming syntax.

The *treatreg* program can be initiated using the following basic syntax:

treatreg depvar [indepvars], treat(depvar_t = indepvars_t) [twostep]

where *depvar* is the outcome variable on which users want to assess the difference between treated and control groups; *indepvars* is a list of variables that users hypothesize would affect the outcome variable; *depvar_t* is the treatment membership variable that denotes intervention condition; *indepvars_t* is the list of variables that users anticipate will determine the selection process; and *twostep* is an optional specification to request an estimation using a two-step consistent estimator. In other words, absence of *twostep* is the default; under the default, Stata estimates the model using a full maximum likelihood. Using notations from the treatment effect model (i.e., Equations 4.8a and 4.8b), *depvar* is *y*, *indepvars* are the vector **x**, and *depvar_t* is *w* in Equation 4.8a, and *indepvars_t* are the vector **z** in Equation 4.8b. By design, **x** and **z** can be the same variables if the user suspects that covariates of selection are also covariates of the outcome regression. Similarly, **x** and **z** can be different variables if the user suspects that covariates of selection are different from covariates of the outcome regression (i.e., **x** and **z** are two different vectors). However, **z** is part of **x**, if the user suspects that additional covariates affect *y* but not *w*, or vice versa, if one suspects that additional covariates affect *w* but not *y*.

The *treatreg* program supports Stata standard functions, such as the Huber-White estimator of variance under the *robust* and *cluster( )* options, as well as incorporating sampling weights into analysis under the *weight* option. These functions are useful to researchers who analyze survey data with complex sampling designs using unequal sampling weights and multistaged stratification. The *weight* option is only available for the maximum likelihood estimation and supports various types of weights, such as sampling weights (i.e., specify *pwieghts = varname*); frequency weights (i.e., specify *fweights = varname*); analytic weights (i.e., specify *aweights = varname*); and importance weights (i.e., specify *iweights = varname*). When the *robust* and *cluster( )* options are specified, Stata follows a convention that does not print model Wald chi-square, because that statistic is misleading in a sandwich correction of standard errors. Various results can be saved for postestimation analysis. You may use either *predict* to save statistics or variables of interest, or *ereturn list* to check scalars, macros, and matrices that are automatically saved.

We now turn to an example (i.e., Section 4.5.1), and we will demonstrate the syntax. We encourage readers to briefly review the study details of the example before moving on to the application of *treatreg*.

To demonstrate the *treatreg* syntax and printed output, we use data from the National Survey of Child and Adolescent Well-Being (NSCAW). As explained in Section 4.5.1, the NSCAW study focused on the well-being of children whose primary caregiver had received treatment for substance abuse problems. For our demonstration study, we use NSCAW data to compare the psychological outcomes of two groups of children: those whose caregivers received substance abuse services (treatment variable *AODSERVE* = 1) and those whose caregivers did not (treatment variable *AODSERVE* = 0). Psychological outcomes were assessed using the Child Behavior Checklist–Externalizing (CBCL-Externalizing) score (i.e., the outcome variable *EXTERNAL3*). Variables entering into the selection equation (i.e., the **z** vector in Equation 4.8b) are *CGRAGE1, CGRAGE2, CGRAGE3, HIGH, BAHIGH, EMPLOY, OPEN, SEXUAL, PROVIDE, SUPERVIS, OTHER, CRA47A, MENTAL, ARREST, PSH17A, CIDI,* and *CGNEED*. Variables entering into the regression equation (i.e., the x vector in Equation 4.8a) are *BLACK, HISPANIC, NATAM, CHDAGE2, CHDAGE3,* and *RA.* Table 4.1 exhibits the syntax and output. Important statistics printed by the output are explained below.

First, *rho* is the estimated ρ in the variance-covariance matrix, which is the correlation between the error ε_i of the regression equation (4.8a) and the error u_i of the selection equation (4.8b). In this example, $\hat{\rho} = -.3603391$, which is estimated by Stata through the inverse hyperbolic tangent of ρ (i.e., labeled as "*/athrho*" in the output). The statistic "*atanh* ρ" is merely a middle step through which Stata obtains estimated ρ. It is the estimated ρ (i.e., labeled as *rho* in the output) that serves an important function.[2] The value of *sigma* is the estimated σ_ε in the above variance-covariance matrix, which is the variance of the regression equation's error term (i.e., variance of ε_i in Equation 4.8a). In this example, $\hat{\sigma}_\varepsilon = 12.1655$, which is estimated by Stata through $\ln(\sigma_\varepsilon)$ (i.e., labeled as "*/lnsigma*" in the output). As with "*atanh* ρ," "*lnsigma*" is a middle-step statistic that is relatively unimportant to users. The statistic labeled "*lambda*" is the inverse Mills ratio, or nonselection hazard, which is the product of two terms: $\hat{\lambda} = \hat{\sigma}_\varepsilon \hat{\rho} = (12.16551)(-.363391) = -4.38371$. Note that this is the statistic Heckman used in his two-step estimator (i.e., $\lambda_i = \phi(c_i)/(1 - \Phi(c_i))$ in Equation 4.7) to obtain a consistent estimation of the first-step equation. In the early days of discussing the Heckman or Heckit models, some researchers, especially economists, assumed that λ could be used to measure the level of selectivity effect, but this idea proved controversial and is no longer widely practiced. The estimated nonselection hazard (i.e., λ) can also be saved as a new variable in the data set for further analysis, if the user specifies *hazard(newvarname)* as a **treatreg** option. Table 4.2 illustrates this specification and prints out the saved hazard (variable *h1*) for the first 10 observations and the descriptive statistics.

Second, because the treatment effect model assumes the level of correlation between the two error terms is nonzero, and because violation of that assumption can lead to estimation bias, it is often useful to test H_0: $\rho = 0$. Stata prints results of a likelihood ratio test against "H_0: $\rho = 0$" at the bottom of the output. This ratio test is a comparison of the joint likelihood of an independent probit model for the selection equation and a regression model on the observed data against the treatment effect model likelihood. Given that $x^2 = 9.47$ ($p < .01$) from Table 4.1, we can reject the null hypothesis at a statistically significant level and conclude that ρ is not equal to 0. This suggests that applying the treatment effect model is appropriate.

Third, the reported model $x^2 = 58.97$ ($p < .0001$) from Table 4.1 is a Wald test of all coefficients in the regression model (except constant) being zero. This is one method to gauge the goodness of fit of the model. With $p < .0001$, the user can conclude that the covariates used in the regression model may be appropriate, and at least one of the covariates has an effect that is not equal to zero.

Table 4.1   Exhibit of Stata *treatreg* Output for the NSCAW Study

```
//Syntax to run treatreg
treatreg external3 black hispanic natam chdage2 chdage3 ra, ///
         treat(aodserv=cgrage1 cgrage2 cgrage3 high bahigh ///
         employ open sexual provide supervis other cra47a ///
         mental arrest psh17a cidi cgneed)

(Output)
Iteration 0:     log likelihood = -5780.7242
Iteration 1:     log likelihood =  -5779.92
Iteration 2:     log likelihood = -5779.9184
Iteration 3:     log likelihood = -5779.9184

Treatment-effects model — MLE                 Number of obs   =      1407
                                              Wald chi2(7)    =     58.97
Log likelihood = -5779.9184                   Prob > chi2     =    0.0000
```

	Coef.	Std. Err.	z	P>|z|	[95% Conf.	Interval]
external3						
black	-1.039336	.7734135	-1.34	0.179	-2.555198	.4765271
hispanic	-3.171652	.9226367	-3.44	0.001	-4.979987	-1.363317
natam	-1.813695	1.533075	-1.18	0.237	-4.818466	1.191077
chdage2	-3.510986	.9258872	-3.79	0.000	-5.325692	-1.696281
chdage3	-3.985272	.7177745	-5.55	0.000	-5.392085	-2.57846
ra	-1.450572	1.068761	-1.36	0.175	-3.545306	.6441616
aodserv	8.601002	2.474929	3.48	0.001	3.75023	13.45177
_cons	59.88026	.6491322	92.25	0.000	58.60798	61.15254
aodserv						
cgrage1	-.7612813	.3305657	-2.30	0.021	-1.409178	-.1133843
cgrage2	-.6835779	.3339952	-2.05	0.041	-1.338197	-.0289593
cgrage3	-.7008143	.3768144	-1.86	0.063	-1.439357	.0377284
high	-.118816	.1299231	-0.91	0.360	-.3734605	.1358286
bahigh	-.1321991	.1644693	-0.80	0.422	-.454553	.1901549
employ	-.1457813	.1186738	-1.23	0.219	-.3783777	.0868151
open	.5095091	.1323977	3.85	0.000	.2500143	.7690039
sexual	-.237927	.2041878	-1.17	0.244	-.6381277	.1622736
provide	.0453092	.1854966	0.24	0.807	-.3182575	.4088759
supervis	.1733817	.1605143	1.08	0.280	-.1412205	.4879839
other	.1070558	.1938187	0.55	0.581	-.272822	.4869335
cra47a	-.0190208	.1213197	-0.16	0.875	-.256803	.2187613
mental	.3603464	.1196362	3.01	0.003	.1258638	.5948289
arrest	.5435184	.1171897	4.64	0.000	.3138308	.7732059
psh17a	.6254078	.1410607	4.43	0.000	.348934	.9018816
cidi	.6945615	.1167672	5.95	0.000	.4657019	.9234211
cgneed	.6525656	.1880198	3.47	0.001	.2840535	1.021078
_cons	-1.759101	.3535156	-4.98	0.000	-2.451979	-1.066223
/athrho	-.3772755	.1172335	-3.22	0.001	-.6070489	-.1475022
/lnsigma	2.498605	.0203257	122.93	0.000	2.458768	2.538443
rho	-.3603391	.1020114			-.5420464	-.1464417
sigma	12.16551	.2472719			11.69039	12.65994
lambda	-4.38371	1.277229			-6.887032	-1.880387

```
LR test of indep. eqns. (rho = 0):   chi2(1) =     9.47    Prob > chi2 = 0.0021
```

Table 4.2   Exhibit of Stata *treatreg* Output: Syntax to Save Nonselection Hazard

```
//To request nonselection hazard or inverse Mills' ratio
treatreg external3 black hispanic natam chdage2 chdage3 ra, ///
         treat(aodserv=cgrage1 cgrage2 cgrage3 high bahigh ///
         employ open sexual provide supervis other cra47a ///
         mental arrest psh17a cidi cgneed) hazard(h1)

(Same output as Table 4.1, omitted)

(Output)

. list h1 in 1/10

          +-----------+
          |        h1 |
          |-----------|
       1. |  -.0496515 |
       2. | -.16962817 |
       3. |  2.0912486 |
       4. |   -.285907 |
       5. | -.11544285 |
          |-----------|
       6. | -.25318141 |
       7. | -.02696075 |
       8. | -.02306203 |
       9. | -.05237761 |
      10. | -.12828341 |
          +-----------+

. summarize h1

    Variable |       Obs        Mean    Std. Dev.       Min        Max
    ---------+-------------------------------------------------------
          h1 |      1407    -4.77e-12    .4633198   -1.517434   2.601461
```

Fourth, interpreting regression coefficients for the regression equation (i.e., the top panel of the output of Table 4.1) is performed in the same fashion as that used for a regression model. The sign and magnitude of the regression coefficient indicate the net impact of an independent variable on the dependent variable: other things being equal, the amount of change observed on the outcome with each one-unit increase in the independent variable. A one-tailed or two-tailed significance test on a coefficient of interest may be estimated using z and its associated p values. However, interpreting the regression coefficients of the selection equation is complicated because the observed w variable takes only two values (0 vs. 1), and the estimation process uses the probability of $w = 1$. Nevertheless, the sign of the coefficient is always meaningful, and significance of the coefficient is important. For example, using the variable *OPEN* (whether a child welfare case was open at baseline: *OPEN* = 1, yes; *OPEN* = 0, no), because the coefficient is positive (i.e., coefficient of *OPEN* = .5095), we know that the sample selection process (receipt or no receipt of services) is positively related to child welfare case status. That is, a caregiver with an open child welfare case was more likely to receive substance abuse services, and this relationship is statistically significant. Thus, coefficients with *p* values less than .05 indicate variables that contribute to selection bias. In this example, we observe eight variables with *p* values of less than .05 (i.e., variables *CGRAGE1, CGRAGE2, OPEN, MENTAL, ARREST, PSH17A, CIDI*, and *CGNEED*). The significance of these variables indicates presence of selection bias and underscores the importance of explicitly considering selection when modeling child outcomes. The eight variables are likely to be statistically significant in a logistic regression using the logit of service receipt (i.e., the logit of *AODSERV*) as a dependent variable and the same set of selection covariates as independent variables.

Fifth, the estimated treatment effect is an indicator of program impact net of observed selection bias; this statistic is shown by the coefficient associated with the treatment membership variable (i.e., *AODSERV* in the current example) in the regression equation. As shown in Table 4.1, this coefficient is 8.601002, and the associated *p* value is .001, meaning that other things being equal, children whose caregivers received substance abuse services had a mean score that was 8.6 units greater than children whose caregivers did not receive such services. The difference is statistically significant at a .001 level.

As previously mentioned, Stata automatically saves scalars, macros, and matrices for postestimation analysis. Table 4.3 shows the saved statistics for the demonstration model (Table 4.1). Automatically saved statistics can be recalled using the command "*ereturn list*."

4.5 Examples

This section describes three applications of the Heckit treatment effect model in social behavioral research. The first example comes from the NSCAW study, and, as in the *treatreg* syntax illustration, it estimates the impact on child well-being of the participation of children's caregivers in substance abuse treatment services. This study is typical of those that use a large, nationally representative survey to obtain observational data (i.e., data generated through a nonexperimental process). It is not uncommon in such studies to use a covariance control approach in an attempt to estimate the impact of program participation.

Our second example comes from a program evaluation that originally included a group randomization design. However, the randomization failed, and researchers were left with a group-design experiment in which treatment assignment was not ignorable. The example demonstrates the use of the

Table 4.3   Exhibit of Stata *treatreg* Output: Syntax to Check Saved Statistics

```
//Syntax to check saved statistics
treatreg external3 black hispanic natam chdage2 chdage3 ra, ///
treat(aodserv=cgrage1 cgrage2 cgrage3 high bahigh ///
employ open sexual provide supervis other cra47a ///
mental arrest psh17a cidi cgneed)
(Same output as Table 4.1, omitted)
ereturn list
(Output)
scalars:
e(rc) =   0
e(ll) =   -5779.918436833443
e(converged) =  1
e(rank) =  28
e(k) =  28
e(k_eq) =  4
e(k_dv) =  2
e(ic) =  3
e(N) =  1407
e(k_eq_model) =  1
e(df_m) =  7
e(chi2) =  58.97266440003305
e(p) =  2.42002594678e-10
e(k_aux) =  2
e(chi2_c) =  9.467229137793538
e(p_c) =  .0020917509015586
e(rho) =  -.3603390977383875
e(sigma) =  12.16551204275612
e(lambda) =  -4.383709633012229
e(selambda) =  1.277228928908404
macros:
e(predict) : "treatr_p"
e(cmd) : "treatreg"
e(title) : "Treatment-effects model—MLE"
e(chi2_ct) : "LR"
e(method) : "ml"
e(diparm3) : "athrho lnsigma, func(exp(@2)*(exp(@1)-exp(-@1))/(exp(@1)+
>  exp(-@1)) ) der( exp(@2)*(1-((exp(@1)-exp(-@1))/(exp(@1)+exp(-@1)))^2) exp(@2)*(
>  exp(@1. ."
e(diparm2) : "lnsigma, exp label("sigma")"
e(diparm1) : "athrho, tanh label("rho")"
e(chi2type) : "Wald"
e(opt) : "ml"
e(depvar) : "external3 aodserv"
e(ml_method) : "lf"
e(user) : "treat_ll"
e(crittype) : "log likelihood"
e(technique) : "nr"
e(properties) : "b V"
matrices:
e(b) :   1 x 28
e(V) :   28 x 28
e(gradient) :  1 x 28
e(ilog) :  1 x 20
e(ml_hn) :  1 x 4
e(ml_tn) :  1 x 4
functions:
e(sample)
```

Heckit treatment effect model to correct for selection bias while estimating treatment effectiveness.

The third example illustrates how to run the treatment effect model after multiple imputations of missing data.

4.5.1 APPLICATION OF THE TREATMENT EFFECT MODEL TO ANALYSIS OF OBSERVATIONAL DATA

Child maltreatment and parental substance abuse are highly correlated (e.g., English et al., 1998; U.S. Department of Health and Human Services [DHHS], 1999). A caregiver's abuse of substances may lead to maltreatment through many different mechanisms. For example, parents may prioritize their drug use more highly than caring for their children and substance abuse can lead to extreme poverty and to incarceration, both of which often leave children with unmet basic needs (Magura & Laudet, 1996). Policymakers have long been concerned about the safety of the children of substance-using parents.

Described briefly earlier, the NSCAW study was designed to address a range of questions about the outcomes of children who are involved in child welfare systems across the country (NSCAW Research Group, 2002). NSCAW is a nationally representative sample of 5,501 children, ages 0 to 14 years at intake, who were investigated by child welfare services following a report of child maltreatment (e.g., child abuse or neglect) between October 1999 and December 2000 (i.e., a multi-wave data collection corresponding to the data employed by this example). The NSCAW sample was selected using a two-stage stratified sampling design (NSCAW Research Group, 2002). The data were collected through interviews conducted with children, primary caregivers, teachers, and child welfare workers. These data contain detailed information on child development, functioning and symptoms, service participation, environmental conditions, and placements (e.g., placement in foster care or a group home). NSCAW gathered data over multiple waves, and the sample represented children investigated as victims of child abuse or neglect in 92 primary sampling units, principally counties, in 36 states.

The analysis for this example uses the NSCAW wave-2 data, or the data from the 18-month follow-up survey. Therefore, the analysis employs one-time-point data that were collected 18 months after the baseline. For the purposes of our demonstration, the study sample was limited to 1,407 children who lived at home (i.e., not in foster care), whose primary caregiver was female, and who were 4 years of age or older at baseline. We limited the study sample to children with female caregivers because females comprised the vast majority (90%) of primary caregivers in NSCAW. In addition, because NSCAW is a large observational database and our research questions focus on the impact of caregivers' receipt of substance abuse services on children's well-being, it is important to model the process of treatment assignment directly; therefore, the heterogeneity of potential causal effects is taken into consideration. In the NSCAW survey, substance abuse treatment was defined using six variables that asked the caregiver or child welfare worker whether the caregiver had received treatment for an alcohol or drug problem at the time of the baseline interview or at any time in the following 12 months.

Our analysis of NSCAW data was guided by two questions: (1) After 18 months of involvement with child welfare services, how were children of care-givers who received substance abuse services faring? and (2) Did children of caregivers who received substance abuse services have more severe behavioral problems than their counterparts whose caregivers did not receive such services?

As described previously, the choice of covariates hypothesized to affect sample selection serves an essential role in the analysis. We chose these variables based on our review of the substance abuse literature through which we determined the characteristics that were most frequently associated with substance abuse treatment receipt. Because no studies focused exclusively on female caregivers involved with child welfare services, we had to rely on literature regarding substance abuse in the general population (e.g., Knight, Logan, & Simpson, 2001; McMahon, Winkel, Suchman, & Luthar, 2002; Weisner, Jennifer, Tam, & Moore, 2001). We found four categories of characteristics: (1) social demographic characteristics (e.g., care-giver's age, *less than 35 years, 35 to 44 years, 45 to 54 years,* and *above 54 years;* caregiver's education, *less than high school degree, high school degree,* and *bachelor's degree or higher;* caregiver's employment status, *employed/not employed,* and whether the caregiver had "trouble paying for basic necessities," which was answered—*yes/no*); child welfare care status—*closed/open;* (2) risks (e.g., care-giver mental health problems—*yes/no;* child welfare care status—*closed/open;* caregiver history of arrest—*yes/no;* and the type of child maltreatment—*physical abuse, sexual abuse, failure to provide, failure to supervise,* and *other*); (3) caregiver's prior receipt of substance abuse treatment (i.e., caregiver alcohol or other drug treatment—*yes/no*); and (4) caregiver's need for alcohol and drug treatment ser-vices (i.e., measured on the World Health Organization's Composite International Diagnostic Interview–Short Form [CIDI-SF] that reports *presence/absence* of need for services and caregiver's self-report of service need—*yes/no*).

The outcome variable is the Achenbach Children's Behavioral Checklist (CBCL/4–18) that is completed by the caregivers. This scale includes scores for externalizing and internalizing behaviors (Achenbach, 1991). A high score on each of these measures indicates a greater extent of behavioral problems. When we con-ducted the outcome regression, we controlled for the following covariates: child's race/ethnicity (Black/non-Hispanic, White/non-Hispanic, Hispanic, and Native American); child's age (4 to 5 years, 6 to 10 years, and 11 and older); and risk assessment by child welfare worker at the baseline (risk absence/risk presence).

Table 4.4 presents descriptive statistics of the study sample. Of 1,407 children, 112 (8% of the sample) had a caregiver who had received substance abuse services, and 1,295 (92% of the sample) had caregivers who had not received services. Of 11 study variables, 8 showed statistically significant differences ($p < .01$) between treated cases (i.e., children whose caregivers had received services) and nontreated cases (i.e., children whose caregivers had not received services). For instance, the following caregivers were more likely to have received treatment services: those with a racial/ethnic minority status, with a positive risk to children, who were currently unemployed, with a current, open child welfare case, investigated for child maltreatment types of failure to provide or failure to supervise, who had trouble paying for basic necessities, with a history of mental health problems, with a history of arrest, with prior receipt of substance abuse treatment, CIDI-SF positive, and those who self-reported needing services. Without controlling for these selection effects, the estimates of differences on child outcomes would clearly be biased.

Table 4.5 presents the estimated differences in psychological outcomes between groups before and after adjustments for sample selection. Taking the externalizing score as an example, the data show that the mean externalizing score for the treatment group at the Wave 2 data collection (Month 18) was 57.96, and the mean score for the nontreatment group at the Wave 2 was 56.92. The unadjusted mean difference between groups was 1.04, meaning that the externalizing score for the treatment group was 1.04 units greater (or worse) than that for the nontreatment group. Using an OLS regression to adjust for covariates (i.e., including all variables used in the treatment effect model, i.e., independent variables used in both the selection equation and the regression equation), the adjusted mean difference is $- 0.08$ units; in other words, the treatment group is 0.08 units lower (or better) than the nontreatment group, and the difference is not statistically significant. These data suggest that the involvement of caregivers in substance abuse treatment has a negligible effect on child behavior. Alternatively, one might conclude that children whose parents are involved in treatment services do not differ from children whose parents are not referred to treatment. Given the high risk of children whose parents abuse substances, some might claim drug treatment to be successful.

Now, however, consider a different analytic approach. The treatment effect model adjusts for heterogeneity of service participation by taking into consideration covariates affecting selection bias. The results show that at the follow-up data collection (Month 18), the treatment group was 8.6 units higher (or worse) than the nontreatment group ($p < .001$). This suggests that both the unadjusted mean difference (found by independent t test) and the adjusted mean difference (found above by regression) are biased because we did not control appropriately for selection bias. A similar pattern is observed for the internalizing score. The findings suggest that negative program impacts may be masked in simple mean differences and even in regression adjustment.

Table 4.4   Sample Description for the Study Evaluating the Impacts of Caregiver's Receipt of Substance Abuse Services on Child Developmental Well-Being

Variable	N	%	% Caregivers Treated (% Service Users)	Bivariate χ^2 Test p Value
Substance-abuse service use				
No	1,295	92.0		
Yes (AODSERV)	112	8.0		
Child's race				
White	771	54.8	7.1	< .000
African American (BLACK)	350	24.9	8.9	
Hispanic (HISPANIC)	219	15.6	5.5	
Native American (NATAM)	67	4.8	20.9	
Child's age				
11+	488	34.7	7.8	.877
4–5 (CHDAGE2)	258	18.3	7.4	
6–10 (CHDAGE3)	661	47.0	8.3	
Risk assessment				
Risk absence	1,212	86.1	2.8	< .000
Risk presence (RA)	195	13.9	40.0	
Caregiver's age				
> 54	27	1.9	11.1	.756
< 35 (CGRAGE1)	804	57.1	7.5	
35–44 (CGRAGE2)	465	33.1	8.8	
45–54 (CGRAGE3)	111	7.9	7.2	
Caregiver's education				
No high school diploma	443	31.5	10.2	.104
High school diploma or equivalent (HIGH)	618	43.9	7.3	
B.A. or higher (BAHIGH)	346	24.6	6.4	

(Continued)

Table 4.4 (Continued)

Variable	N	%	% Caregivers Treated (% Service Users)	Bivariate χ^2 Test p Value
Caregiver's employment status				
Not employed	682	48.5	10.1	.004
Employed (EMPLOY)	725	51.5	5.9	
Child welfare case status				
Closed	607	43.1	3.8	< .000
Open (OPEN)	800	56.9	11.1	
Maltreatment type				
Physical abuse	375	26.7	5.3	.002
Sexual abuse (SEXUAL)	256	18.2	3.9	
Failure to provide (PROVIDE)	231	16.4	10.0	
Failure to supervise (SUPERVIS)	353	25.9	11.6	
Other (OTHER)	192	13.7	9.4	
Trouble paying for basic necessities				
No	988	70.2	6.4	.001
Yes (CRA47A)	419	29.8	11.7	
Caregiver mental health				
No problem	1,030	73.2	5.2	< .000
Mental health problem (MENTAL)	377	26.8	15.7	
Caregiver arrest				
Never arrested	959	68.2	4.1	< .000
Arrested (ARREST)	448	31.8	16.3	
AOD treatment receipt				
No treatment	1,269	90.2	5.5	< .000
Treatment (PSH17A)	138	9.8	30.4	
CIDI-SF				
Absence	1,005	71.4	4.1	< .000
Presence (CIDI)	402	28.6	17.7	
Caregiver report of need				
No	1,348	95.8	6.8	< .000
Yes (CGNEED)	59	4.2	35.6	

NOTES:

1. Reference group is shown next to the variable name.

2. Variable name in capital case is the actual name used in programming syntax.

Table 4.5   Differences in Psychological Outcomes Before and After Adjustments of Sample Selection

Group and Comparison	Outcome Measures: CBCL Scores	
	Externalizing	Internalizing
Mean (*SD*) of outcome 18 months after baseline		
Children whose caregivers received services (*n* = 112)	57.96 (11.68)	54.22 (12.18)
Children whose caregivers did not receive services (*n* = 1,295)	56.92 (12.29)	54.13 (11.90)
Unadjusted mean difference[a]	1.04	0.09
Regression-adjusted mean (*SE*) difference[a]	−0.08 (1.40)	−2.05 (1.37)
Adjusted mean (*SE*) difference controlling sample selection	8.60 (2.47)***	7.28 (2.35)**

a. Independent *t* tests on mean differences or *t* tests on regression coefficients show that none of these mean differences are statistically significant.

**p < .01, ***p < .001, two-tailed test.

4.5.2 EVALUATION OF TREATMENT EFFECTS FROM A PROGRAM WITH A GROUP RANDOMIZATION DESIGN

The "Social and Character Development" (SACD) program was jointly sponsored by the U.S. Department of Education and the Centers for Disease Control and Prevention. The SACD intervention project was designed to assess the impact of schoolwide social and character development education in elementary schools. Seven proposals to implement SACD were chosen through a peer review process, and each of the seven research teams implemented different SACD programs in elementary schools across the country. At each of the seven sites, schools were randomly assigned to receive either an intervention program or a control curriculum, and one cohort of students was followed from third grade (beginning in fall 2004) through fifth grade (ending in spring 2007). A total of 84 elementary schools were randomized to intervention and control at seven sites: Illinois (Chicago); New Jersey; New York (Buffalo, New York City, and Rochester); North Carolina; and Tennessee.

Using site-specific data (as opposed to data collected across all seven sites), this example reports preliminary findings from an evaluation of the SACD program implemented in North Carolina (NC). The NC intervention was also known as the Competency Support Program, which included a skills-training curriculum, *Making Choices,* designed for elementary school students. The primary goal of the *Making Choices* curriculum was to increase students' social competence and reduce their aggressive behavior. During their third-grade year, the treatment group received 29 *Making Choices* classroom lessons, and 8 follow-up classroom lessons in each of the fourth and fifth grades. In addition, special in-service training for classroom teachers in intervention schools focused on the risks of peer rejection and social isolation, including poor academic outcomes and conduct problems. Throughout the school year, teachers received consultation and support (2 times per month) in providing the *Making Choices* lessons designed to enhance children's social information processing skills. In addition, teachers could request consultation on classroom behavior management and social dynamics.

The investigators designed the Competency Support Program evaluation as a group randomization trial. The total number of schools participating in the study within a school district was determined in advance, and then schools were randomly assigned to treatment conditions within school districts; for each treated school, a school that best matched the treated school on academic yearly progress, percentage of minority students, and percentage of students receiving free or reduced-price lunch was selected as a control school (i.e., data collection only without receiving intervention). Over a 2-year period, this group randomization procedure resulted in a total of 14 schools (Cohort 1, 10 schools; Cohort 2, 4 schools) for the study: Seven received the Competency Support Program intervention, and seven received routine curricula. In this example, we focus on the 10 schools in Cohort 1.

As it turned out—and is often the case when implementing randomized experiments in social behavioral sciences—the group randomization did not work out as planned. In some school districts, as few as four schools met the study criteria and were eligible for participation. When comparing data from the 10 schools, the investigators found that the intervention schools differed from the control schools in significant ways: The intervention schools had lower academic achievement scores on statewide tests (Adequate Yearly Progress [AYP]); a higher percentage of students of color; a higher percentage of students receiving free or reduced-price lunches; and lower mean scores on behavioral composite scales at baseline. These differences were statistically significant at the .05 level using bivariate tests and logistic regression models. The researchers were confronted with the failure of randomization. Had these selection effects not been taken into consideration, the evaluation of the program effectiveness would be biased.

The evaluation used several composite scales that proved to have good psychometric properties. Scales from two well-established instruments were used for the evaluation: (1) the Carolina Child Checklist (CCC) and (2) the Interpersonal Competence Scale–Teacher (ICST). The CCC is a 35-item teacher questionnaire that yields factor scores on children's behavior, including social contact ($\alpha = .90$), cognitive concentration ($\alpha = .97$), social competence ($\alpha = .90$), and social aggression ($\alpha = .91$). The ICST is also a teacher questionnaire. It uses 18 items that yield factor scores on children's behavior, including aggression ($\alpha = .84$), academic competence ($\alpha = .74$), social competence ($\alpha = .75$), internalizing behavior ($\alpha = .76$), and popularity ($a = .78$).

Table 4.6 presents information on the sample and results of the Heckit treatment effect model used to assess change scores in the fifth grade. The two outcome measures used in the treatment effect models included the ICST Social Competence Score and the CCC Prosocial Behavior Score, which is a subscale of CCC Social Competence. On both these measures, high scores indicate desirable behavior. The dependent variable employed in the treatment effect model was a change score; that is, a difference of an outcome variable (i.e., ICST Social Competence or CCC Prosocial Behavior) at the end of the spring semester of the fifth grade minus the score at the beginning of fall semester of the fifth grade. Though "enterers" (students who transfer in) are included in the sample and did not have full exposure, most students in the intervention condition received *Making Choices* lessons during the third, fourth, and fifth grades. Thus, if the intervention was effective, then we would expect to observe a higher change (i.e., greater increase on the measured behavior) for the treated students than the control group students.

Before evaluating the treatment effects revealed by the models, we need to highlight an important methodological issue demonstrated by this example: the control of clustering effects using the Huber-White sandwich estimator of variance. As noted earlier, the Competency Support Program implemented in North Carolina used a group randomization design. As such, students were nested within schools, and students within the same school tended to exhibit similar behavior on outcomes. When analyzing this type of nested data, the analyst can use the option of *robust cluster* (•) in **treatreg** to obtain an estimation of robust standard error for each coefficient. The Huber-White estimator only corrects standard errors and does not change the estimation of regression coefficients. Thus, in Table 4.6 we present one column for the "Coefficient," along with two columns of estimated standard errors: one under the heading of "*SE*" that was estimated by the regular specification of **treatreg**, and the other under the heading of "Robust *SE*" that was estimated by the robust estimation of **treatreg**. Syntax that we used to create this analysis specifying control of clustering effect is shown in a note to Table 4.6.

As Table 4.6 shows, the estimates of "Robust *SE*" are different from those of "*SE*," which indicates the importance of controlling for the clustering effects. As a consequence of adjusting for clustering, conclusions of significance testing using "Robust *SE*" are different from those using "*SE*." Indeed, many covariates included in the selection equation are significant under "Robust *SE*" but not under "*SE*". In the following discussion, we focus on "Robust *SE*" to explore our findings.

The main evaluation findings shown in Table 4.6 are summarized below. First, selection bias appears to have been a serious problem because many

(Text continued on page 120)

Table 4.6　Estimated Treatment Effect Models of Fifth Grade's Change on ICST Social Competence Score and on CCC Prosocial Behavior Score

Predictor Variable	Descriptives % or M (SD)	Change on ICST Social Competence			Change on CCC Prosocial Behavior		
		Coefficient	SE	Robust SE	Coefficient	SE	Robust SE
Regression equation							
Age	7.90 (.50)	0.035	.0766	.0697	0.076	.0833	.1058
Gender female (Ref.[a] male)	53.35%	0.026	.0741	.0584	0.034	.0806	.0862
Race							
Black (Ref. Other)	27.70%	−0.125	.1810	.1858	−0.100	.1969	.1452
White	57.73%	−0.194	.1754	.1527	−0.190	.1907	.1662
Hispanic	9.62%	−0.070	.2093	.2286	0.068	.2275	.1217
Primary caregiver's education	5.51 (2.02)	0.026	.0229	.0330	0.062	.0248*	.0400
Income-to-needs ratio	170.55 (109.05)	0.000	.0004	.0004	0.000	.0005	.0004
Primary caregiver full-time employed (Ref. other)	56.85%	0.043	.0777	.0846	0.037	.0845	.0961
Father's presence at home (Ref. absence)	77.25%	−0.006	.0975	.1206	0.091	.1060	.0671

Predictor Variable	Descriptives % or M (SD)	Change on ICST Social Competence			Change on CCC Prosocial Behavior		
		Coefficient	SE	Robust SE	Coefficient	SE	Robust SE
Intervention (Ref. control)	40.23%	0.170	.0935+	.0941+	0.203	.1004*	.0723**
Constant		−0.505	.6629	.6078	−1.082	.7208	.8119
Selection equation							
School AYP Composite Score 2005	68.11 (9.58)	−0.350	.0421***	.1033**	−0.353	.0432***	.1070**
School's % of minority 2005	52.10 (14.51)	−0.146	.0319***	.0513**	−0.150	.0328***	.0545**
School's % of free lunch 2005	47.46 (9.98)	−0.122	.0249***	.0523*	−0.122	.0253***	.0525*
School's pupil-to-teacher ratio 2005	15.69 (1.57)	−1.214	.1596***	.2824***	−1.224	.1644***	.2972***
Age	7.90 (0.50)	0.361	.2708	.1400*	0.334	.2720	.1391*
Gender female (Ref. male)	53.35%	0.112	.2677	.1150	0.122	.2704	.01011
Race							
Black (Ref. other)	27.70%	−0.702	.5672	.4118+	−0.761	.5711	.3864*
White	57.73%	−1.119	.5478*	.5424*	−1.187	.5537*	.5173*
Hispanic	9.62%	−0.732	.6795	.1851***	−0.725	.6826	.1728***

(Continued)

Table 4.6 (Continued)

Predictor Variable	Descriptives % or M (SD)	Change on ICST Social Competence			Change on CCC Prosocial Behavior		
		Coefficient	SE	Robust SE	Coefficient	SE	Robust SE
Primary caregiver's education	5.51 (2.02)	0.070	.0835	.0176***	0.085	.0828	.0117***
Income-to-needs ratio	170.55 (109.05)	0.001	.0016	.0009	0.001	.0016	.0009
Primary caregiver full-time employed (Ref. other)	56.85%	−0.258	.2688	.0634***	−0.306	.2689	.0710***
Father's presence at home (Ref. absence)	77.25%	−0.045	.3175	.0898	−0.043	.3235	.0829
Baseline ICSTAGG—aggression	2.54 (1.51)	0.278	.1658+	.0766***	0.279	.1694	.0788***
Baseline ICSTACA—academic competence	5.26 (1.68)	0.161	.1058	.0583**	0.154	.1073	.0592**
Baseline ICSTINT—internalizing behavior	3.26 (1.16)	0.164	.1337	.0419***	0.188	.1345	.0389***
Baseline CCCCON—cognitive concentration	3.43 (1.01)	−0.308	.2151	.0992**	−0.302	.2183	.0931**

Predictor Variable	Descriptives % or M (SD)	Change on ICST Social Competence			Change on CCC Prosocial Behavior		
		Coefficient	SE	Robust SE	Coefficient	SE	Robust SE
Baseline CCCSTACT—social contact	3.83 (0.82)	−0.410	.1927*	.1076***	−0.416	.1977*	.1129***
Baseline CCCRAGG—relational aggression	3.97 (0.87)	0.844	.2700**	.1020***	0.832	.2778**	.1107***
Constant		49.884	6.7632***	14.2766***	50.593	7.0000***	14.9670**
Rho		0.206	.1510	.0914	0.083	.1481	.0608
Sigma		0.672	.0258	.0433	0.731	.0279	.0313
Lambda		0.139	.1022	.0693	0.061	.1084	.0455
Wald test of ρ = 0: χ^2 ($df = 1$)		1.71		4.80*		.31	1.85
Number of students	343						
Number of schools (clusters)	10						

NOTES:

Syntax to create the results of estimates with robust standard errors for the "Change on ICST Social Competence":

```
treatreg icstsc_ age Femalei Black White Hisp PCEDU IncPovL ///
    PCempF Father, treat(INTSCH=AYP05Cs pmin05 freel ///
    puptch05 age Femalei Black White Hisp PCEDU ///
    IncPovL PCempF Father icstagg icstaca icstint ///
    ccccon cccstact cccragg)robust cluster(school)
```

a. Ref. stands for reference group.

*$p < .05$, **$p < .01$, ***$p < .001$, +$p < .1$, two-tailed test.

variables included in the selection equation were statistically significant. We now use the analysis of the ICST Social Competence score as an example. All school-level variables (i.e., school AYP composite test score, school's percentage of minority students, school's percentage of students receiving free lunch, and school's pupil-to-teacher ratio) in 2005 (i.e., the year shortly after the intervention was completed) distinguished the treatment schools from the control schools. Students' race and ethnicity compositions were also different between the two groups, meaning that the African American, Hispanic, and Caucasian students are less likely than other students to receive treatment. The sign of the primary caregiver's education variable in the selection equation was positive, which indicated that primary caregivers of students from the intervention group had higher education than their control group counterparts ($p < .001$). In addition, primary caregivers of the treated students were less likely to have been employed full-time than were their control group counterparts. All behavioral outcomes at baseline were statistically different between the two groups, which indicated that treated students were rated as more aggressive ($p < .001$), had higher academic competence scores ($p < .01$), exhibited more problematic scores on internalizing behavior ($p < .001$), demonstrated lower levels of cognitive concentration ($p < .001$), displayed lower levels of social contact with prosocial peers ($p < .001$), and showed higher levels of relational aggression ($p < .001$). It is clear that without controlling for these selection effects, the intervention effect would be severely biased.

Second, we also included students' demographic variables and caregivers' characteristics in the regression equation based on the consideration that they were covariates of the outcome variable. This is an example of using some of the covariates of the selection equation in the regression equation (i.e., the x vector is part of the z vector, as described in Section 4.4). Results show that none of these variables were significant.

Third, our results indicated that the treated students had a mean increase in ICST Social Competence in the fifth grade that was 0.17 units higher than that of the control students ($p < 0.1$) and a mean increase in CCC Prosocial Behavior in the fifth grade that was 0.20 units higher than that of the control students ($p < .01$). Both results are average treatment effects of the sample that can be generalized to the population, although the difference on ICST Social Competence only approached significance ($p < .10$). The data showed that the Competency Support Program produced positive changes in students' social competence, which was consistent with the study's focus on social information processing skills. Had the study analysis not used the Heckit treatment effect model, the intervention effects would have been biased and inconsistent. An independent sample t test confirmed that the mean differences on both change scores were statistically significant at a .000 level, with inflated mean differences.

The *t* test showed that the intervention group had a mean change score on ICST Social Competence that was 0.25 units higher than the control group (instead of 0.17 units higher as shown by the treatment effect model) and a mean change score on CCC Prosocial Behavior that was 0.26 units higher than the control group (instead of 0.20 units higher as shown by the treatment effect model).

Finally, the null hypothesis of zero ρ, or zero correlation between the errors of the selection equation and the regression equation, was rejected at a significance level of .05 for the ICST Social Competence model, but it was not rejected for the CCC Prosocial Behavior model. This indicates that the assumption of nonzero ρ may be violated by the CCC Prosocial Behavior model. It suggests that the selection equation of the CCC Prosocial Behavior model may not be adequate, a topic that we will address in Chapter 8.

4.5.3 RUNNING THE TREATMENT EFFECT MODEL AFTER MULTIPLE IMPUTATIONS OF MISSING DATA

Missing data are nearly always a problem in research, and missing values represent a serious threat to the validity of inferences drawn from findings. Increasingly, social science researchers are turning to *multiple imputation* to handle missing data. Multiple imputation, in which missing values are replaced by values repeatedly drawn from conditional probability distributions, is an appropriate method for handling missing data when values are not missing completely at random (Little & Rubin, 2002; Rubin, 1996; Schafer, 1997). The following example illustrates how to analyze a treatment effect model based on multiply imputed data sets after missing data imputation using Rubin's rule for inference of imputed data. Given that this book is not focused on missing data imputation, we ignore the description about methods of multiply imputation. Readers are directed to the references mentioned above to find full discussion of multiple imputation. In this example, we attempt to show the method analyzing the treatment effect model based on multiply imputed data sets to generate a combined estimation of *treatreg* within Stata.

The Stata programs we recommend to fulfill this task are called *mim* and *mimstack*; both were created by John C. Galati at U.K. Medical Research Council and Patrick Royston at Clinical Epidemiology and Biostatistics Unit, the United Kingdom (Galati, Royston, & Carlin, 2009). Stata users may use the commands *findit mim* and *findit mimstack* within Stata with a Web-aware environment to search the programs and then install them by following the online instructions. The *mimstack* command is used for stacking a multiply imputed data set into the format required by *mim*, and *mim* is a prefix command for working with multiply imputed data sets to estimate the required model such as *treatreg*.

The commands to conduct a combined *treatreg* analysis look like the following:

mimstack, m(#) sortorder(varlist) istub(string) [nomj0 clear]

mim, cat(fit): **treatreg** depvar [indepvars], treat(depvar_t = indepvars_t)

where *m* specifies the number of imputed data sets, *sortorder* specifies a list of one or more variables that uniquely identify the observations in each of the data sets to be stacked, *istub* specifies the filename of the imputed data files to be stacked with the name specified in *string, nomj0* specifies that the original nonimputed data are not to be stacked with the imputed data sets, *clear* allows the current data set to be discarded, *mim, cat(fit)* informs that the program to be estimated is a regression model, and *treatreg* and its following commands are specifications one runs based on a single data set (i.e., data file without multiple imputation).

For the example depicted in Section 4.5.2, we had missing data on most independent variables. Using multiple imputation, we generated 50 imputed data files. Analysis shows that with 50 data sets, the imputation achieved a relative efficiency of 99%. The syntax to run a *treatreg* model analyzing outcome variable CCC Social Competence change score *ccscomch* using 50 data files is shown in the lower panel of Table 4.7.

In this *mimstack* command, *id* is the ID number used in all 50 files that uniquely identifies observations within each data set; *g3scom* is the common-portion name of the 50 files (i.e., the 50 imputed data files are named as g3scom1, g3scom2, . . . , and g3scom50); *nomj0* indicates that the original nonimputed data set was not used; and *clear* allows the program to discard the current data set once estimation of the current model is completed. In the above *mim* command, *cat(fit)* informs Stata that the combined analysis (i.e., *treatreg*) is a regression-type model; *treatreg* specifies the treatment effect model as usual, where the outcome variable for the regression equation is *ccscomch*, the independent variables for the regression equation are *ageyc, fmale, blck, whit, hisp, pcedu, ipovl, pcemft,* and *fthr,* the treatment membership variable is *intbl,* and the independent variables included in the selection equation are *ageyc, fmale, blck, whit, hisp, pcedu, ipovl, pcemft, fthr, dicsaca2,* and *dicsint2*. The *treatreg* model also estimates robust standard error to control for clustering effect where the variable identifying clusters is *schbl*.

Table 4.7 is an exhibition of the combined analysis invoked by the above commands. Results of the combined analysis are generally similar to those produced by a single-file analysis, but with an important difference: The combined analysis does not provide *rho, sigma,* and *lambda,* but instead shows *athrho* and *lnsigma* based on 50 files. Users may examine *rho, sigma,* and *lambda* by checking individual files to assess these statistics, particularly

Table 4.7   Exhibit of Combined Analysis of Treatment Effect Models Based on Multiple Imputed Data Files

```
Multiple-imputation estimates (treatreg)              Imputations =      50
Treatment-effects model — MLE                         Minimum obs =     590
Using Li-Raghunathan-Rubin estimate of VCE matrix     Minimum dof =   817.2
```

raggrch in~1	Coef.	Std. Err.	t	P>|t|	[95% Conf. Int.]		M.df
raggrch							
ageyc	-.084371	.041684	-2.02	0.043	-.16617	-.002573	995.5
fmale	-.025434	.081485	-0.31	0.755	-.185336	.134467	997.8
blck	-.108327	.190727	-0.57	0.570	-.482599	.265944	997.7
whit	-.128004	.225938	-0.57	0.571	-.571373	.315366	997.4
hisp	-.08513	.170175	-0.50	0.617	-.419073	.248813	995.9
pcedu	-.016804	.025657	-0.65	0.513	-.067152	.033544	987.0
ipovl	.000269	.000273	0.99	0.324	-.000267	.000806	817.2
pcemft	.008156	.111237	0.07	0.942	-.21013	.226442	995.6
fthr	-.04736	.080869	-0.59	0.558	-.206057	.111336	977.9
intbl	.580029	.427241	1.36	0.175	-.258367	1.41843	996.7
_cons	.71825	.457302	1.57	0.117	-.179138	1.61564	994.6
intbl							
ageyc	-.023355	.136161	-0.17	0.864	-.29055	.243841	996.4
fmale	.036754	.120963	0.30	0.761	-.200618	.274125	996.9
blck	.107904	.511518	0.21	0.833	-.89587	1.11168	997.5
whit	-.779496	.463681	-1.68	0.093	-1.6894	.130406	997.6
hisp	-.652384	.621296	-1.05	0.294	-1.87158	.566814	997.5
pcedu	.077337	.057483	1.35	0.179	-.035467	.190141	975.1
ipovl	-.000278	.001028	-0.27	0.787	-.002294	.001739	967.1
pcemft	-.21878	.138034	-1.58	0.113	-.489655	.052095	985.5
fthr	-.038307	.244794	-0.16	0.876	-.518684	.442069	986.9
dicsaca2	.052524	.064487	0.81	0.416	-.074021	.17907	996.5
dicsint2	.117797	.048969	2.41	0.016	.021701	.213892	991.5
_cons	-.384643	1.25788	-0.31	0.760	-2.85305	2.08376	994.7
athrho							
_cons	-.386324	.377802	-1.02	0.307	-1.1277	.355055	996.5
lnsigma							
_cons	-.372122	.122431	-3.04	0.002	-.612373	-.13187	997.5

Syntax to create the above results:
```
mimstack, m(50) sortorder("id") istub(g3scom) clear nomj0

mim, cat(fit): treatreg ccscomch ageyc fmale blck whit hisp ///
                pcedu ipovl pcemft fthr,treat(intbl=ageyc ///
                fmale blck whit hisp pcedu ipovl pcemft fthr ///
                dicsaca2 dicsint2) robust cluster(schbl)
```

if these statistics are consistent across files. If the user does not find a consistent pattern of these statistics across files, then the user will need to further investigate relations between the imputed data and the treatment effect model.

4.6 Conclusions

In 2000, the Nobel Prize Review Committee named James Heckman as a corecipient of the Nobel Prize in Economics in recognition of "his development of theory and methods for analyzing selective samples" (Nobel Prize Review Committee, 2000). This chapter reviews basic features of the Heckman sample selection model and its related models, including the treatment effect model and instrumental variables model. The Heckman model was invented at approximately the same time that statisticians started to develop the propensity score matching models, which we will examine in the next chapter. The Heckman model emphasizes modeling structures of selection bias rather than assuming mechanisms of randomization work to balance data between treated and control groups. However, surprisingly the Heckman sample selection model shares an important feature with the propensity score matching model: It uses a two-step procedure to model the selection process first and then uses the conditional probability of receiving treatment to control for bias induced by selection in the outcome analysis. Results show that the Heckman model, particularly its revised version called the treatment effect model, is useful in producing improved estimates of average treatment effects, especially when the causes of selection processes are known and are correctly specified in the selection equation.

To conclude this chapter, we share a caveat in running Heckman's treatment effect model. That is, the treatment effect model is sensitive to model "misspecification." It is well established that when the Heckman model is misspecified (i.e., when the predictor or independent variables are incorrect or omitted), particularly when important variables causing selection bias are not included in the selection equation, and when the estimated correlation between errors of the selection equation and the regression equation (i.e., the estimated ρ) is zero, then results of the treatment effect model are biased. The *Stata Reference Manual* (StataCorp, 2003) correctly states that

> the Heckman selection model depends strongly on the model being correct; much more so than ordinary regression. Running a separate probit or logit for sample inclusion followed by a regression, referred to in the literature as the two-part model (Manning, Duan, & Rogers, 1987)—not to be confused with Heckman's two-step procedure—is an especially attractive alternative if

the regression part of the model arose because of taking a logarithm of zero values. (p. 70)

Kennedy (2003) argues that the Heckman two-stage model is inferior to the selection model or treatment effect model using maximum likelihood because the two-stage estimator is inefficient. He also warns that in solving the omitted-variable problem, the Heckman procedure introduces a measurement error problem, because an estimate of the expected value of the error term is employed in the second stage. Finally, it is not clear whether the Heckman procedure can be recommended for small samples.

In practice, there is no definite procedure to test conditions under which the assumptions of the Heckman model are violated. As a consequence, sensitivity analysis is recommended to assess the stability of findings under the stress of alternative violations of assumptions. In Chapter 8, we present results of a Monte Carlo study that underscore this point. The Monte Carlo study shows that the Heckman treatment effect model works better than other approaches when ρ is indeed nonzero, and it works worse than other approaches when ρ is zero.

Notes

1. You could consider a set of omitted variables. Under such a condition, the model would use multiple instruments. All omitted variables meeting the required conditions are called *multiple instruments*. However, for simplicity of exposition, we omit the discussion of this kind of IV approach. For details of the IV model with multiple instruments, readers are referred to Wooldridge (2002, pp. 90–92).

2. The relation between *atanh* ρ and ρ is as follows:

$$atanh\,\rho = \frac{1}{2}\ln\left(\frac{1+\rho}{1-\rho}\right)\text{ or }-.3772755 = \frac{1}{2}\left(\frac{1+(-.3603391)}{1-(-.3603391)}\right),$$

using data of Table 4.1.

5

Propensity Score Matching and Related Models

This chapter describes three propensity score models, including propensity score matching, propensity score stratification (subclassification), and propensity score weighting. Although these models employ slightly different statistical theories and algorithms, and are sometimes described as addressing different analytical questions, all the models stem from Rosenbaum and Rubin's (1983) seminal work that defined a propensity score as the conditional probability of assignment to a particular treatment given a vector of observed covariates. When applied appropriately, these models can help solve the problem of selection bias and provide valid estimates of average treatment effects (ATEs) as well as average treatment effects for the treated (ATTs). In this chapter, we also introduce recent advances in propensity score matching. These include *generalized boosted regression* and *optimal matching*.

Section 5.1 provides an overview of propensity score models. The overview conceptualizes the modeling process as either a two-step or a three-step sequenced analysis. Section 5.2 reviews key propositions and corollaries derived and proved by Rosenbaum and Rubin (1983, 1984, 1985). The purpose of this review is to address two key questions: (1) How do propensity score models balance data? and (2) How do propensity score models solve the problem of dimensionality that plagued the classical matching algorithm? Section 5.3 focuses on Step 1 of the analysis: the specification of a logistic regression model and the search for a best set of conditioning variables that optimizes estimates of propensity scores. We review the procedure of generalized boosted regression in this section. Section 5.4 focuses on the second analytic step, that is, *resampling* based on estimated propensity scores. We then review various types of matching algorithms, including greedy matching and optimal matching. Section 5.5 focuses on the third step: postmatching analysis. We review various methods that follow matching, including the calculation of an overall mean and a variance of the treatment effect following subclassification, general procedures for multivariate analysis following greedy matching, procedures for conducting the Hodges-Lehmann aligned rank test of the treatment effect following optimal full matching or optimal variable matching, and regression adjustment (i.e., regressing difference scores of outcome on difference scores of covariates) following optimal pair matching. Section 5.6 illustrates an approach that omits matching: multivariate analysis using propensity scores as sampling weights, sometimes called *propensity score weighting*. Section 5.7 is an extension of the binary treatment-condition model: propensity scores with doses of treatment. Section 5.8 summarizes key features of computing software packages that can be used to run most models described in the chapter. Section 5.9 presents examples of selected models. Section 5.10 summarizes and concludes the discussion.

5.1 Overview

In 1983, Rosenbaum and Rubin published a seminal paper on propensity score analysis. That paper articulated the theory and application principles for a variety of propensity score models. Ever since this work, the propensity score method has grown at a rapid pace and moved in various directions for refinement. New models, such as the application of generalized boosted regression, have been developed to refine logistic regression and, in turn, to refine estimation of propensity scores. Other innovations include new models developed to refine matching algorithms and include optimal matching that applies developments in operations research (i.e., network flow theory) to matching. Additional new models, such as propensity score weighting, have been developed to combine propensity scores and conventional statistical methods.

Novice users of propensity score methods are often puzzled by new terminologies and seemingly different techniques. It is easy to get lost when first encountering the propensity score literature. To chart the way, we summarize the process of propensity score analysis in Figure 5.1 as either a two-step or a three-step analytic process.

We first examine propensity score models as a three-step analytic process.

Step 1: Seek the best conditioning variables or covariates that are speculated to be causing an imbalance between treated and control groups. A rigorous propensity score modeling always begins with estimation of the conditional probability of receiving treatment. Data analysts fulfill the task by estimating a logistic regression model (or similar model such as probit regression or discriminant analysis) or multinomial logit model to analyze the effects of multiple doses of treatment. The objective of analysis at this stage is to identify the observed

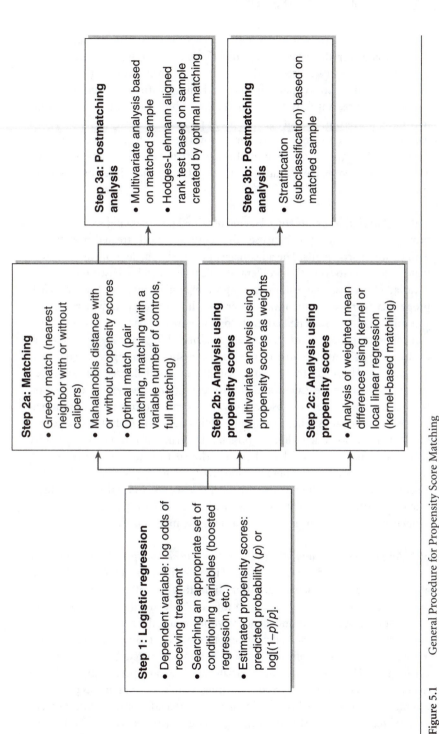

Figure 5.1 General Procedure for Propensity Score Matching

129

covariates affecting selection bias and further to specify a functional form of the covariates for the propensity score model. Ideally, we seek an optimal estimate of propensity scores. By definition, a propensity score is a conditional probability of a study participant receiving treatment given observed covariates; hence, not only treated participants but also control participants may have nonzero propensity scores. More precisely, the propensity score is a balancing score representing a vector of covariates. In this context, a pair of treated and control participants sharing a similar propensity score are essentially viewed as comparable, even though they may differ on values of specific covariates.

Step 2a: Matching or resampling. Having obtained the balancing scores (i.e., the propensities), the analyst then uses the scores to match treated participants with control participants. The advantage of using the single propensity score is that it allows us to solve the problem of failure in matching based on multiple covariates. Because the common support region formed by the estimated propensity scores does not always cover the whole range of study participants, you might not find matched controls for some treated participants, and some control participants may never be used; thus, matching typically leads to a loss of study participants. Because of this characteristic, matching is referred to as resampling. Although the original sample is not balanced on observed covariates between treatment and control conditions, the resample based on propensity scores balances observed covariates and controls for selection bias on observed measures. The key objective at this stage is to make the two groups of participants as much alike as possible in terms of estimated propensity scores. Various algorithms have been developed to match participants with similar propensity scores. These include greedy matching, Mahalanobis metric distance matching with or without propensity scores, and optimal matching. These algorithms differentially deal with the loss of participants whose propensity scores may be so extreme as to make matching difficult.

Step 3a: Postmatching analysis based on the matched samples. In principle, the new sample developed in Step 2a corrects for selection bias (on observed covariates) and violations of statistical assumptions that are embedded in multivariate models (such as the assumption embedded in regression about independence between an independent variable and the regression equation's error term). With this matched new sample, the analyst can perform multivariate analysis as is normally done using a sample created by a randomized experiment. However, most multivariate analyses are permissible only for matched samples created by greedy matching. With matched samples created by optimal matching, special types of analyses are needed. These include a special type of regression adjustment (i.e., regressing difference scores of outcomes between treated and control participants on difference scores of covariates) for a sample created by optimal pair matching; a special type of regression adjustment (i.e., regressing difference scores of the Hodges-Lehmann aligned rank of the outcome variable on difference scores of the aligned rank of covariates) for a sample created by either optimal variable matching or optimal full matching; and a test of the ATE using the Hodges-Lehmann aligned rank statistic for samples created by optimal full or variable matching.

Step 3b: Postmatching analysis using stratification of the propensity scores. Researchers could also perform stratification of the estimated propensity scores without conducting multivariate modeling; this stratification would also be conducted in a way similar to that in which researchers analyze treatment effects with data generated by randomized experiments, that is, by comparing the mean difference of an outcome between treatment and control conditions within a stratum, and then generating a mean and variance for the overall sample to gauge the sample ATE and its statistical significance.

As indicated in Figure 5.1, propensity score models are also used in two-step analytic processes. Models of this type use almost identical methods to estimate propensity scores and share the same Step 1 features as the three-step models. But the two-step models skip resampling (i.e., matching). They use propensity scores in a different way. For two-step models, the main features of Step 2 are shown below.

Step 2b: Multivariate analysis using propensity scores as sampling weights. As indicated earlier, this method (Hirano & Imbens, 2001; Robins & Ronitzky, 1995; Rosenbaum, 1987) does not resample the data, and therefore, avoids undesirable loss of study participants. Use of propensity scores as weights is analogous to the reweighting procedures used in survey sampling, where adjustments are made for observations on the basis of the probabilities for inclusion in a sample (McCaffrey, Ridgeway, & Morral, 2004). Propensity score weighting not only overcomes the problem of loss of sample participants but also offers two kinds of estimates for treatment effects: the ATE and the ATT.

Step 2c: Analysis of weighted mean differences using kernel or local linear regression (i.e., the kernel-based matching estimator developed by Heckman et al., 1997, and Heckman, Ichimura, & Todd, 1998). This method conducts a "latent" matching and combines weighting and outcome analysis into one step using nonparametric regression (i.e., either a tri-cubic kernel smoothing technique or a local linear regression). Given that this method is categorically different from Rosenbaum and Rubin's models, we describe kernel-based matching in Chapter 7.

Based on this overall process, we now move to the details to review key statistical theories, modeling principles, practice problems, and solutions for each step of the propensity score analysis approach.

5.2 The Problem of Dimensionality and the Properties of Propensity Scores

With complete data, Rosenbaum and Rubin (1983) defined the propensity score for participant i ($i = 1, \ldots, N$) as the conditional probability of assignment to a particular treatment ($W_i = 1$) versus nontreatment ($W_i = 0$) given a vector of observed covariates, x_i:

$$e(x_i) = pr(W_i = 1 | X_i = x_i).$$

The advantage of the propensity score in matching or stratification is its reduction of dimensions: The vector X may include many covariates, which represent many dimensions, and the propensity approach reduces all this dimensionality to a one-dimensional score. In conventional matching, as the number of matching variables increases, the researcher is challenged by the difficulty of finding a good match from the control group for a given treated participant. Rosenbaum (2002b) illustrated this with p covariates: Even if each covariate is a binary variable, there will be 2^p possible values of x. Suppose $p = 20$ covariates, then $2^{20} = 1,048,576$, or more than a million possible values of x. With a sample of hundreds or even thousands of participants, it is likely that many participants will have unique values of x and, therefore, can neither find matches from the control condition nor be used as a match for any treated case. Exact matching in this context often results in dropping cases and, in the presence of a large number of covariates or exceptional variation, may become infeasible.

The propensity score $e(x_i)$ is a balancing measure (so called, *the coarsest score*) that summarizes the information of vector x_i in which each x covariate is a *finest score*. Rosenbaum and Rubin (1983) derived and proved a series of theorems and corollaries showing the properties of propensity scores. The most important property is that a coarsest score can sufficiently balance differences observed in the finest scores between treated and control participants. From Rosenbaum and Rubin (1983) and Rosenbaum (2002b), the properties of propensity scores include the following:

1. Propensity scores balance observed differences between treated and control participants in the sample. Rosenbaum (2002b, p. 298) showed that a treated and control participant with the same value of the propensity score have

the same distribution of the observed covariate **X**. This means that in a stratum or matched set that is homogeneous on the propensity score, treated and control participants may have differing values for **X** (i.e., if two participants have the same propensity score, they still could differ on an observed covariate such as gender, if gender—the finest score—is included in the **X** vector), but the differences will be chance differences rather than systematic differences.

2. Treatment assignment and the observed covariates are conditionally independent given the propensity score, that is,

$$\mathbf{x}_i \perp w_i | e(\mathbf{x}_i).$$

This property links the propensity score to the assumption regarding strongly ignorable treatment assignment. In other words, conditional on the propensity score, the covariates may be considered independent of assignment to treatment. Therefore, for observations with the same propensity score, the distribution of covariates should be the same across the treated and control groups. Furthermore, this property means that, conditional on the propensity score, each participant has the same probability of assignment to treatment, as in a randomized experiment.

3. If the strongly ignorable treatment assignment assumption holds and $e(\mathbf{x}_i)$ is a balancing score, then the expected difference in observed responses to the two treatment conditions at $e(\mathbf{x}_i)$ is equal to the ATE at $e(\mathbf{x}_i)$. This property links the propensity score model to the counterfactual framework and shows how the problem of not observing outcomes for the treated participants under the control condition (a problem discussed in Section 2.2) can be resolved. It follows that the mean difference of the outcome variable between treated and control participants for all units with the same value of the propensity score is an unbiased estimate of the ATE at that propensity score. That is,

$$E[E(Y_1, |e(\mathbf{x}_i), W_i = 1) - E(Y_0, |e(\mathbf{x}_i), W_i = 0)] = E[Y_1 - Y_0|e(\mathbf{x}_i)].$$

4. Rosenbaum and Rubin (1983, p. 46) derived corollaries to justify three key approaches using the propensity scores. These corollaries form the foundation for all models described in this chapter.

    a. *Pair matching:* The expected difference in responses of treatment and control units in a matched pair with same value of propensity score $e(\mathbf{x})$ equals the ATE at $e(\mathbf{x})$, and the mean of matched pair differences obtained by this two-step sampling process is unbiased for the ATE $\tau = E(Y_1|W = 1) - E(Y_0|W = 0) = E[Y_1 - Y_0|e(\mathbf{x})].$

    b. *Subclassification of propensity scores* such that all units within a stratum have the same $e(\mathbf{x})$ and at least one unit in the stratum receives each treatment condition—the expected difference in treatment mean equals the ATE at that value of $e(\mathbf{x})$, and the weighted average of such differences is unbiased for the treatment effect $\tau = E(Y_1|W=1) - E(Y_0|W=0)$.

c. *Covariance adjustment:* Assume that the treatment assignment is strongly ignorable at the balancing score $e(\mathbf{x})$, and the conditional expectation of Y_t $(t=0,1)$ is linear: $E[Y_t|W=t, e(\mathbf{x})] = \alpha_t + \beta_t e(\mathbf{x})$, then the estimator $(\hat{\alpha}_1 - \hat{\alpha}_0) + (\hat{\beta}_1 - \hat{\beta}_0)e(x)$ is conditionally unbiased given $e(\mathbf{x}_i)$ $(i=1, \ldots, n)$ for the treatment effect at $e(\mathbf{x})$, namely $E[Y_1 - Y_0|e(\mathbf{x})]$, if $\hat{\alpha}_t$ and $\hat{\beta}_t$ are conditionally unbiased estimators of α_t and β_t, such as least squares estimators.

In the above descriptions, the propensity score $e(\mathbf{x})$ is defined as a predicted probability of receiving treatment in a sample where treatment assignment is nonignorable. Typically, this value is a predicted probability saved from an estimated logistic regression model. In practice, Rosenbaum and Rubin (1985) suggested using the logit of the predicted probability as a propensity score (i.e., $\hat{q}(x) = \log[(1 - \hat{e}(x))/\hat{e}(x)]$), because the distribution of $\hat{q}(x)$ approximates to normal. Note that in the literature, the quantity $\hat{q}(x)$ is also called an *estimated propensity score*, though $\hat{q}(x)$ differs from $\hat{e}(x)$ as given by the previous equation. In this chapter and elsewhere, except when explicitly noted, we will follow the convention of referring to $\hat{e}(x)$ as the *propensity score*. Readers should keep in mind that, in practice, a logit transformation of $\hat{e}(x)$ (i.e., $\hat{q}(x)$) may be used, and $\hat{q}(x)$ has distributional properties that may make it more desirable than $\hat{e}(x)$.

These theorems and corollaries established the foundation of the propensity score matching approach and many related procedures. For instance, the property of $x_i \perp w_i|e(\mathbf{x}_i)$ leads to a procedure often used to check whether estimated propensity scores successfully remove imbalance on observed covariates between treated and control groups. The procedure is as follows: (a) the analyst conducts a bivariate test (i.e., a Wilcoxon rank-sum test— also known as the Mann-Whitney two-sample statistic—an independent sample t test, or a one-way ANOVA, for a continuous covariate; or a chi-square test for a categorical covariate) using treatment condition as a grouping variable before matching; if the bivariate test shows significant difference between treated and control groups on a covariate, then the analyst needs to control the covariate by including the covariate in the model estimating propensity scores; (b) after matching on the propensity scores, the analyst performs similar bivariate tests with some adjustments (e.g., instead of using a Wilcoxon rank-sum test the analyst may use a Wilcoxon matched pairs signed-rank test or absolute standardized difference in covariate means); if the postmatching bivariate tests are nonsignificant, then we may conclude that the propensity score has successfully removed group differences on the observed covariates; and (c) if the postmatching bivariate tests show significant differences, the model predicting propensity scores should be reconfigured and rerun until the matching successfully removes all significant imbalances.

Implied in our use above, the covariates in the vector **X** are called *conditioning* or *matching* variables. A correct specification of covariates in the Step 1 model is crucial to the propensity score approach because the final estimation of the treatment effect is sensitive to this specification (Rubin, 1997). Many studies show that the choice of conditioning variables can make a substantial difference in the overall performance of propensity score analysis (e.g., Heckman, Ichimura, Smith, & Todd, 1998; Heckman et al., 1997; Lechner, 2000).

A *correct specification of the conditioning model* predicting propensity scores has two aspects: One is to include the correct variables in the model; that is, researchers should include important covariates that have theoretical relevance. To do this, we typically rely on substantive information and prior studies about predictors of receiving treatment. The other is to specify correctly the functional form of conditioning variables. This may involve introducing polynomial and interaction terms. The dilemma the analyst faces is that there is no definitive procedure or test available to provide guidance in specifying a best propensity score model, and theory often provides weak guidance as to how to choose and configure conditioning variables (Smith & Todd, 2005). This dilemma prompted the development of promising methods such as generalized boosted regression for searching best propensity scores. The next section describes the development of propensity scores and strategies for dealing with the specification problem.

5.3 Estimating Propensity Scores

As mentioned earlier, several methods for estimating the conditional probability of receiving treatment using a vector of observed covariates are available. These methods include logistic regression, the probit model, and discriminant analysis. Of these methods, this book describes only logistic regression because it is the prevailing approach. A closely related method is *Mahalanobis metric distance*, which was invented prior to methods for propensity score matching (Cochran & Rubin, 1973; Rubin, 1976, 1979, 1980a). A Mahalanobis metric distance per se is not a model-based propensity score. However, the Mahalanobis metric distance serves a similar function as a propensity score and is an important statistic used in greedy matching, optimal matching, and multivariate matching. Accordingly, we will examine the Mahalanobis metric distance in Sections 5.4.1 and 5.4.2 and in Chapter 6.

5.3.1 BINARY LOGISTIC REGRESSION

The conditional probability of receiving treatment when there are two treatment conditions (i.e., treatment vs. control) is estimated using binary logistic regression. Denoting the binary treatment condition as W_i ($W_i = 1$, if a study case is in the treatment condition, and $W_i = 0$, if the case is in the control condition) for the ith case ($i = 1, \ldots, N$), the vector of conditioning variables as X_i, and the vector of regression parameters as β_i, a binary logistic regression depicts the conditional probability of receiving treatment as follows:

$$P(W_i | X_i = x_i) = E(W_i) = \frac{e^{x_i \beta_i}}{1 + e^{x_i \beta_i}} = \frac{1}{1 + e^{-x_i \beta_i}}. \qquad (5.1)$$

This is a nonlinear model, meaning that the dependent variable W_i is not a linear function of the vector of conditioning variables x_i. However, by using an appropriate link function such as a logit function, we can express the model as a generalized linear model (McCullagh & Nelder, 1989). Although W_i is not a linear function of x_i, its transformed variable through the logit function (i.e., the natural logarithm of odds or $\log\{P(W_i)/[1 - P(W_i)]\}$) becomes a linear function of x_i:

$$\log_e \left(\frac{P}{1 - P} \right) = x_i \beta_i,$$

where P denotes $P(W_i)$.

Model 5.1 is estimated with the maximum likelihood estimator. To ease the exposition, we now assume that there are only two conditioning variables x_1 and x_2. The log likelihood function of Model 5.1 with two conditioning variables can be expressed as follows:

$$\log_e \ell(\beta_0, \beta_1, \beta_2) = \sum_{i=1}^{n} W_i(\beta_0 + \beta_1 x_{1i} + \beta_2 x_{2i})$$
$$- \sum_{i=1}^{n} \log_e [1 + \exp(\beta_0 + \beta_1 x_{1i} + \beta_2 x_{2i})]. \qquad (5.2)$$

The partial derivative of log ℓ with respect to β maximizes the likelihood function. In practice, the problems are seldom solved analytically, and we often rely on a numerical procedure to find estimates of β. Long (1997, pp. 56–57) described three numerical estimators: the Newton-Raphson method, the scoring method, and the B-triple-H (BHHH) method. Typically, a numerical method involves the following steps: (a) insert starting values (i.e., "guesses") of β_0, β_1, and β_2 in the right-hand side of Equation 5.2 to obtain a first guess of log ℓ; (b) insert a different set of β_0, β_1, and β_2 into the right-hand side equation to obtain a second guess of log ℓ; by comparing the new log ℓ with the old one, the analyst knows the direction for trying the next set of β_0, β_1, and β_2; the process from Step (a) to Step (b) is called an *iteration*; (c) replicate the above process several times (i.e., running several iterations) until the largest value of log ℓ is obtained (i.e., the maximum log likelihood function) or until the difference in log ℓ between two iterations is no longer greater than a predetermined criterion value, such as 0.000001.

Estimated values of $\hat{\beta}_0, \hat{\beta}_1,$ and $\hat{\beta}_2$ (i.e., $\hat{\beta}_0, \hat{\beta}_1,$ and $\hat{\beta}_2$) are logistic regression coefficients at which the likelihood of reproducing sample observations is maximized. Using these estimated coefficients and applying Equation 5.1 (i.e., replacing $\beta_0, \beta_1,$ and β_2 with $\hat{\beta}_0, \hat{\beta}_1,$ and $\hat{\beta}_2$), the analyst obtains the predicted probability of receiving treatment (i.e., estimated propensity score) for each sample participant i.

As in running OLS regression or other multivariate models, we must be sensitive to the nature of the data at hand and the possibility of violations of assumptions. Routine diagnostic analyses, such as tests of multicollinearity, tests of influential observations, and sensitivity analyses should be used to assess the fit of the final model to the data. A number of statistics have been developed to assess the goodness of fit of the model. Unfortunately, none of these statistics indicates whether the estimated propensity scores are representative of the true propensity scores. Notwithstanding, meeting requirements embedded in these fit statistics is a minimum requirement or starting point.

Details of goodness-of-fit indices for the logistic regression model can be found in textbooks on logistic regression or limited dependent variable analysis (e.g., Kutner, Nachtsheim, & Neter, 2004; Long, 1997). Here, we summarize a few indices and include cautionary statements for their use.

1. *Pearson chi-square goodness-of-fit test:* This test detects major departures from a logistic response function. Large values of the test statistic (i.e., those associated with a small or significant p value) indicate that the logistic response function is not appropriate. However, it is important to note that the test is not sensitive to small departures (Kutner et al., 2004).

2. *Chi-square test of all coefficients:* This test is a likelihood ratio test and analogous to the F test for linear regression models. We can perform a chi-square test using the log likelihood ratio, as follows:

   Model chi-square = 2 log likelihood of the full model −
   $\qquad\qquad\qquad$ 2 log likelihood of the model with intercept only

   If the Model chi-square $> \chi^2$ $(1 - \alpha,$ $df =$ number of conditioning variables), then we reject the null hypothesis stating that all coefficients except the intercept are equal to zero. As a test of models estimated by the maximum likelihood approach, a large sample is required to perform the likelihood ratio test, and this test is problematic when the sample is small.

3. *Hosmer-Lemeshow goodness-of-fit test:* This test first classifies the sample into small groups (e.g., g groups) and then calculates a test statistic using the Pearson chi-squares from the $2 \times g$ tables of observed and estimated expected frequencies. A test statistic that is less than χ^2 $(1 - \alpha,$ $df = g - 2)$ indicates a good model fit. The Hosmer-Lemeshow test is sensitive to sample size. That is,

in the process of reducing the data through grouping, we may miss an important deviation from fit due to a small number of individual data points. Hence, we advocate that, before concluding that a model fits, an analysis of the individual residuals and relevant diagnostic statistics be performed (Hosmer & Lemeshow, 1989, p. 144).

4. *Pseudo R^2:* Because the logistic regression model is estimated by a non-least-squares estimator, the common linear measure of the proportion of the variation in the dependent variable that is explained by the predictor variables (i.e., the coefficient of determination R^2) is not available. However, several pseudo R^2s for the logistic regression model have been developed by analogy to the formula defining R^2 for the linear regression model. These pseudo R^2s include Efron's, McFadden's, adjusted McFadden's, Cox and Snell's, Nagelkerke/Cragg and Uhler's, McKelvey and Zavoina's, count R^2, and adjusted count R^2. In general, a higher value in a pseudo R^2 indicates a better fit. However, researchers should be aware of several limitations of pseudo R^2 measures and interpret their findings with caution. The UCLA Academic Technology Services (2008) provides a detailed description of each of these pseudo R^2s and concludes as follows:

> Pseudo R-squares cannot be interpreted independently or compared across datasets: They are valid and useful in evaluating multiple models predicting the same outcome on the same dataset. In other words, a pseudo R-squared statistic without context has little meaning. A pseudo R-square only has meaning when compared to another pseudo R-square of the same type, on the same data, predicting the same outcome. In this situation, the higher pseudo R-square indicates which model better predicts the outcome.

Given these basic concepts in logistic regression, how do we optimize the estimates of propensity scores in the context of observational studies? It is worth underscoring a key point mentioned previously: A good logistic regression model that meets routine requirements and standards is a necessary but insufficient condition for arriving at the best propensity scores.

5.3.2 STRATEGIES TO SPECIFY A CORRECT MODEL PREDICTING PROPENSITY SCORES

A principal question then arises: What defines the "best" logistic regression? The answer is simple: We need propensity scores that balance the two groups on the observed covariates. By this criterion, a best logistic regression is a model that leads to estimated propensity scores that best represent the true propensity scores. The challenge is that the true propensity scores are unknown, and therefore, we must seek methods to measure the fit between the estimated and the unknown true scores.

The literature on propensity score matching is nearly unanimous in its emphasis on the importance of including in models carefully chosen and appropriate conditioning variables in the correct functional form. Simulation and replication studies have found that estimates of treatment effects are sensitive to different specifications of conditioning variables. For example, Smith and Todd (2005) found that using more conditioning variables may exacerbate the common support region problem—an issue we describe in detail later. In general, a sound logistic regression model should minimize the overall sample prediction error. That is, it should minimize the overall sample difference between the observed value W_i and $P(W_i = 1)$. McCaffrey et al. (2004) developed a procedure using generalized boosted modeling (GBM) that seeks the best balance of the two groups on observed covariates. This procedure is described in Section 5.3.4. The algorithm McCaffrey et al. invoked altered the GBM criterion in such a way that iterations stop only when the sample average standardized absolute mean difference (ASAM) in the covariates is minimized. Suffice it to say that a best logistic regression model should take covariate balance into consideration, and in practice outside of propensity score estimation, this may or may not be a crucial concern in running logistic regression. Strategies for fitting the propensity score model are summarized below.

1. Rosenbaum and Rubin (1984, 1985) described a procedure that used high-order polynomial terms and/or cross-product interaction terms in the logistic regression through repeatedly executing the following tasks: running logistic regression, matching, bivariate tests of covariate balances based on the matched data, and rerunning logistic regression if covariate imbalances remain. As mentioned earlier, it is a common practice to run bivariate analysis before and after matching. Using a Wilcoxon rank-sum (Mann-Whitney) test, t test, chi-square, or other bivariate method, we test whether the treated and control groups differ on covariates included in the logistic regression. The propensity score matching approach aims to achieve approximately the same distribution of each covariate between the two groups. Nonetheless, even after matching, significant differences may remain between groups. When these differences exist, the propensity score model can be reformulated or the analyst can conclude that the covariate distributions did not overlap sufficiently to allow the subsequent analysis to adjust for these covariates (Rubin, 1997). In rerunning the propensity score model, we may include either a square term of the covariate that shows significance after matching or a product of two covariates if the correlation between these two covariates is likely to differ between the groups (Rosenbaum & Rubin, 1984).

2. Rosenbaum and Rubin (1984) further recommended applying stepwise logistic regression to select variables. Note that data-driven approaches determine the inclusion or exclusion of conditioning variables based on a Wald statistic (or t statistic) and its associated p value. Thus, the estimated model that results contains only those variables that are significant at a predetermined

level. Rosenbaum (2002b) suggested a similar rule of thumb: The logistic regression model should include all pretreatment covariates whose group differences met a low threshold for significance, such as $|t| > 1.5$.

3. Eichler and Lechner (2002) used a variant of a measure suggested by Rosenbaum and Rubin (1985), which was based on standardized differences between the treatment and matched comparison groups in terms of means of each variable in x, squares of each variable in x, and first-order interaction terms between each pair of variables in x.

4. Dehejia and Wahba (1999) used a procedure similar to the method recommended by Rosenbaum and Rubin (1984), but they added stratification to determine the use of higher-order polynomial terms and interactions. They first fit a logistic regression model specifying main effects only, and then they stratify the sample by estimated propensity scores. Based on the stratified sample, they test for differences in the means and standard deviations of covariates between treated and control groups within each stratum. If significant differences remain, they add higher-order polynomial terms and interactions. The process continues until no significant differences are observed within strata.

5. Hirano and Imbens (2001) developed a procedure that fully relies on a statistical criterion to seek conditioning predictors in logistic regression and then predictors in a follow-up outcome regression. Their search for predictor variables for both the logistic regression and the follow-up regression model is, to some extent, similar to the stepwise method, but it tests a range of models by using different cutoff values of the t statistic. Because this method is innovative, important, and deserving of scrutiny, we describe it in a separate section below.

In sum, a careful selection of conditioning variables and a correct specification of the logistic regression are crucial to propensity score matching. Although scholars in the field have suggested a variety of rules and approaches, there is no definitive procedure of which we are aware. Because the selection of conditioning variables affects both balance on propensity scores and the final estimate of the treatment effect, every effort must be made to ensure that the estimate of propensity scores has considered all substantively relevant factors and used observed data in a way that is not sensitive to specification errors. The method described below illustrates these points.

5.3.3 HIRANO AND IMBENS'S METHOD FOR SPECIFYING PREDICTORS RELYING ON PREDETERMINED CRITICAL t VALUES

Hirano and Imbens's (2001) study was innovative in several aspects: It treated estimated propensity scores as sampling weights and conducted propensity score weighting analysis; it combined propensity score weighting and regression adjustment; it demonstrated the importance of carefully searching predictors for both logistic regression and outcome regression; and it tested the sensitivity of specifications of critical t (i.e., statistical decisions) to the targeted outcome of estimates of treatment effectiveness. In the following description, we focus on their method using predetermined critical t values and put aside the methodology of propensity score weighting. We deal with the weighting procedure in Section 5.6.

The problem of variable selection encountered by Hirano and Imbens (2001) warrants a detailed description. A total of 72 covariates was available for use in analysis: Some or all may be used in the logistic regression, and some or all may be used in the outcome regression. Hirano and Imbens viewed the inclusion of variables as a classic subset selection problem in regression (e.g., in the sense of Miller, 1990). With 72 variables, it seemed that any predetermined rule for selecting covariates into either equation (i.e., logistic regression or outcome regression) was subjective and would affect the results of the other model. Therefore, instead of determining which variables to include in the equations using theoretical guidance or rules derived empirically, Hirano and Imbens ran a range of possible models using different values of the t statistic (i.e., a critical t whose value determines whether a covariate should be entered into an equation). The authors then showed estimated treatment effects under all possible combinations of models.

Hirano and Imbens used the following steps to estimate the treatment effects:

1. Denoting the critical t value for inclusion of a covariate in the logistic regression as t_{prop}, and the critical t value for inclusion of a covariate in the outcome regression as t_{reg}, they considered all pairs with t_{prop} and t_{reg} in the set of $\{0, 1, 2, 4, 8, 16, \infty\}$.

2. They ran 72 *simple* logistic regression models under a given value of t_{prop}. Each time, they used only one of the 72 covariates in the model. Suppose $t_{prop} = 2$. Under this critical value, if an estimated regression coefficient had observed $t < 2$, then the covariate was ruled out; otherwise, it was retained and was included in the final model of logistic regression under $t_{prop} = 2$.

3. After running all 72 simple logistic regressions under $t_{prop} = 2$, Hirano and Imbens found that only a portion of the 72 variables were significant individually. These variables then became the covariates chosen for a logistic regression predicting propensity scores under $t_{prop} = 2$.

4. Hirano and Imbens then ran outcome regressions in a fashion similar to Steps 2 and 3. That is, they ran 72 *simple* regressions, with each model containing only one covariate. A combined regression using all covariates that were individually significant under $t_{reg} = 2$ followed.

5. The authors then replicated Steps 2 to 4 for all other critical t values; that is, for $t_{prop} = 0, 1, 4, 8, 16$, and ∞, and for $t_{reg} = 0, 1, 4, 8, 16$, and ∞.

6. Finally, Hirano and Imbens calculated treatment effects under conditions defined and produced by all pairs of t_{prop} and t_{reg}.

In essence, Hirano and Imbens's approach is a sensitivity analysis, that is, it tests how sensitive the estimated treatment effect is to different specifications of the logistic regression and outcome regression. Under a critical value for both t_{prop} and $t_{reg} = \infty$ (i.e., a scenario requiring an extremely large observed t for its inclusion), if none of the 72 covariates is used in the logistic regression and in the outcome regression, then the estimated treatment effect is a mean difference between the treated and control patients without any control of covariates. At the other extreme, when a critical value for both t_{prop} and $t_{reg} = 0$ (i.e., no restriction is imposed on the entering criterion, because any covariate could have an observed t value greater than 0), all 72 covariates are used in both logistic regression and outcome regression. Under this scenario, the estimated treatment effect is a *stringent* estimation. The number of covariates used in scenarios of other paired t values (i.e., 1, 2, 4, 8, 16) varies, and a high value of critical t typically leads to a model using few covariates. Using this setup, Hirano and Imbens presented a table showing estimated treatment effects under all combinations of t_{prop} and t_{reg}. The table includes a total of 49 cells, which represent seven rows for t_{prop} and seven columns for t_{reg}, and correspond to the seven values in the predetermined set of critical t. The key finding is that there is a great range of variation among estimates of treatment effects for certain scenarios but not for others. Specifically, Hirano and Imbens found that the ranges of estimated treatment effects were the smallest for $t_{prop} = 2$ and for $t_{reg} = 2$. The range of treatment effects under $t_{prop} = 2$ is $(-.062, -.053)$ and under $t_{reg} = 2$ is $(-.068, -.061)$. Based on this finding, the authors concluded that the true treatment effect was around $-.06$. This estimated effect was further verified and confirmed by a different estimation using a *bias-adjusted matching estimator.*[1]

The crucial feature of Hirano and Imbens's (2001) approach is flexibility in selecting covariates and specifying models. "By using a flexible specification of the propensity score, the sensitivity to the specification of the regression function is dramatically reduced" (p. 271). Hirano and Imbens (2001) concluded their work by stating,

> Estimation of causal effects under the unconfoundedness assumption can be challenging where the number of covariates is large and their functional relationship to the treatment and outcome are not known precisely. By flexibly estimating both the propensity score and the conditional mean of the outcome given the treatment and the covariates one can potentially guard against misspecification in a relatively general way. Here we propose a simple rule for deciding on the specification of the propensity score and the

regression function. This rule only requires the specification of two readily interpretable cutoff values for variable selection, and is therefore relatively easy to implement and interpret. However, more work needs to be done to understand its properties, and also to investigate alternative approaches to variable selection in similar problems. (pp. 273–274)

5.3.4 GENERALIZED BOOSTED MODELING

Seeking a best logistic regression model may take a completely different route. One of the problems we have seen with logistic regression is specifying an unknown functional form for each predictor. If specifying functional forms can be avoided, then the search for a best model involves fewer subjective decisions and, therefore, may lead to a more accurate prediction of treatment probability. Generalized boosted modeling (GBM), also known as generalized boosted regression, is a method that offers numerous advantages and appears to be promising in solving the variable specification problem. McCaffrey et al. (2004) first applied the GBM approach to the estimation of propensity scores and developed a special software program for the R statistical environment.

GBM is a general, automated, data-adaptive algorithm that fits several models by way of a regression tree, and then merges the predictions produced by each model. As such, GBM can be used with a large number of pretreatment covariates to fit a nonlinear surface and predict treatment assignment. GBM is one of the latest prediction methods that have been made popular in the machine-learning community as well as mainstream statistics research (Ridgeway, 1999). From a statistical perspective, the breakthrough in applying boosting to logistic regression and exponential family models was made by Friedman, Hastie, and Tibshirani (2000). They showed that an exponential loss function used in a machine-learning algorithm, *AdaBoost*, was closely related to the Bernoulli likelihood. From this basis, Friedman et al. developed a new boosting algorithm that finds a classifier to directly maximize a Bernoulli likelihood. Prediction models that use a modern regression tree approach are known as GBMs.

Details of the GBM approach may be found in Ridgeway (1999), Friedman (2002), and a recent paper by Mease, Wyner, and Buja (2007). Here, we follow McCaffrey et al. (2004) to highlight application issues in procedures using GBM to estimate propensity scores.

First, GBM does not provide estimated regression coefficients such as $\hat{\beta}_0, \hat{\beta}_1$, and $\hat{\beta}_2$ as we normally have with a maximum likelihood estimator. The key feature and advantage of the regression tree method is that the analyst does not need to specify functional forms of the predictor variables. As McCaffrey et al. (2004) pointed out, trees handle continuous, nominal, ordinal, and missing independent variables, and they capture nonlinear and interaction effects. A useful property of trees is that they are invariant to one-to-one transformations of the independent variables. Thus, "whether we use age, log(age), or age^2 as a participant's attribute, we get exactly the same propensity score adjustments" (McCaffrey et al., 2004, p. 408). This property explains why uncertainty about a correct functional form for each predictor variable is no longer an issue when GBM is used. Because of this, GBM does not produce estimated regression coefficients. Instead, it provides *influence*, which is the percentage of log likelihood explained by each input variable. The percentages of influence for all predictor variables sum to 100%. For instance, suppose there are three predictor variables: age (x_1), gender(x_2), and a pretreatment risk factor (x_3). A GBM output may show that the influence of x_1 is 20%, of x_2 is 30%, and of x_3 is 50%. Based on this output, the analyst can conclude that the pretreatment risk factor makes the largest contribution to the estimated log-likelihood function, followed by gender and age.

Second, to further reduce prediction error, the GBM algorithm follows Friedman's (2002) suggestion to use a random subsample in the estimation. In some software programs, this subsample is labeled as *training data*. Friedman (2002) suggested subsampling 50% of the observations at each iteration. However, programs use different specifications for the subsample size. For instance, the default of training data used by Stata *boost* is 80%.

Third, McCaffrey et al. (2004, pp. 408–409) provided a detailed description of how GBM handles interaction terms. Based on their experiments, they recommended a maximum of four splits for each simple tree used in the model, which allows all four-way interactions between all covariates to be considered for optimizing the likelihood function at each iteration. To reduce variability, GBM also requires using a *shrinkage coefficient*. McCaffrey et al. suggested using a value of .0005, relatively small shrinkage, to ensure a smooth fit.

Finally, it is noteworthy that McCaffrey et al. suggested stopping the algorithm at the number of iterations that minimized the ASAM in the covariates.[2] This is a recommendation particularly directed toward users of GBM who wish to develop propensity scores. As previously mentioned, the GBM procedure stops the algorithm at the number of iterations that minimize prediction errors. Thus, an optimal estimation of propensity scores that minimizes prediction error may not best balance the sample treated and control groups on the observed covariates. Based on their experience, McCaffrey et al. observed that the ASAM decreases initially with each additional iteration and reaches a minimum, and afterward the ASAM increases with additional iterations. For this reason, McCaffery et al. suggest stopping when ASAM is minimized.

5.4 Matching

After propensity scores are estimated, the next step of analysis often entails matching treated to control participants based on the estimated propensity scores (i.e., to proceed to the Step 2a tasks shown in Figure 5.1). Alternatively, it is possible to skip matching and move to analyzing outcome data using propensity scores as sampling weights (i.e., to proceed to the Step 2b tasks shown in Figure 5.1) or calculate a weighted mean difference in the outcome variable using nonparametric regression (i.e., to proceed to the Step 2c tasks shown in Figure 5.1). This section discusses the various methods of matching pertaining to Step 2a. The section is divided into three topics. The first describes conventional greedy matching and its related methods of matching with or without Mahalanobis metric distance. The second describes *optimal matching*, a method that overcomes some of the shortcomings of greedy matching. The third topic highlights key features of the *fine balance* procedure.

5.4.1 GREEDY MATCHING

The core idea of matching, after obtaining estimated propensity scores, is to create a new sample of cases that share approximately similar likelihoods of being assigned to the treatment condition. Perhaps the most common matching algorithm is the so-called greedy matching. It includes Mahalanobis metric matching, Mahalanobis metric matching with propensity scores, nearest neighbor matching, caliper matching, nearest neighbor matching within caliper, and nearest available Mahalanobis metric matching within a caliper defined by the propensity score. All methods are called greedy matching. Following D'Agostino (1998) and Smith and Todd (2005), we summarize the major features of greedy algorithms below.

1. Mahalanobis Metric Matching

The Mahalanobis metric matching method was invented prior to propensity score matching (Cochran & Rubin, 1973; Rubin, 1976, 1979, 1980a). To apply this method, we first randomly order study participants, and then calculate the distances between the first treated participant and all controls, where the distance, $d(i, j)$, between a treated participant i and a nontreated participant j is defined by the Mahalanobis distance:

$$d(i, j) = (\mathbf{u} - \mathbf{v})^T \mathbf{C}^{-1}(\mathbf{u} - \mathbf{v}), \tag{5.3}$$

where $\mathbf{u}$ and $\mathbf{v}$ are values of the matching variables for treated participant i and nontreated participant j, and $\mathbf{C}$ is the sample covariance matrix of the matching variables from the full set of nontreated participants. The nontreated participant, j, with the minimum distance $d(i, j)$ is chosen as the match for treated participant i, and both participants are removed from the pool. This process is repeated until matches are found for all treated participants. Because

Mahalanobis metric matching is not based on a one-dimensional score, it may be difficult to find close matches when many covariates are included in the model. As the number of covariates increases, the average Mahalanobis distance between observations increases as well. This relationship is a drawback that may be overcome by using two methods that combine the Mahalanobis metric matching with propensity scores (see below). Parenthetically, it is worth noting that C is defined somewhat differently by different researchers, though these different definitions all refer to the method as Mahalanobis metric matching. For instance, D'Agostino (1998) defines C as the sample covariance matrix of the matching variables from the set of control participants, while Abadie et al. (2004) define C as the sample covariance matrix of matching variables from both sets of the treated and control participants.

2. Mahalanobis Metric Matching Including the Propensity Score

This procedure is performed exactly as described above for Mahalanobis metric matching, with an additional covariate, the estimated propensity score $\hat{q}(x)$. The other covariates are included in the calculation of the Mahalanobis distance.

3. Nearest Neighbor Matching

P_i and P_j are the propensity scores for treated and control participants, respectively, I_1 is the set of treated participants, and I_0 is the set of control participants. A neighborhood $C(P_i)$ contains a control participant j (i.e., $j \in I_0$) as a match for treated participant i (i.e., $i \in I_1$), if the absolute difference of propensity scores is the smallest among all possible pairs of propensity scores between i and j, as

$$C(P_i) = \min_{j} ||P_i - P_j||, \quad j \in I_0. \tag{5.4}$$

Once a j is found to match to i, j is removed from I_0 without replacement. If for each i there is only a single j found to fall into $C(P_i)$, then the matching is nearest neighbor pair matching or 1-to-1 matching. If for each i there are n participants found to fall into $C(P_i)$, then the matching is 1-to-n matching.

4. Caliper Matching

In the above matching, there is no restriction imposed on the distance between P_i and P_j, as long as j is a nearest neighbor of i in terms of the estimated propensity score. By this definition, even if $||P_i - P_j||$ is large (i.e., j is very different from i on the estimated propensity score), j is still considered as a match to i. To overcome shortcomings of erroneously choosing j, we select j as a match for i, only if the absolute distance of propensity scores between the two participants meets the following condition:

$$||P_i - P_j|| < \varepsilon, \quad j \in I_0, \tag{5.5}$$

where ε is a prespecified tolerance for matching or a caliper. Rosenbaum and Rubin (1985) suggested using a caliper size of a quarter of a standard deviation of the sample estimated propensity scores (i.e., $\varepsilon \leq .25\sigma_p$, where σ_p denotes standard deviation of the estimated propensity scores of the sample).

5. Nearest Neighbor Matching Within a Caliper

This method is a combination of the two approaches described above. We begin with randomly ordering the treated and nontreated participants. We then select the first treated participant i and find j as a match for i, if the absolute difference of propensity scores between i and j falls into a predetermined *caliper* ε, and is the smallest among all pairs of absolute differences of propensity scores between i and other js within the caliper. Both i and j are then removed from consideration for matching and the next treated participant is selected. The size of the caliper is determined by the investigator but typically is set as $\varepsilon \leq .25\sigma_p$. Nearest neighbor matching within a caliper has become popular because multivariate analyses using the matched sample can be undertaken, if the sample is sufficiently large.

6. Nearest Available Mahalanobis Metric Matching Within Calipers Defined by the Propensity Score

This method combines Mahalanobis distance and nearest neighbor within caliper matching into a single approach. The treated participants are randomly ordered, and the first treated participant is selected. All nontreated participants within a predetermined caliper of the propensity score $\hat{q}(x)$ are then selected, and Mahalanobis distances, based on a smaller number of covariates (i.e., covariates without $\hat{q}(x)$), are calculated between these participants and the treated participant. One j is chosen as a match for i among all candidates, if the chosen j has the shortest Mahalanobis distance from i. The closest nontreated participant and treated participant are then removed from the pool, and the process repeated. According to Rosenbaum and Rubin (1985), this method produces the best balance between the covariates in the treated and nontreated groups, as well as the best balance of the covariates' squares and cross-products between the two groups.

All greedy matching algorithms described above share a common characteristic: Each divides a large decision problem (i.e., matching) into a series of smaller, simpler decisions each of which is handled optimally. Each makes those decisions one at a time without reconsidering early decisions as later ones are made (Rosenbaum, 2002b).

As such, users of greedy matching typically encounter a dilemma between incomplete matching and inaccurate matching. Taking nearest neighbor matching within caliper as an example, we often have to make a decision between choices such as the following: while trying to maximize exact matches, cases may be excluded due to incomplete matching; or while trying to maximize cases, more inexact matching typically results (Parsons, 2001). Neither of the above decisions is optimal. Within the framework of greedy matching, we often wind up recommending running different caliper sizes, checking the sensitivity of results to different calipers, and choosing a method that seems to be best afterward.

Greedy matching is criticized also because it requires a sizable common support region to work. When we define the logit in Figure 5.2 as a propensity score (i.e., $\hat{q}(x) = \log[(1 - \hat{e}(x))/\hat{e}(x)]$) and set a routine common support region, we see that greedy matching excludes participants because treated cases fall outside the lower end of the common support region (i.e., those who have low logit) and nontreated cases fall outside the upper end of the common support region (those who have high logit). These participants simply have no matches. The common support region is sensitive to different specifications of the Step 1 model used to predict propensity scores, because logistic regressions

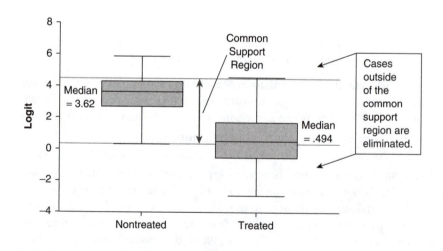

Figure 5.2   Illustration of Common Support Region Using Hypothetical Data ($\hat{q}(x) = \log[(1 - \hat{e}(x))/\hat{e}(x)]$)

with different predictor variables and/or functional forms produce different common support regions. To solve the problem within the conventional framework of propensity score matching, the recommended procedure is for the analyst to test different models and conduct sensitivity analyses by varying the size of the common support region.

Addressing these limitations led to the development of optimal matching, which has proven to have numerous advantages over greedy matching. However, before dismissing greedy matching, we want to emphasize that despite its limitations (e.g., it requires a large sample size and loses study participants because of a narrowed common support region under some settings), greedy matching, particularly its nearest neighbor matching within caliper, has unique advantages. One of these advantages is its permission of subsequent multivariate analysis of almost any kind that allows researchers to evaluate causal effects as they do with randomized experiments. Because of this unique flexibility, greedy matching is widely applied by researchers from a range of disciplines.

5.4.2 OPTIMAL MATCHING

Although the application of optimal matching to propensity score analysis has a history of only about 10 years, the application has grown rapidly and fruitfully primarily for two reasons: the use of network flow theory to optimize matching and the availability of fast computing software packages that makes the implementation feasible. From Hansen (2007), matched adjustment requires analysts to articulate a distinction between desirable and undesirable potential matches, and then to match treated and control participants in such a way as to favor the more desirable pairings. As such, the second task (i.e., matching itself) is less statistical in nature, but completing the matching task well can substantially improve the power and robustness of matched inference (Hansen, 2004; Hansen & Klopfer, 2006). Rosenbaum (2002b, pp. 302–322) offers a comprehensive review of the theory and application principles of optimal matching. Hansen developed an *optmatch* that performs optimal matching in *R* and is available free with *R*. Hansen's *optmatch* package is by far the fastest matching package. Ming and Rosenbaum (2001) proposed using *SAS Proc Assign*, which is a reasonable alternative for individuals who prefer programming in SAS. Haviland, Nagin, and Rosenbaum (2007) provided an excellent example of applying optimal matching to analysis of group-based trajectories, and their accessible work outlined important concerns and strategies for conducting optimal matching. Furthermore, Haviland et al. (2007) present the material about optimal matching in an accessible way. The central ideas of optimal matching are described below.

As we said earlier, all greedy matching algorithms share a common characteristic: Each method divides a large decision-making problem into a series of smaller, simpler decisions each of which is handled optimally. Decisions are made one at a time without reconsidering early decisions as later ones are made. In this sense, greedy matching is not *optimal*.

Taking a numerical example, let's consider creating two matched pairs from four participants with the following propensity scores: .1, .5, .6, and .9. A greedy matching would first pick up the second and third participants to form the first pair, because their propensity score distance is smallest and the two participants look most similar (i.e., $|.5 - .6| = .1$) among the four; next, the greedy matching would use the first and last participants to form the second matched pair. By doing so, the total distance on propensity scores from the two pairs are $|.5 - .6| + |.1 - .9| = .9$. An optimal matching, described in this section, would form the following two pairs: Use the first and second participants to form the first pair, and use the third and fourth participants to form the second pair. By doing so, none of the two pairs created by the optimal matching is better than the first pair created by the greedy matching, because the distance for each pair is larger than .1. However, the total distance from the optimal matching is $|.1 - .5| + |.6 - .9| = .7$, which is better than the total distance of the greedy matching (i.e., .9). It is from this example that we see the importance of conducting optimal matching.

To facilitate discussion, we first introduce the notation used by Rosenbaum (2002b). Initially, we have two sets of participants: The treated participants are in a set A and the controls are in a set B, with $A \cap B = \varnothing$. The initial number of treated participants is $|A|$ and the number of controls is $|B|$, where $|\bullet|$ denotes the number of elements of a set.

For each $a \in A$ and each $b \in B$, there is a distance, δ_{ab} with $0 \leq \delta_{ab} \leq \infty$. The distance measures the difference between a and b in terms of their observed covariates, such as their difference on propensity scores or Mahalanobis metrics. Matching is a process to develop S strata $(A_1, \ldots, A_s; B_1, \ldots, B_s)$ consisting of S nonempty, disjoint participants of A and S nonempty, disjoint subsets of B, so that $|A_s| \geq 1$, $|B_s| \geq 1$, $A_s \cap A_{s'} = \varnothing$ for $s \neq s'$, $B_s \cap B_{s'} = \varnothing$ for $s \neq s'$, $A_1 \cup \cdots \cup A_S \subseteq A$, and $B_1 \cup \cdots \cup B_S \subseteq B$.

By this definition, a matching process produces S matched sets, each of which contains $|A_1|$ and $|B_1|$, $|A_2|$ and $|B_2|$, . . . , and $|A_s|$ and $|B_s|$. Note that, by definition, within a stratum or matched set, treated participants are similar to controls in terms of propensity scores. Depending on the structure (i.e., the ratio of the number of treated participants to control participants within each stratum) the analyst imposes on matching, we can classify matching into the following three types:

1. *Pair matching:* Each treated participant matches to a single control or a stratification of $(A_1, \ldots, A_s; B_1, \ldots, B_s)$ in which $|A_s| = |B_s| = 1$ for each s.

2. *Matching using a variable ratio or variable matching:* Each treated participant matches to, for instance, at least one and at most four controls. Formally, this is a stratification whose ratio of $|A_s|:|B_s|$ varies.

3. *Full matching:* Each treated participant matches to one or more controls, and similarly each control participant matches to one or more treated participants. Formally, this is a stratification of $(A_1, \ldots, A_s; B_1, \ldots, B_s)$ in which the *minimum of* $(|A_s|, |B_s|) = 1$ for each s.

Optimal matching is the process of developing matched sets $(A_1, \ldots, A_s; B_1, \ldots, B_s)$ with size of (α, β) in such a way that the total sample distance of propensity scores is minimized. Formally, optimal matching minimizes the total distance Δ defined as

$$\Delta = \sum_{s=1}^{S} \omega(|A_s|, |B_s|)\delta(A_s, B_s), \qquad (5.6)$$

where $\omega(|A_s|,|B_s|)$ is a weight function. Rosenbaum (2002b) defined three choices among weight functions: (1) the proportion of α treated participants who fall in set s or $\omega(|A_s|,|B_s|) = |A_s| / \alpha$; (2) the proportion of the β control participants who fall in set s or $\omega(|A_s|,|B_s|) = |B_s| / \beta$; and (3) the proportion of the sum of treated and control participants who fall in set s or $\omega(|A_s|,|B_s|) = (|A_s| + |B_s|) / (\alpha + \beta)$. For each of the three weight functions, the sum of weights equals to one, and the total distance Δ is truly a weighted average of the distance $\delta(A_s B_s)$. The actual choice of weight function in an application is not so important. Of greater importance is that optimal matching develops matched sets (i.e., the challenge is to create S sets and identify which controls are matched to which treated participants) in such a way that the matching optimizes or minimizes the total distance for a given data set and prespecified structure.

How does optimal matching accomplish this goal? Suffice it to say that this method achieves the goal by using a network flow approach (i.e., a topic in operations research) to matching. Rosenbaum (2002b) provides a detailed description of the optimal matching method. A primary feature of network flow is that it concerns the cost of using b for a as a match, where a *cost* is defined as the effect of having the pair of (a, b) on the total distance defined by Equation 5.6. Standing in sharp contrast to greedy matching, optimal matching identifies matched sets in such a way that the process aims to optimize the total distance, and decisions made later take into consideration decisions made earlier. Indeed, later decisions may alter earlier ones.

From an application perspective, the structure imposed on optimal matching (i.e., whether you want to run a 1-to-1 pair matching, a matching with a constant ratio of treated to control participants, a variable matching with specifications of the minimum and maximum number of controls for each treated participant, or a full matching) affects both the level of bias reduction and efficiency. In this context, *efficiency* is defined as the reciprocal of the variance, and therefore, a high level of efficiency is associated with a low variance. Haviland et al. (2007) review this topic in detail and distill from it two practice implications: (1) there are substantial gains in bias reduction from discarding some controls, yet there is little loss in efficiency from doing so, provided multiple controls are matched to each treated participant and (2) there are substantial gains in bias reduction from permitting the number of matched controls to vary from one treated participant to another, yet there is little loss in efficiency from doing so if imbalance is not extreme. From this, they draw three general principles for making decisions about matching structure:

1. Having two controls for each treated participant is more efficient than matched pairs (i.e., a 1-to-1 match).

2. A large number of controls yields negligible gains in efficiency.

3. Some variation in the number of matched controls among all strata S does not greatly harm efficiency.

In practice, the selection of a matching structure should be based on the number of treated participants and controls. Sometimes the decision is implied by the structure of the data. For example, assume you want to evaluate data generated by a quasi-experimental design or a randomized experiment in which randomization has failed and that the number of treated participants is close to the number of controls. Given this scenario, a 1-to-1 pair matching is probably the only choice. On the other hand, suppose you have conducted a case-cohort design (e.g., a design described by Joffe & Rosenbaum, 1999) and the ratio of control to treated participants is large, perhaps on the order of 3:1 or 4:1. Under such conditions, there is a range of possible choices to specify the structure of optimal matching, and the choice of structure will have an impact on bias reduction and efficiency. A common practice under such conditions is to test different structures and then compare results (i.e., estimates of treatment effects) among matching schemes.

Methods in this field of matching are rapidly evolving and we recommend also consulting the literature. What seems a good choice now may not be the best choice 2 or 3 years from now. Hansen (2004), for example, found that in the context of a specific application, variable matching with a specific structure worked best; that is, each treated participant was matched to at least $.5(1 - \hat{P}) / \hat{P}$ controls and at most $2(1 - \hat{P}) / \hat{P}$ controls, where $\hat{P}$ represents the proportion of treated participants in the sample. We cannot be sure that a variable matching approach with such a structure will continue to be a best choice in other data sets; however, it should clearly be explored as a viable option.

Finally, it is important to note that pair matching generated by an optimal-matching algorithm is different from the pair matching generated by greedy matching, particularly so for a 1-to-1 or 1-to-n nearest neighbor matching within caliper. The primary reason for these differences is that greedy matching is not an optimal process, and decisions made earlier affect the level of optimizations accomplished later. On the other hand, even though the researcher might use an optimal-matching algorithm such as *optmatch* to conduct a pair matching, Rosenbaum (2002b, pp. 310–311) showed that such pair matching is not generally optimal, especially when compared with full matching based on the same data.

5.4.3 FINE BALANCE

The matching procedures implemented by the greedy and optimal algorithms share a key feature: Treated participants are matched with control participants on a single propensity score to balance a large number of covariates. An innovative method called *fine balance* does not require individually matching on the propensity score (Rosenbaum, Ross, & Silber, 2007). Because this method is quite new, we briefly highlight its main ideas here.

Fine balance refers to exactly balancing a nominal variable, often a variable with many discrete categories, without trying to match individuals on this variable. Fine balance employs a principle similar to the network optimization algorithm (see Section 5.4.2) to create a patterned distance matrix, which is passed to a subroutine that optimally pairs the rows and columns of the matrix. In an illustrative example, Rosenbaum et al. (2007) addressed the problem of matching on a nominal variable with 72 categories (i.e., study participants had 9 possible years of diagnosis and 8 geographic locations, which were considered substantively important for achieving an exact balance). Using the fine balance method, *an exact balance* on the 72 categories and *close individual matches* on a total of 61 covariates based on a propensity score are obtained. The key idea of fine balancing is that the nominal variable is balanced exactly at every level. Rosenbaum et al. used *SAS Proc Assign* to implement the fine balance strategy. As a matching tool, fine balance is used in conjunction with other matching tools, such as propensity scores, minimum distance matching, or Mahalanobis metric distance. It exemplifies the rapid growth of the field. In practice, the choice is not which one tool to use, but rather it has become whether or not to apply a particular tool in conjunction with other tools. For a discussion of when fine balance is implied, see Rosenbaum et al. (2007).

5.5 Postmatching Analysis

This section describes postmatching procedures (i.e., Step 3a or 3b tasks shown in Figure 5.1) for a three-step analysis of propensity scores. Because methods applied at Step 2 for matching vary, methods of postmatching analysis also vary. We describe six methods: (1) multivariate analysis after greedy matching, (2) stratification after greedy matching, (3) checking covariate imbalance before and after optimal matching, (4) outcome analysis for an optimally matched sample using the Hodges-Lehmann aligned rank test, (5) regressing the difference scores of an outcome on difference scores of covariates based on an optimal pair-matched sample, and (6) regression using the Hodges-Lehmann aligned ranks of both outcomes and covariates based on an optimally matched sample.

5.5.1 MULTIVARIATE ANALYSIS AFTER GREEDY MATCHING

The property described in 4(c) under Section 5.2 (i.e., corollary 4.3 of Rosenbaum & Rubin, 1983) is the theoretical justification for conducting multivariate analysis after greedy matching. The impetus for developing propensity scores and matching is that observational data are usually not balanced, and hence, we cannot assume that treatment assignment is ignorable. After matching on the estimated propensity scores, at least the sample is balanced on observed covariates (between treated and control participants), and therefore, we can perform multivariate analyses and undertake covariate adjustments for the matched sample as is done in randomized experiments. In theory, regression, or any regression-type models, may be used at this stage to estimate ATEs by using a dichotomous explanatory variable indicating treatment conditions. Many studies have used this approach to adjust and estimate ATEs. For instance, following a caliper matching, Morgan (2001) conducted a regression analysis to estimate the impact of Catholic schools on learning. Smith (1997) conducted a variance-components analysis (also known as a hierarchical linear regression) based on a matched sample generated by a random order, nearest available pair-matching method. In a sample of Medicare eligible patients, he sought to estimate the effects of an organizational innovation on mortality within hospitals. Guo et al. (2006) conducted a survival analysis (i.e., Kaplan-Meier product limit estimates) after nearest neighbor matching within a caliper. They were interested in estimating the impact of caregivers' use of substance abuse services on the hazard rate of child maltreatment rereport.

5.5.2 STRATIFICATION AFTER GREEDY MATCHING

After greedy matching, stratification on the estimated propensity scores and comparison of treated and control participants within stratum is a common analytic strategy. The justification for conducting stratification following greedy matching is provided in 4(b) of Section 5.2 (i.e., corollary 4.2 of Rosenbaum and Rubin, 1983). The process is intuitive: sort the sample by estimated propensity scores in an ascending order, divide the sample into five strata using quintiles of the estimated propensity scores, calculate mean difference of outcome and variance of difference between treated and control participants within each stratum, estimate the mean difference (ATE) for the whole sample (i.e., all five strata), and test whether or not the sample difference on outcome is statistically significant.

The treatment effect for the whole sample is the average of the five subclass-specific differences of the mean responses in the two treatment conditions, as

$$\hat{\delta} = \sum_{k=1}^{K} \frac{n_k}{N} \left[\overline{Y}_{0k} - \overline{Y}_{1k} \right], \tag{5.7}$$

where k indexes the propensity score subclasses, N is the total number of participants, n_k is the number of participants in the kth subclass, and $\overline{Y}_{0k}, \overline{Y}_{1k}$ are the mean responses corresponding to the two treatment groups in the kth subclass. The variance of this estimate is calculated using the following formula:

$$\text{Var}(\hat{\delta}) = \sum_{k=1}^{K} \left(\frac{n_k}{N} \right)^2 \text{Var}[\overline{Y}_{0k} - \overline{Y}_{1k}]. \tag{5.8}$$

Note that in the above equation, $\text{Var}[\overline{Y}_{0k} - \overline{Y}_{1k}]$ denotes the variance of the mean difference, and the variance of the difference between two independent random variables is the *sum* of the variances. Using $z^* = \hat{\delta}/SE(\hat{\delta})$, we can perform a significance test of a nondirectional (i.e., perform a two-tailed test) or a directional hypothesis (i.e., perform a one-tailed test).

To illustrate the calculation of the ATE of the sample and its significance test, we use an example provided by Perkins, Tu, Underhill, Zhou, and Murray (2000). Based on propensity score stratification, Perkins et al. (2000) reported means and standard errors of an outcome variable as in Table 5.1.

Table 5.1   Estimating Overall Treatment Effect After Stratification

| Stratum | No. of Patients | Mean Outcome | | | Standard Error | |
		Treatment 1	Treatment 2	Difference	Treatment 1	Treatment 2
Subclass 1	1,186	0.0368	0.0608	−0.024	0.0211	0.0852
Subclass 2	1,186	0.035	0.0358	−0.0008	0.0141	0.0504
Subclass 3	1,186	0.0283	0.0839	−0.0556	0.0083	0.0288
Subclass 4	1,186	0.0653	−0.0106	0.0759	0.0121	0.0262
Subclass 5	1,186	0.0464	0.0636	−0.0172	0.0112	0.0212
Total	5,930					

SOURCE: Perkins, Tu, Underhill, Zhou, and Murray (2000, table 2). Reprinted by permission of John Wiley & Sons, Ltd.

Applying Equation 5.7 to these data, the sample ATE is

$$\hat{\delta} = \sum_{k=1}^{K} \frac{n_k}{N} \left[\overline{Y}_{0k} - \overline{Y}_{1k} \right]$$

$$= \frac{1186}{5930}(-.024) + \frac{1186}{5930}(-.0008) + \frac{1186}{5930}(-.0556)$$

$$+ \frac{1186}{5930}(.0759) + \frac{1186}{5930}(-.0172) = -.00434.$$

Applying Equation 5.8 to these data, the variance and standard error of the ATE are

$$\mathrm{Var}(\hat{\delta}) = \sum_{k=1}^{K} \left(\frac{n_k}{N} \right)^2 \mathrm{Var}\left[\overline{Y}_{0k} - \overline{Y}_{1k} \right]$$

$$= \left(\frac{1186}{5930} \right)^2 \left[(.0211)^2 + (.0852)^2 \right]$$

$$+ \left(\frac{1186}{5930} \right)^2 \left[(.0141)^2 + (.0505)^2 \right] + \left(\frac{1186}{5930} \right)^2 \left[(.0083)^2 + (.0288)^2 \right]$$

$$+ \left(\frac{1186}{5930} \right)^2 \left[(.0121)^2 + (.0262)^2 \right] + \left(\frac{1186}{5930} \right)^2 \left[(.0112)^2 + (.0212)^2 \right]$$

$$= .000509971.$$

$$SE(\hat{\delta}) = \sqrt{\mathrm{Var}(\hat{\delta})} = \sqrt{.000509971} = .023.$$

Because $(-.0043/.023) = -.1887$, the mean difference between treatment groups for the whole sample (i.e., the average sample treatment effect) is not statistically significant at the level of $\alpha = .05$.

5.5.3 COMPUTING INDICES OF COVARIATE IMBALANCE

It is often desirable to check covariate balance before and after optimal matching. Haviland et al. (2007) developed the *absolute standardized difference in covariate means*, d_X for use before matching and d_{Xm} for use after matching. This measure is similar to ASAM in the literature.

Before matching, d_X is used to check imbalance on covariate X. It is estimated using the following formula:

$$d_X = \frac{|M_{Xt} - M_{Xp}|}{S_X} ,;$$ (5.9)

where M_{Xt} and M_{Xp} are the means of X for treated and potential control groups, respectively. Denoting the standard deviations of the treated and potential control groups as S_{Xt} and S_{Xp}, we compute the overall standard deviation as

$$S_X = \sqrt{\left(S_{Xt}^2 + S_{Xp}^2\right)/2}.$$

The value d_{Xm} for the level of imbalance on covariate X after matching is estimated by

$$d_{Xm} = \frac{|M_{Xt} - M_{Xc}|}{S_X}.$$ (5.10)

In this equation, subscript c denotes the control group, and M_{xc} denotes the unweighted mean of means of the covariate X for the controls matched to treated participants. This covariate X can be computed by the following method: after matching, each treated participant i in stratum s is matched to m_{si} controls, $j = 1, \ldots, m_{si}$. The number of treated participants in stratum s is n_s, and the total number of treated participants in the whole sample is n_+. The values of a covariate X have a subscript t or c for treated or control group, a subscript s for the stratum, a subscript i to identify the treated participant, and a subscript j for controls to distinguish the m_{si} controls matched to treated participant i. Thus, X_{csij} denotes the value of X for the jth control who matches to treated participant i, $j = 1, \ldots, m_{si}$. Denoting $M_{csi\bullet}$ the mean of the m_{si} values of the covariate X for the controls matched to treated participant i, and M_{Xc} the unweighted mean of these means, we have

$$M_{csi\bullet} = \frac{1}{m_{si}} \sum_{j=1}^{m_{si}} X_{csij} \text{ and } M_{Xc} = \frac{1}{n_+} \sum_{s=1}^{S} \sum_{i=1}^{n_s} M_{csi\bullet}; \tag{5.11}$$

d_X and d_{Xm} can be interpreted as the difference between treated and control groups on X in terms of the standard deviation unit of X. Note that d_X and d_{Xm} are standardized measures that can be compared with each other. Typically, one expects to have $d_X > d_{Xm}$, because the need to correct for imbalance before matching is greater, and the sample balance should improve after matching. Taking the data reported by Haviland et al. (2007, table 4, p. 256) as an example, before optimal matching, the d_X of the covariate "peer-rated popularity" is 0.47, meaning that the treated and control groups are almost half a standard deviation apart on peer-rated popularity, whereas, after optimal matching, the d_{Xm} of the same covariate is 0.18, meaning that the difference between the two groups is 18% of a standard deviation for peer-rated popularity; and indeed, matching improves balance.

An illustrative example of computing d_X and d_{Xm} will be presented in Section 5.9.2.

5.5.4 OUTCOME ANALYSIS USING THE HODGES-LEHMANN ALIGNED RANK TEST AFTER OPTIMAL MATCHING

After optimal matching, we usually want to estimate the ATE and perform a significance test. In this section, we describe these procedures for a matched sample created by full matching or variable matching. Methods for a sample created by optimal pair matching are described in the next section. The sample ATE may be assessed by a weighted average of the mean differences between treated and control participants of all matched sets, using an equation similar to Equation 5.7, as

$$\hat{\delta} = \sum_{i=1}^{b} \frac{n_i + m_i}{N} \left[\overline{Y}_{0i} - \overline{Y}_{1i} \right],$$

where i indexes the b matched strata, N the total number of sample participants, n_i the number of treated participants in the ith stratum, and m_i the number of controls in the ith stratum, and $\overline{Y}_{0i}, \overline{Y}_{1i}$ the mean responses corresponding to the control and treated groups in the ith stratum.

The significance test of the ATE may be performed by the Hodges-Lehmann aligned rank test (Hodges & Lehmann, 1962). Lehmann (2006, pp. 132–141) described this test in detail. Its major steps include the following:

1. Compute the mean of the outcome for each matched stratum i, then create a centering score for each participant by subtracting the stratum's mean from the observed value of the outcome.

2. Sort the whole sample by the centering scores in an ascending order, and then rank the scores; the ranked score is called *aligned rank* and is denoted as k_{ij} ($j = 1, \ldots, N_i$), where i indicates the ith stratum, j the jth observation within the ith stratum, and N_i the total number of participants in the ith stratum.

3. For each stratum i, compute

$$k_{i\bullet} = \frac{k_{i1} + \cdots + k_{iN_i}}{N_i} \text{ and } E\left(\hat{W}_s^{(i)}\right) = n_i k_{i\bullet},$$

$$\text{and Var}\left(\hat{W}_s^{(i)}\right) = \frac{m_i n_i}{N_i(N_i - 1)} \sum_{j=1}^{N_i} (k_{ij} - k_{i\bullet})^2.$$

4. Across strata, calculate

$$\hat{W}_s = \sum \hat{W}_s^{(i)}, \ E\left(\hat{W}_s\right) = \sum E\left(\hat{W}_s^{(i)}\right), \text{ and Var}\left(\hat{W}_s\right) = \sum \text{Var}\left(\hat{W}_s^{(i)}\right),$$

where $\hat{W}_s^{(i)}$ is the sum of the aligned ranks for the *treated* participants within the ith stratum. Note that the subscript s in all the above equations indicates treatment participants.

5. Finally, calculate the following test statistic z^*:

$$z^* = \left[\hat{W}_s - E\left(\hat{W}_s\right)\right] / \sqrt{\text{Var}(\hat{W}_s)}. \tag{5.12}$$

The z^* statistic follows a standard normal distribution. Using z^*, the analyst can perform a significance test of a nondirectional hypothesis (i.e., perform a two-tailed test) or a directional hypothesis (i.e., perform a one-tailed test).

Social and behavioral sciences researchers and, indeed, policymakers, are often interested in effect sizes. An exact measure of the size of the treatment effect under the current setting is yet to be developed. However, we recommend using d_{Xm} defined by Equation 5.10 as an approximation of effect size for postmatching analysis. The d_{Xm} statistic, according to Haviland et al. (2007), is similar to Cohen's d. To approximate an effect size, simply calculate d_{Xm} for the outcome variable after matching.

An example of outcome analysis after optimal full or variable matching is presented in Section 5.9.3.

5.5.5 REGRESSION ADJUSTMENT BASED ON SAMPLE CREATED BY OPTIMAL PAIR MATCHING

After obtaining a matched sample using optimal pair matching, an ATE is estimated using a special type of *regression adjustment* developed by Rubin

(1979). The basic concept of regression adjustment is straightforward: regressing the difference scores between treated and control participants on the outcome variable on the difference scores between treated and control participants on the covariates.

Following Rubin (1979), we write $\alpha_i + W_i(\mathbf{X})$ to denote the expected value of the outcome variable Y given a covariate matrix $\mathbf{X}$ in the population P_i, where i denotes treatment condition ($i = 1$, treated; and $i = 0$, control). The difference in expected values of Y for P_1 and P_0 units with the same value of $\mathbf{X}$ is $\alpha_1 - \alpha_0 + W_1(\mathbf{X}) - W_0(\mathbf{X})$. When P_1 and P_0 represent two treatment populations such that the variables in $\mathbf{X}$ are the only ones that affect Y and have different distributions in P_1 and P_0, then this difference is the effect of the treatment at X. If $W_1(\mathbf{X}) = W_0(\mathbf{X}) = W(\mathbf{X})$ for all $\mathbf{X}$, the response surfaces are then parallel, and $\alpha_1 - \alpha_0$ is the treatment effect for all values of the covariates $\mathbf{X}$.

The regression adjustment is undertaken as follows:

1. Take the difference scores on the outcome variable Y between treated and control participants $Y = Y_1 - Y_0$.

2. Take the difference scores on the covariate matrix $\mathbf{X}$ between treated and control participants $\mathbf{X} = \mathbf{X}_1 - \mathbf{X}_0$.

3. Regressing Y on X, we obtain the following estimated regression function:

$$Y = \hat{\alpha} + \mathbf{X}'\hat{\beta}$$

$\hat{\alpha}$ then is the estimated ATE. We can use the observed t statistic and p value associated with $\hat{\alpha}$ to perform a significance test (two-tailed or one-tailed).

An example of outcome analysis after optimal pair matching is presented in Section 5.9.4.

5.5.6 REGRESSION ADJUSTMENT USING HODGES-LEHMANN ALIGNED RANK SCORES AFTER OPTIMAL MATCHING

The method described in Section 5.5.4 is bivariate, that is, it analyzes ATE on the outcome between treated and control participants where control of covariates is obtained through optimal matching. Analysts could also evaluate ATE in conjunction with covariance control (i.e., regression adjustment) to conduct a multivariate analysis, which is the procedure suggested by Rosenbaum (2002a, 2002b). In essence, this method is a combination of the Hodges-Lehmann aligned rank method and regression adjustment. The central idea of this approach is summarized as follows: create aligned rank scores of the outcome variable; create aligned rank scores for each of the covariates; then regress (perhaps using robust regression) the aligned rank scores of the outcome on the aligned rank scores of covariates; the residuals are then ranked from 1 to N (i.e., the total number of participants in the sample) with average ranks for ties. By following this procedure, the sum of the ranks of the residuals for treated participants becomes the test statistic (i.e., Equation 5.12). At this stage, you may use the Hodges-Lehmann aligned rank test to discern whether the treatment effect is statistically significant after the combination of optimal matching and regression adjustment.

5.6 Propensity Score Weighting

Propensity scores may be used without matching, a process that reduces the analysis to two steps. This section describes one such method: multivariate analysis using propensity scores as sampling weights (i.e., proceeding to Step 2b shown in Figure 5.1). Details of this method can be found in Rosenbaum (1987), Hirano and Imbens (2001), Hirano, Imbens, and Ridder (2003), and McCaffrey et al. (2004). We summarize the main features of the analysis below.

In this section, we define propensity scores as the estimated probabilities of receiving treatment; that is, we use $\hat{e}(x)$, rather than the logit:

$$\hat{q}(x) = \log[(1 - \hat{e}(x))/\hat{e}(x)]$$

Propensity score weighting aims to reweight treated and control participants to make them representative of the population of interest, as in Horvitz-Thompson estimators (Horvitz & Thompson, 1952) for stratified sampling. The crucial element of analysis is the development of weights based on the estimated propensity scores. Different types of weights are used, depending on whether an average treatment effect (ATE) or the average treatment effect for the treated (ATT) is desired.

1. For estimating ATE, we define weights as follows:

$$\omega(W, x) = \frac{W}{\hat{e}(x)} + \frac{1 - W}{1 - \hat{e}(x)}. \tag{5.13}$$

By this definition, when $W = 1$ (i.e., a treated participant), Equation 5.13 becomes $\omega(W, x) = 1/\hat{e}(x)$; and when $W = 0$ (i.e., a control), Equation 5.13 becomes $\omega(W, x) = \hat{e}(x)/(1 - \hat{e}(x))$.

2. For estimating the ATT, we define weights as follows:

$$\omega(W, x) = W + (1 - W)\frac{\hat{e}(x)}{1 - \hat{e}(x)}. \tag{5.14}$$

By this definition, when $W = 1$ (i.e., a treated participant), Equation 5.14 becomes $\omega(W, x) = 1$; and when $W = 0$ (i.e., a control), Equation 5.14 becomes $\omega(W, x) = \hat{e}(x)/(1 - \hat{e}(x))$.

In summary, if we denote $\hat{e}(x)$ simply as P, the weight for a treated participant is $1/P$, and for a control participant is $[1/(1 - P)]$ when estimating ATE; the weight for a treated participant is 1 and for a control participant is $[P/(1 - P)]$ when estimating ATT. After creating these weights, we can simply use them in multivariate analysis. Most software packages allow users to specify the name of a weight variable in procedures of multivariate analysis. The analysis then is analogous to multivariate modeling that incorporates sampling weights.

Although weighting with propensity scores is a creative idea and is easy to implement, recent studies suggest that it may have limitations. Focusing on the weighting procedure to estimate ATE, Freedman and Berk (2008) conducted a series of data simulations. They found that propensity score weighting was optimal only under three circumstances: (1) study participants were independent and identically distributed, (2) selection was exogenous, and (3) the selection equation was properly specified (i.e., with correct predictor variables and functional forms). When these conditions are not observed, they found that weighting is likely to increase random error in the estimates. Indeed, it appears to bias the estimated standard errors downward, even when selection mechanisms are well understood. Moreover, in some cases, weighting may increase the bias in estimated causal parameters. Given these findings, Freedman and Berk recommend that if investigators have a good causal model, it may be better just to fit the model without weights; if the causal model is improperly specified, there can be significant problems in retrieving the situation by weighting, and in such a circumstance investigators should exercise caution in implementing the weighting procedure. Freedman and Berk warn that it rarely makes sense to use the same set of covariates in the outcome equation and the selection equation that predicts propensity scores. Reflecting on both the developmental nature of the field and the uncertainty surrounding the validity of emerging procedures, Kang and Schafer (2007) recently showed that the use of inverse probabilities as weights is sensitive to misspecification of the propensity score model when some estimated propensities are small. Caution seems warranted in the use of propensity scores as sampling weights.

5.7 Modeling Doses of Treatment

All methods described thus far concern a binary treatment condition, that is, a treated group compared with a control group. However, in practice, there may be more than two conditions. This often happens when we want to estimate the impact of treatment dosage. For example, in a school-based study, a researcher may have accurately recorded the number of minutes of student exposure to an intervention, and therefore, in addition to receiving or not receiving treatment (i.e., a zero minute of dosage), it may be informative to test the hypothesis that an increase in exposure to treatment results in improved outcomes. Other examples of treatment dosage include the length of exposure to intervention-related advertisements (such as participants' frequencies of seeing antidrug commercials on TV, hearing them on radio, or reading antidrug ads in newspapers or magazines), the number of times using services or visiting doctors, the number of treatment sessions attended, and the like.

The analysis of treatment dosage with propensity score matching may be generalized in two directions. In the first direction, one estimates a single scalar propensity score using ordered logistic regression, and then matches on the scalar propensity score and proceeds as for two treatment groups (Joffe & Rosenbaum, 1999). In the second direction, one estimates propensity score for each level of treatment dosage (i.e., if there are five treatment conditions defined by differential doses, one estimates five propensity scores for each participant) using a multinomial logit regression, and then defines the inverse of a particular estimated propensity score as sampling weight to conduct a multivariate analysis of outcome (Imbens, 2000; Rosenbaum, 1987). Statistical theories underlying both methods are discussed by Imai and Van Dyk (2004). In this section, we summarize the statistical principles and application procedures involved in each of the two methods.

1. *Modeling Doses With a Single Scalar Balancing Score Estimated by an Ordered Logistic Regression.* In general, matching with multiple doses of treatment is an extension of the propensity score matching under a binary condition. This method was originally proposed by Joffe and Rosenbaum (1999). Lu, Zanutto, Hornik, and Rosenbaum (2001) review details of the method with illustrations. When moving from a binary condition to multiple treatments, matching essentially requires the creation of matched pairs in such a way that high- and low-dose groups have similar or balanced distributions of observed covariates. However, balancing covariates with propensity scores under the condition of multiple doses raises three considerations or, perhaps, complications (Lu et al., 2001).

First, under the multiple-doses condition, the original definition of a propensity score (i.e., it is a conditional probability of receiving treatment, or a single scalar score, given observed covariates) is no longer applicable. Because there are multiple doses, each participant now can have multiple propensity scores, and each score compares one dose with another. Indeed, the second method of modeling multiple doses defines propensity scores in this way and estimates multiple propensity scores. Joffe and Rosenbaum's (1999) method uses a *single* scalar balancing score and shows that such a score exists only for certain models. These include McCullagh's (1980) ordered logistic regression and a Gaussian multiple linear regression model *with errors of constant variance.* The key application issue is the choice of a statistical model that estimates the propensity score, and the recommended model is the ordered logistic regression

(Joffe & Rosenbaum, 1999; Lu et al., 2001). Using an OLS regression to estimate the propensity score is problematic, because such a model assumes errors of constant variance, but in practice error variances typically vary by levels of treatment dosage and heteroscedasticity is likely to be present.

Second, under the multiple-doses condition, one needs to redefine the distance between a treated case and a control case in optimization of matching. In this situation, the goal is to identify pairs that are similar in terms of observed covariates but very different in terms of dosage. Hence, the distance must measure both the similarity in terms of covariates and the differences in terms of doses.

And finally, under the multiple-doses condition, the matching algorithm employed is also different from that employed in the binary condition. In the network flow literature of operations research, matching one group to another disjoint group (i.e., under the binary condition) is called a *bipartite* matching problem. But matching under the condition of multiple doses is considered matching *within a single group* and is called *nonbipartite* matching. Because of this difference, the optimization algorithms one uses in the binary condition, such as the algorithms of optimal matching described in Section 5.4.2, are no longer applicable.

These three complexities have led to the development of new features of matching with multiple doses. The matching procedure using a single scalar score is summarized below.

Step 1: Develop a single scalar score based on an ordered logistic regression. We first run an ordered logistic regression in the form of McCullagh (1980). The distribution of doses Z_k for a sample of K participants ($k = 1, 2, \ldots, K$), given observed covariates $\mathbf{x}_k$, is modeled as

$$\log\left(\frac{\Pr(Z_k \geq d)}{\Pr(Z_k < d)}\right) = \theta_d + \boldsymbol{\beta}'\mathbf{x}_k, \text{ for } d = 2, 3, 4, 5,$$

assuming there are five treatment doses being modeled. This model compares the probability of a response greater than or equal to a given category ($d = 2, \ldots,$ 5) to the probability of a response less than this category, and the model is composed of $d - 1$ parallel linear equations. In the linear part of the model, θ_d is called a "cutoff" value to be used in calculating predicted probabilities of falling into each of the five responses, with the probability of the omitted response equaling 1 minus the cumulative probability of falling into all of the other four categories. Note that there are four θ_d values for five ordered responses and five model-based predicted probabilities for each participant k. None of these quantities is a single scalar. The crucial feature of Joffe and Rosenbaum's model is that it defines $\hat{\boldsymbol{\beta}}'\mathbf{x}_k$ as the estimated propensity score, or $e(\mathbf{x}_k) = \hat{\boldsymbol{\beta}}'\mathbf{x}_k$, because the distribution of doses given covariates depends on the observed covariates only through $\hat{\boldsymbol{\beta}}'\mathbf{x}_k$, and the observed covariates $\mathbf{x}$ and the doses Z are conditionally independent given the scalar $\hat{\boldsymbol{\beta}}'\mathbf{x}_k$. Under this setup, $e(\mathbf{x}_k) = \hat{\boldsymbol{\beta}}'\mathbf{x}_k$ is a single balancing score, and the maximum likelihood estimate $\hat{\boldsymbol{\beta}}'\mathbf{x}_k$ is used after running the ordered logistic regression in the matching.

Step 2: Calculate distance between participants k and k', where k ≠ k'. Recall that matching is an optimization to minimize the sample total distances between treated and control participants. This goal remains the same for the multiple-doses condition, but the distance formula is revised. Lu et al. (2001) provide the following equation to calculate the distance under the multiple-doses condition:

$$\Delta(\mathbf{x}_k, \mathbf{x}_{k'}) = \frac{\left(\hat{\boldsymbol{\beta}}'\mathbf{x}_k - \hat{\boldsymbol{\beta}}'\mathbf{x}_{k'}\right)^2 + \varepsilon}{(Z_k - Z_{k'})^2}, \tag{5.15}$$

where $\hat{\boldsymbol{\beta}}'\mathbf{x}_k$ and $\hat{\boldsymbol{\beta}}'\mathbf{x}_{k'}$ are the estimated propensity scores, and Z_k and $Z_{k'}$ are dose values ($= 1, 2, \ldots, d$, if there are d doses) for k and k', respectively. The main aspect of Equation 5.15 is ε, which is a vanishingly small but strictly positive number ($\varepsilon > 0$). The constant ε is a formal device signifying how perfect matches on covariates or doses will be handled. It is ε that makes the calculation of distances for the multiple-doses condition differ from that for the binary condition. Thus ε serves two functions. It specifies that (1) if participants k and k' have the same dose, $Z_k = Z_{k'}$, then the distance between them is ∞ even if they have identical observed covariates $\mathbf{x}_k = \mathbf{x}_{k'}$, and the distance between them is 0, and (2) when two participants have identical observed covariates $\mathbf{x}_k = \mathbf{x}_{k'}$ and the distance between them is 0, the dose distance $\Delta(x_k, x_{k'})$ will be smaller as the difference in dose $(Z_k - Z_{k'})^2$ increases.

Step 3: Conduct nonbipartite pair matching using the distances so defined. For a sample of K participants, each participant k has a distance from each of the remaining participants in the sample on the estimated propensity scores. The researcher then conducts an optimal pair matching in such a way that the total distance associated with all matched pairs is minimized. Each of the resultant pairs then contains one high-dose participant and one low-dose participant, because $\Delta(x_k, x_{k'}) = \infty$ if $Z_k = Z_{k'}$, which is forbidden by Equation 5.15. It is worth noting that the optimal matching under the current context is the so-called nonbipartite matching, which is different from the bipartite optimal matching described in Section 5.4.2. Therefore, one has to use special software programs to conduct the matching. Specifically, the

R program *optmatch* developed by Hansen (2007) conducts bipartite matching, and therefore, should *not* be used under the current context. Lu et al. (2001) used a revised algorithm based on Derigs's (1988) program. An alternative strategy is to use *SAS Proc Assign* described in Ming and Rosenbaum (2001).

Step 4: Check covariate balance after matching. Having obtained matched pairs, the next step involves checking covariate balance between high- and low-dose participants to see how well the propensity score matching performed, that is, whether or not high- and low-dose participants are comparable in terms of observed covariates. The balance check is straightforward; that is, you may calculate mean differences and conduct independent sample *t* tests between the high- and low-dose participants on each observed covariate. One hopes that at this stage all *t* tests would show nonsignificant differences between the high- and low-dose participants. If significant differences remain, you may return to the ordered logistic regression and matching steps to change specifications, and rerun the previous analyses.

Step 5: Evaluate the impact of treatment doses on the outcome. At this final stage, the impact of treatment doses on outcome differences is estimated. Because multiple doses are modeled, not only is it possible to evaluate treatment effectiveness, but it is also possible to assess the degree to which dosage plays a role in affecting the outcome difference. The outcome evaluation pools together all pairs showing the same contrasts of high and low doses, calculates mean difference on the outcome variable between the high- and low-dose participants, and conducts a Wilcoxon signed-rank test to evaluate whether or not the difference is statistically significant. For an illustration, see Lu et al. (2001).

2. *Modeling Doses With Multiple Balancing Scores Estimated by a Multinomial Logit Model.* Imbens (2000) proposes to estimate multiple balancing scores by using a multinomial logit model, and then conduct an outcome analysis that employs the inverse of a specific propensity score as sampling weight. This method requires fewer assumptions and is easier to implement. In contrast to Joffe and Rosenbaum's approach, Imbens's method could be used with several unordered treatments. This approach has two steps.

Step 1: Estimate the generalized propensity scores by using the multinomial logit model. Imbens first defines the conditional probability of receiving a particular level of the treatment dose given the observed covariates as the generalized propensity score, which can be estimated by the multinomial logit model. Suppose there are *d* treatment doses; then each participant will have *d* generalized propensity scores, and under the current context, propensity scores are multiple and are no longer a scalar function.

Step 2: Conduct outcome analysis by following the process of propensity score weighting. Next, the researcher calculates the inverse of a specific generalized propensity score and defines the inversed propensity score as a sampling weight to be used in outcome analysis (i.e., analysis with propensity score weighting). Denoting $e(\mathbf{x}_{k,d}) = \mathrm{pr}(D = d|X = \mathbf{x})$ as the generalized propensity score of receiving treatment dose d for participant k with observed covariates $\mathbf{x}$, then the inverse of the generalized propensity score (i.e., $1/e(\mathbf{X}_{k,d})$) is defined as a sampling weight for participant k. Note that even though each participant has multiple propensity scores obtained from the multinomial logit model, there is only one such score that is used and defined in the propensity score weighting analysis: It is the predicted probability for participant k to fall into the d dose category that is used, and the inverse of this score is defined as the weight in the outcome analysis.

After creating the weight, we simply use it in multivariate analyses evaluating outcome differences. Most software packages allow users to specify the name of a weight variable in procedures of multivariate analysis. The analysis then is analogous to multivariate modeling that incorporates sampling weights, as described in Section 5.6. In the outcome analysis, a set of $d - 1$ dummy variables is created, with one dose category omitted as a reference group. These dummy variables are specified as predictor variables in the outcome analysis. They signify the impact of doses on the outcome variable. The p value associated with the coefficient for each dummy variable indicates the dosewise statistical significance and can be used in hypothesis testing. An example illustrating Imbens's method is presented in Section 5.9.6.

Readers who are interested in both methods will find details of modeling features in the references listed above. However, in our opinion, conducting an efficacy subset analysis is an efficient and viable alternative to modeling doses of treatment. We describe efficacy subset analysis in Chapter 6.

5.8 Overview of the Stata and R Programs

Currently, no commercial software package offers procedures for implementing propensity score matching. In SAS, Lori Parsons (2001) developed several macros (e.g., the GREEDY macro does nearest neighbor within caliper matching). There are several user-developed programs available in Stata and R that allow users to undertake most tasks described in this chapter. Based on our experience with Stata, we found *psmatch2* (Leuven & Sianesi, 2003); *boost* (Schonlau, 2007); *imbalance* (Guo, 2008b); and *hodgesl* (Guo, 2008a) to be especially useful. In the following section, we provide an overview of the main features of these programs. In addition, the overview includes the R program *optmatch* (Hansen, 2007), which we found to be a comprehensive program for conducting optimal matching. To our knowledge, there is no current package available in either Stata or R that can be used to conduct the nonbipartite matching.

To obtain from the Internet a user-developed program in Stata, you may use the *findit* command followed by the name of the program (e.g., *findit psmatch2*), and then follow online instructions to install the program. All user-developed programs contain a help file that offers basic instructions for running the program.

The *psmatch2* program implements full Mahalanobis matching and a variety of propensity score matching methods (e.g., greedy matching and propensity score matching with nonparametric regression). Table 5.2 exhibits the syntax and output of three *psmatch2* examples: nearest neighbor matching within a caliper of $.25\sigma_p$, Mahalanobis matching without propensity scores, and Mahalanobis matching with propensity scores. In our nearest neighbor matching example, we used propensity scores (named *logit3*) estimated by a foreign program. In other words, we input propensity scores into *psmatch2*. This *psmatch2* step is often needed if users want to use programs such as *R-gbm* or other software packages to estimate propensity scores. Alternatively, users can specify the names of the conditioning variables and let the program estimate the propensity scores directly.

A few cautionary statements about running *psmatch2* are worth mentioning. When one treated case is found, several nontreated cases—each of which has the same value of propensity score—may be tied. In a 1-to-1 match, identifying which of the tied cases was the matched case depends on the order of the data. Thus, it is important to first create a random variable, and then sort data using this variable. To guarantee consistent results from session to session, users must control for the seed number by using a *set seed* command. For nearest neighbor and Mahalanobis matching, the literature (e.g., D'Agostino, 1998) has suggested the use of nonreplacement. That is, once a treated case is matched to one nontreated case, both cases are removed from the pool. Nonreplacement can be done in nearest neighbor matching in *psmatch2* by using *noreplacment descending*. However, this command does not work for Mahalanobis matching. In the matched sample created by *psmatch2* Mahalanobis, it is possible that one control case can be used as a match for several treated cases. To perform nonreplacement for Mahalanobis, users need to examine matched data carefully, keep one pair of the matched and treated cases in the data set, and delete all pairs that used the matched control more than once.

The *boost* program estimates boosted regression for the following link functions: Gaussian (normal), logistic, and Poisson. Table 5.3 exhibits the syntax

Table 5.2   Exhibit of Stata *psmatch2* Syntax and Output Running Greedy
               Matching and Mahalanobis Metric Distance

```
//Chapter 5 Example of Running psmatch2
clear
cd "D:\Sage\ch5"
use nscaw1, replace
summarize logit3

(Output)
. summarize logit3

Variable |     Obs       Mean     Std. Dev.      Min        Max
---------+------------------------------------------------------------
logit3 |    2758    3.063446   1.604109   -2.976967   5.905981

summarize logit3
set seed 1000
generate x=uniform()
sort x

// Nearest neighbor within caliper (.25*SD=.401)

psmatch2 aodserv, pscore(logit3) caliper(0.401) ///
         noreplacement descending
sort _id
g match=nscawid[_n1]
g treat=nscawid if _nn==1
drop if treat==.
list match treat in 1/10
summarize match treat

(Output)
. list match treat in 1/5
        +-----------+-----------
        |  match     treat |
        |-----------|----------
   1. | 200451    201910 |
   2. | 202521    203574 |
   3. | 203896    200013 |
   4. | 205200    204683 |
   5. | 203070    204690 |
        |-----------|----------

. summarize match treat
Variable |     Obs    Mean    Std. Dev.     Min       Max
---------+------------------------------------------------------------
match |   245   202706.7   1549.137   200009   205488
treat |   245   202713.2   1657.963   200007   205475
```

(Continued)

Table 5.2 (Continued)

```
clear
use nscaw1, replace
set seed 1000
generate x=uniform()
sort x

// Mahalanobis without propensity score

psmatch2 aodserv, mahal(married high bahigh poverty2 ///
        poverty3 poverty4 poverty5 employ open black ///
        hispanic natam chdage1 chdage2 chdage3 cgrage1 ///
        cgrage2 cgrage3 cra47a mental arrest psh17a ///
        sexual provide supervis other ra cidi cgneed)
sort _id
generate match=nscawid[_n1]
generate treat=nscawid if _n1 !=.
drop if treat==.
list match treat in 1/10
summarize match treat

  (Output)
. list match treat in 1/10
       +---------+---------+
       |  match     treat  |
       |---------|---------|
  1.   | 200174    202656  |
  2.   | 200740    200280  |
  3.   | 205482    201873  |
  4.   | 202589    203963  |
  5.   | 201406    200809  |
       |---------|---------|
  6.   | 204334    204305  |
  7.   | 200463    204621  |
  8.   | 202592    205035  |
  9.   | 202224    205089  |
 10.   | 202437    203528  |
       +---------+---------+

. summarize match treat

Variable |      Obs        Mean    Std. Dev.        Min        Max
---------+---------------------------------------------------------
  match  |      298      202741    1598.967     200042     205500
  treat  |      298    202708.6    1675.858     200001     205475
clear
use nscaw1, replace
set seed 1000
generate x=uniform()
sort x
```

```
// Mahalanobis with propensity score

psmatch2 aodserv, mahal(married high bahigh poverty2 ///
         poverty3 poverty4 poverty5 employ open black ///
         hispanic natam chdage1 chdage2 chdage3 cgrage1 ///
         cgrage2 cgrage3 cra47a mental arrest psh17a sexual
///
         provide supervis other ra cidi cgneed) add
pscore(logit3)
sort _id
generate match=nscawid[_n1]
generate treat=nscawid if _n1 !=.
drop if treat==.
list match treat in 1/10
summarize match treat

(Output)

. list match treat in 1/10
     +-----------------+
     |  match    treat |
     |-----------------|
  1. | 200174   202656 |
  2. | 200740   200280 |
  3. | 205482   201873 |
  4. | 202589   203963 |
  5. | 201406   200809 |
     |-----------------|
  6. | 204334   204305 |
  7. | 200463   204621 |
  8. | 202592   205035 |
  9. | 202224   205089 |
 10. | 202437   203528 |
     +-----------------+

. summarize match treat

Variable |     Obs       Mean    Std. Dev.     Min       Max
---------+--------------------------------------------------
   match |     298     202741    1598.967    200042    205500
   treat |     298   202708.6    1675.858    200001    205475
```

and output of *boost*. Following *boost*, the analyst specifies the names of the dependent variable and independent variables used in the regression model, the name of link function *distribution (logistic)*, and other specifications of the model. In our example, we specified a maximum number of iterations of 1,000, a training data set of 80%, the name of saved predicted probability as *p*, a maximum of four interactions allowed, a shrinkage coefficient of .0005, and a request for showing influence of each predictor variable in the output.

Table 5.3   Exhibit of Stata *boost* Syntax and Output Running Propensity Score Model Using GBM

```
//Chapter 5 Example of running boost
cd "D:\Sage\ch5\"
use chpt5_2, replace
//create propensity scores using boost
program boost_plugin, plugin using("D:/Data/boost.dll")
gen x=uniform()
sort x
set seed 1000
boost kuse pcg_adc age97 mratio96 pcged97 black, ///
   distribution(logistic) maxiter(1000)
trainfraction(0.8) ///
   pred(p) inter(4) shrink(.0005)influence

(Output)

influence
Distribution=logistic
predict=p
Trainfraction=.8 Shrink=.0005 Bag=.5 maxiter=1000
Interaction=4
Fitting ...
Assessing Influence ...
Predicting ..

bestiter= 1000
Test R2= .1444125
trainn= 802
Train R2= .19706948
Influence of each variable (Percent):
4.0252428 pcg_adc
.3782218 age97
84.372294 mratio96
7.0719709 pcged97
4.1522708 black
```

Information about running *gbm* in R and the program developed by McCaffrey et al. (2004) can be found at http://dx.doi.org/10.1037/1082-989X.9.4.403.supp.

The *imbalance* program is used to produce the covariate imbalance statistics d_x and d_{xm} developed by Haviland et al. (2007), and *hodgesl* is a program that performs the Hodges-Lehmann aligned rank test. Both programs are available at this book's companion Web site. A feature of *hodgesl* is that it saves the mean of the outcome variable and the number of participants for treated and control groups within each matched set for future analysis.

To conduct propensity score weighting and to model the impacts of multiple doses of treatment in Stata, users need to create weight variables first using data management commands, and then specifying the name of the weight variable using *pweight* when running a multivariate model such as regression.

To run *optmatch*, the analyst needs to first install the free statistical software package R from www.r-project.org. After installing and starting R, to obtain the *optmatch* package, go to the "Packages" menu in R, choose the "Load package" feature, select *optmatch* from the list, type *library(optmatch)* at the R command prompt, and type *help(fullmatch)* for instructions. Table 5.4 exhibits syntax and the output of running *optmatch* for creating the propensity scores within the program (i.e., by using R's *glm* function). The example also shows how to perform a full matching, to request a calculation of mean distance and total distance based on the matched sets, and finally to request output showing the stratum structure (i.e., the number of matched sets associated with all possible ratios of treated to control participants after matching).

Table 5.4   Exhibit of R Syntax and Output Running Logistic Regression and Full Matching

```
#optmatch using glm to create propensity scores and then optmatch
set.seed(10)
setwd ("D:/Sage/ch5")
library(foreign)
cds <- read.dta("chpt5_2.dta")
attach(cds)
#logistic regression
lcds <- glm(kuse ~ pcg.adc + age97 + mratio96 + pcged97 +
  black, family = binomial, data=cds)
summary(lcds)

(R output)
Results of Logistic regression
Deviance Residuals:
Min             1Q          Median            3Q            Max
-2.1750       -0.5974        -0.2311         0.4896        4.1454
```

(Continued)

Table 5.4 (Continued)

```
Coefficients:
              Estimate    Std. Error    z value       Pr(>|z|)
(Intercept)    3.05915      0.73939       4.137     3.51e-05 ***
pcg.adc        0.19130      0.04620       4.140     3.47e-05 ***
age97          0.06468      0.03214       2.013     0.0441 *
mratio96      -1.05239      0.10424     -10.096     < 2e-16 ***
pcged97       -0.26944      0.05830      -4.622     3.80e-06 ***
black          0.80379      0.20179       3.983     6.80e-05 ***
--
Signif. codes:  0 '***' 0.001 '**' 0.01 '*' 0.05 '.' 0.1 ' ' 1

(Dispersion parameter for binomial family taken to be 1)

    Null deviance: 1176.31  on 1002  degrees of freedom
Residual deviance:  748.02  on  997  degrees of freedom
AIC: 760.02
Number of Fisher Scoring iterations: 6

library(optmatch)
#create propensity scores based on the logistic regression
pdist <- pscore.dist(lcds)
#run full match
fm <- fullmatch(pdist)
(fm.d <- matched.distances(fm,pdist,pres=TRUE))
unlist(fm.d,max)
mean(unlist(fm.d))
sum(unlist(fm.d))
stratumStructure(fm)
(Output omitted ...)
Results of Full Matching
> mean(unlist(fm.d))
[1] 0.1168186
> sum(unlist(fm.d))
[1] 99.64628
> stratumStructure(fm)
stratum treatment:control ratios
17:1 14:1 13:1 8:1 7:1 6:1 5:1 4:1 3:1 2:1 1:1 1:2 1:3 1:4 1:5 1:6
   1    1    1   1   2   3   2   4   7  15  45  16   8  12   9   3
1:7 1:9 1:11 1:13 1:15 1:16 1:18 1:19 1:26 1:33 1:36 1:42 1:46 1:54 1:92
   4   2    1    2    1    1    1    1    1    1    1    1    1    1    1
```

5.9 Examples

We now present examples to illustrate the various models described in this chapter. Each example is based on analyses of observational data from recent studies. These examples represent substantive interests in three topic areas. The first study is an evaluation of caregivers' use of substance abuse services on the hazard rate of child maltreatment. Technically, the study focuses on a sample of families referred to child welfare and the outcome is defined as a subsequent report—a rereport—of child maltreatment after referral. This study used a large nationally representative sample and longitudinal data to assess the causal effect of substance abuse services on child welfare outcomes, which is an issue with important implications for both policymakers and child welfare workers (Section 5.9.1). The second study is a causal study of the impacts of poverty and multigenerational dependence on welfare on children's academic development. This study also used a nationally representative sample and longitudinal data to test important hypotheses derived from theoretical models (Sections 5.9.2 to 5.9.7). The third study is an evaluation of a school-based intervention aimed at promoting children's social competence and decreasing their aggressive behavior. This intervention was originally designed as using a group randomization trial, but in practice the randomization did not work (Section 5.9.8).

Methodologically, these examples illustrate most models depicted in this chapter. Section 5.9.1 demonstrates greedy matching followed by a survival analysis. Section 5.9.2 demonstrates optimal matching and evaluation of covariate imbalance before and after matching. Section 5.9.3 demonstrates the Hodges-Lehmann aligned rank test after optimal matching. Section 5.9.4 demonstrates regression adjustment after optimal pair matching. Section 5.9.5 demonstrates propensity score weighting. Section 5.9.6 demonstrates modeling doses of treatment. Section 5.9.7 deals with the issue of model comparison and demonstrates conclusions the analyst may draw based on findings using the same data but different models (i.e., a comparison of models depicted by Sections 5.9.2 to 5.9.5). And Section 5.9.8 compares the GBM algorithm developed by Rand Corporation to a user-developed GBM algorithm available in Stata.

5.9.1 GREEDY MATCHING AND
SUBSEQUENT ANALYSIS OF HAZARD RATES

This study uses sample and conditioning variables similar to those used in the example in Section 4.5.1. It analyzes a subsample of 2,758 children from the panel data of the National Survey of Child and Adolescent Well-Being (NSCAW). The primary interest of the study is whether caregivers' use of substance abuse services reduces the likelihood of having a rereport of child maltreatment. Thus, the dependent variable is the timing of a maltreatment rereport 18 months after the baseline interview; study participants who did not have a rereport at the end of the 18-month window are defined as censored. As described in Section 4.5.1, the study subsample was limited to children who lived at home (e.g., they were not in foster care) and whose primary caregiver was female. The study was limited to female caregivers because they constitute the vast majority (90%) of primary caregivers in NSCAW.

We conducted a three-step analysis. At Step 1, we used conditioning variables to develop propensity scores. At Step 2, we used nearest neighbor matching within caliper and Mahalanobis metric matching to create various matched samples. At Step 3, because the timing to first maltreatment rereport involves censoring (i.e., we knew only the timing of rereport within an 18-month window and those who did not have rereport by the end of the study window are censored), we conducted a Kaplan-Meier product limit analysis to assess differences on survivor functions between the treated participants (i.e., caregivers who receive substance abuse treatment between baseline and the 18th month) and controls (those who did not receive such services in the same period). The matched samples are 1-to-1 match (i.e., each treated case matches to only one nontreated case in the resamples).

The 1-to-1 match for the rereport analysis was a "three by two by two" design. That is, we used *three* logistic regression models (i.e., each model specified a different set of conditioning variables to predict the propensity scores of receiving treatment), *two* matching algorithms (i.e., nearest neighbor within caliper and Mahalanobis), and *two* matching specifications (i.e., for nearest neighbor we used two different specifications on caliper size and for Mahalanobis we used one with and one without propensity score as a covariate to calculate the Mahalanobis metric distances). Hence, we tested a total of 12 matching schemes. The design using multiple matching schemes was directly motivated by the need to compare results among varying methods and to test the sensitivity of study findings to various model assumptions. We defined the logit or $\log[(1 - \hat{e}(x))/\hat{e}(x)]$ rather than the predicted probability $\hat{e}(x)$ as propensity score, because the logit is approximately normally distributed.

Table 5.5 presents sample descriptive statistics and three logistic regression models. Among the 2,758 children, 10.8% had female caregivers who received substance abuse treatment and the remaining 89.2% had female caregivers who did not receive treatment services. Bivariate chi-square tests showed that most variables were statistically significant ($p < .05$) before matching, indicating that the covariate distributions were not sufficiently overlapped between the treated and nontreated participants in the original sample. Clearly, the treatment group, overall, had many more problems with substance abuse and greater exposure to risk-related conditions.

The three logistic regression models differ in the following ways: Logistic 1 contains all predetermined covariates except four variables measuring service needs; Logistic 2 adds the four service need variables; and Logistic 3 drops the variable "Child welfare worker (CWW) report of need for service" because we

(*Text continued on page 181*)

Table 5.5 Sample Description and Logistic Regression Models Predicting Propensity Scores (Example 5.9.1)

| | | | | | | *B* | |
Variable	N	%	% Caregivers Treated (Service Users)	Bivariate χ^2 Test	Logistic 1	Logistic 2	Logistic 3
Marital status							
Not married	1,926	69.8	11.5	.085			
Married (MARRIED)	832	30.2	9.3		.055	.397	.180
Education							
No degree	926	33.6	13.1	.005			
High school diploma or equivalent (HIGH)	1,232	44.7	10.6		−.161	.078	−.210
B.A. or higher (BAHIGH)	600	21.8	7.8		−.424*	−.064	−.253
Poverty							
< 50%	623	22.6	13.5	.023			
50% to < 100% (POVERTY2)	898	32.6	11.7		−.146	−.206	−.088
100% to < 150% (POVERTY3)	503	18.2	8.0		−.329	−.220	−.221
150% to < 200% (POVERTY4)	339	12.3	8.9		−.125	−.099	.095
> 200% or more (POVERTY5)	395	14.3	9.9		.095	−.277	.011
Employment							
Not employed	1,424	51.6	13.6	< .0001			
Employed (EMPLOY)	1,334	48.4	7.9		−.175	−.323	−.162

(Continued)

Table 5.5 (Continued)

Variable	N	%	% Caregivers Treated (Service Users)	Bivariate χ^2 Test	Logistic 1	Logistic 2	Logistic 3
Case status							
Closed	1,211	43.9	5.7	< .0001			
Open (OPEN)	1,547	56.1	14.8		.807***	.167	.509**
Child's race							
White	1,504	54.5	9.8	.010			
African American (BLACK)	706	25.6	12.2		.167	-.367	.021
Hispanic (HISPANIC)	404	14.7	9.7		.298	-.089	.332
Native American (NATAM)	144	5.2	18.1		.882**	.769*	.819**
Child's Age							
11+	559	20.3	6.8	< .0001			
0–2 (CHDAGE1)	937	34.0	16.9		1.088***	.739*	1.027***
3–5 (CHDAGE2)	452	16.4	9.1		.341	.223	.358
6–10 (CHDAGE3)	810	29.4	7.5		.233	.179	.190
Caregiver's age							
> 54	43	1.6	18.6	.313			
< 35 (CGRAGE1)	1,904	69.0	10.4		-1.225**	-.925	-1.210*
35–44 (CGRAGE2)	653	23.7	11.3		-.754	-.460	-.813
45–54 (CGRAGE3)	158	5.7	12.0		-.719	-.421	-.410

178

Variable	N	%	% Caregivers Treated (Service Users)	Bivariate χ^2 Test	B Logistic 1	Logistic 2	Logistic 3
Trouble paying for basic necessities							
No	1,911	69.3	9.2	< .0001			
Yes (CRA47A)	847	30.7	14.5		−.059	−.195	−.083
Caregiver mental health							
No problem	2,014	73.0	7.4	< .0001			
Mental health problem (MENTAL)	744	27.0	20.2		.734***	.203	.633***
Caregiver arrest							
Never arrested	1,837	66.6	6.1	< .0001			
Arrested (ARREST)	921	33.4	20.2		1.034***	.767***	.858***
AOD treatment receipt							
No treatment	2,469	89.5	8.1	< .0001			
Treatment (PSH17A)	289	10.5	33.9		1.366***	.358	.630**
Maltreatment type							
Physical abuse	681	24.7	7.8	< .0001			
Sexual abuse (SEXUAL)	356	12.9	3.7		−.667*	−.422	−.422

(Continued)

179

Table 5.5 (Continued)

Variable	N	%	% Caregivers Treated (Service Users)	Bivariate χ^2 Test	B Logistic 1	B Logistic 2	B Logistic 3
Failure to provide (PROVIDE)	596	21.6	17.0		.440*	−.191	.276
Failure to supervise (SUPERVIS)	764	27.7	11.9		.108	−.202	−.020
Other (OTHER)	361	13.1	11.1		.211	−.569	.034
Risk assessment							
Risk absence	2,284	82.8	4.3	< .0001			
Risk presence (RA)	474	17.2	42.2			1.388***	2.026***
"CIDI-SF"							
Absence	1,958	71.0	6.1	< .0001			
Presence (CIDI)	800	29.0	22.4			.971***	.912***
Caregiver report of need—No	2,635	95.5	9.3	< .0001			
Yes (CGNEED)	123	4.5	43.1			1.109**	1.210***
CWW report of need for service							
No	2,425	87.9	3.1	< .0001			
Yes (CWWREP)	333	12.1	67.0			3.398***	
Constant of the logistic regression					−3.084***	−3.933***	−3.799***

SOURCE: Guo, S., Barth, R. P., & Gibbons, C. (2006). Table 1, p. 372–373. Reprinted with permission from Elsevier.

NOTE: Reference group is shown next to the variable name.

*$p < .05$, **$p < .01$, ***$p < .001$.

determined that this variable was highly correlated with the actual delivery of treatment and therefore was not an appropriate covariate for matching.

Table 5.6 describes the 12 matching schemes and numbers of participants for the resamples: Schemes 1 to 4 were based on Logistic 1, Schemes 5 to 8 were based on Logistic 2, and Schemes 9 to 12 were based on Logistic 3. Within each set of schemes using the same logistic regression, the first two schemes used nearest neighbor matching within caliper (i.e., one used a caliper size that is a quarter of the standard deviation of the propensity scores or $.25\sigma_p$, and the other employed a more restrictive or narrowed caliper of 0.1), and the next two schemes used Mahalanobis metric matching (i.e., one did not use and the other used the propensity score as a matching covariate). The use of different caliper sizes shows the dilemma we encountered in matching: while a wide caliper results in more matches and a larger sample (i.e., $N_{\text{Scheme 1}} > N_{\text{Scheme 2}}$, $N_{\text{Scheme 5}} > N_{\text{Scheme 6}}$, and $N_{\text{Scheme9}} > N_{\text{Scheme10}}$), inexact matching may occur as indicated by large distances on the propensity score between the treated and nontreated cases. We included both caliper sizes in the analysis to test the sensitivity of findings to varying caliper sizes. Note that the sample sizes were the smallest when using Schemes 5 and 6: Using the same nearest neighbor within a caliper of $.25\sigma_p$, the sample size dropped from 564 for Scheme 1 (based on Logistic 1) to 328 for Scheme 5 (based on Logistic 2), which indicated that adding the four need variables greatly restricted successful matching and reduced sample size. Yet further runs indicated that the resample sizes were most sensitive to the inclusion of the variable "CWW report of need." Logistic 3 retained three need variables from Logistic 2, but dropped "CWW report of need," which increased the sample size from 328 for Scheme 5 to 490 for Scheme 9.

Because different matching schemes produce different resamples, it is important to check covariate distributions after matching and to examine sensitivity of the results to different resampling strategies. Table 5.7 presents this information. Among these 12 matching schemes, there were only two matching methods (Schemes 5 and 6) that successfully removed all significant differences of covariate distributions between treated and nontreated groups. However, because of the problem of nontrivial reduction in sample size noted earlier, these matching methods did not produce resamples representative of the original sample.

All schemes using Mahalanobis matching (Schemes 3, 4, 7, 8, 11, and 12) could not remove significant differences between treated and nontreated groups. Testing the schemes this way suggests that the Mahalanobis approach may not be a good method for matching that involves a large number of matching covariates, such as is the case for this study. Furthermore, using the propensity score as an additional matching variable (Schemes 4, 8, and 12) did not help.

(Text continued on page 184)

Table 5.6   Description of Matching Schemes and Resample Sizes (Example 5.9.1)

Scheme	Description of Matching Method	N of the New sample Treated	Control
1. Nearest 1-1	Propensity scores predicted by logistic 1, Nearest 1-to-1 using caliper = .311 (.25 σ)	282	282
2. Nearest 1-2	Propensity scores predicted by logistic 1, Nearest 1-to-1 using caliper = .1	281	281
3. Mahalanobis 1	Covariates used in the calculation of the Mahalanobis distances same as logistic 1	265	265
4. Mahalanobis 1 with p-score added	Mahalanobis 1 with propensity score added, Propensity scores predicted by logistic 1	265	265
5. Nearest 2-1	Propensity scores predicted by logistic 2, Nearest 1-to-1 using caliper = .490 (.25 σ)	164	164
6. Nearest 2-2	Propensity scores predicted by logistic 2, Nearest 1-to-1 using caliper = .1	163	163
7. Mahalanobis 2	Covariates used in the calculation of the Mahalanobis distances same as logistic 2	182	182
8. Mahalanobis 2 with p-score added	Mahalanobis 2 with propensity score added, Propensity scores predicted by logistic 2	182	182
9. Nearest 3-1	Propensity scores predicted by logistic 3, Nearest 1-to-1 using caliper=.401 (.25 σ)	245	245
10. Nearest 3-2	Propensity scores predicted by logistic 3, Nearest 1-to-1 using caliper=.1	245	245
11. Mahalanobis 3	Covariates used in the calculation of the Mahalanobis distances same as logistic 3	235	235
12. Mahalanobis 3 with p-score added	Mahalanobis 3 with propensity score added, Propensity scores predicted by logistic 3	235	235

SOURCE: Guo, S., Barth, R. P., & Gibbons, C. (2006). Table 2, p. 375. Reprinted by permission from Elsevier.

Table 5.7   Results of Sensitivity Analyses (Example 5.9.1)

| | | Results of Survival Analysis on Timing of Rereport | | |
| | | 85th Percentile of Survivor Function in Months (Kaplan-Meier Estimation) | | *p Value Testing Group Difference (Wilcoxon)* |
Scheme	*Covariate Distributions Did Not Overlap Sufficiently: Covariates Significant After Matching p < .05*	*Treated*	*Control*	
Original Sample or All (n = 2,723)[a]	BAHIGH, POVERTY3, EMPLOY, OPEN, NATAM, CHDAGE1, CHDAGE3, CRA47A, MENTAL, ARREST, PSH17A, SEXUAL, PROVIDE, RA, CIDI, CGNEED, CWWREP	7.6	13.6	< .0001
1. Nearest 1-1	RA, CIDI, CGNEED, CWWREP	7.4	9.5	.48
2. Nearest 1-2	RA, CIDI, CGNEED, CWWREP	7.6	10.2	.33
3. Mahalanobis 1	OPEN, MENTAL, ARREST, RA, CIDI, CGNEED, CWWREP	7.6	12.1	.01
4. Mahalanobis 1 with *p*-score added	OPEN, MENTAL, RA, CIDI, CGNEED, CWWREP	90% 5.2	90% 9.2	.01
5. Nearest 2-1		7.4	9.5	.36
6. Nearest 2-2		8.8	9.5	.45
7. Mahalanobis 2	CGRAGE2, MENTAL, ARREST, PSH17A, RA, CIDI, CWWREP	7.8	9.9	.41

(Continued)

Table 5.7 (Continued)

| | | Results of Survival Analysis on Timing of Rereport | | |
| | | 85th Percentile of Survivor Function in Months (Kaplan-Meier Estimation) | | p Value Testing Group Difference (Wilcoxon) |
Scheme	Covariate Distributions Did Not Overlap Sufficiently: Covariates Significant After Matching p < .05	Treated	Control	
8. Mahalanobis 2 with p-score added	CGRAGE1, CGRAGE2, MENTAL, PSH17A, RA, CIDI, CWWREP	7.3	9.9	.44
9. Nearest 3-1	CWWREP	7.6	12.7	.02
10. Nearest 3-2	CWWREP	7.6	12.7	.02
11. Mahalanobis 3	CGRAGE1, CGRAGE2, MENTAL, ARREST, RA, CIDI, CWWREP	90% 5.3	90% 7.6	.08
12. Mahalanobis 3 with p-score added	CGRAGE1, CGRAGE2, MENTAL, ARREST, RA CIDI, CWWREP	90% 6.0	90% 7.6	.06

SOURCE: Guo, S., Barth, R. P., & Gibbons, C. (2006). Table 3, p. 375. Reprinted by permission from Elsevier.

a. Thiry-five study participants were eliminated from the analysis because of missing data.

Among the rest of the matching schemes that used nearest neighbor within calipers, only Schemes 9 and 10 successfully removed the significant differences between groups, although the variable CWWREP remains significant in these samples. Both schemes were based on Logistic 3, which excludes CWWREP as a matching covariate. This exclusion was defined because a closer look at the distribution of CWWREP indicated that the CWWREP variable was highly correlated with the dependent variable of the logistic regression, that is, the dichotomous variable for receipt of substance abuse services. In the original sample, 95.5% of the non–service users had a zero need for service use identified by child welfare workers, whereas 74.8% of the service users had an identified need. This is almost certainly likely to have occurred because child welfare workers who observed phenomena such as positive drug tests are more likely to conclude that a caregiver is involved with substance abuse treatment. The presence of a high correlation between CWWREP and the dependent variable of logistic regression prevents the use of CWWREP as a conditioning variable. Thus, we conclude that among the three logistic regressions, Logistic 3 is the best.

Table 5.7 also presents results of the survival analysis, that is, the 85th percentile of survivor function associated with each scheme and significance tests on the null hypothesis about equal survivor functions between groups. We report the 90th percentile (instead of 85th percentile) for Schemes 4, 11, and 12, because the proportion of survivors for groups in these schemes is greater than 85% by the end of the study window. In this analysis, the 85th percentile indicates the number of months it takes for the remaining 15% of study participants to have a maltreatment rereport, and the smaller the number the sooner the rereport and the greater the risk of maltreatment recurrence. As the statistics show, for the original sample of 2,723 children, it took 7.6 months for 15% of the children whose caregivers used substance abuse services to have a rereport, while it took 13.6 months for 15% of the nontreated children to have a rereport, and the difference is statistically significant ($p < .0001$). Thus, children whose caregivers used substance abuse services were more likely to have a child maltreatment rereport than children whose caregivers did not use the substance abuse treatment services.

All matching schemes showed differences in survivor functions in the same direction; that is, the treated group had a higher hazard for rereport than the nontreated group. This finding is consistent across different matching methods, indicating that children of service users have a higher likelihood of being rereported than children of non–service users. The remaining question is whether this difference is statistically significant. Because we know that the methods of nearest neighbor matching within caliper using Logistic 3 (i.e., Schemes 9 and 10) are the only ones that meet the assumption about ignorable treatment assignment on the covariates, and the group difference on survivor functions is statistically significant in these schemes (i.e., p value is .02), we can conclude that the difference between groups is statistically significant. Children of substance abuse service users appear to live in an environment that elevates risk for maltreatment and as compared with the children of caregivers who were non–service users, warrants continued protective supervision.

Figure 5.3 is a graphic representation of the survivor curves comparing the original sample with the resample of Scheme 9. The figure shows that (a) the gap between treated and nontreated groups was slightly wider for the resample of Scheme 9 than the original sample between months 8 and 12 and (b) by the end of the study window, the proportion of children remaining in a "no rereport" state was slightly higher in the original sample than in the Scheme 9 resample. The analysis based on the original sample without controlling for heterogeneity of service receipt masked the fact that substance abuse treatment may be a marker for greater risk and an indication for the continued involvement of families with child welfare services.

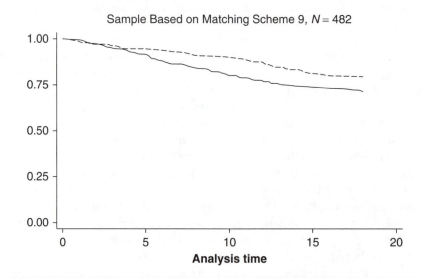

Figure 5.3   Survivor Functions: Percentage Remaining No Rereport (Example 5.9.1)

SOURCE: Guo, S., Barth, R. P., & Gibbons, C. (2006). Figure 1, p. 377. Reprinted by permission from Elsevier.

In sum, the propensity score matching analysis of the rereport risk enabled an analysis of observational data, when experimental data were unavailable or could not be made available, that provides evidence that substance abuse treatment is not generating safety for children of service users. Additional confirmatory analyses and discussion of these findings are available elsewhere (Barth, Gibbons, & Guo, 2006).

5.9.2 OPTIMAL MATCHING

We now present a study investigating intergenerational dependence on welfare and its relation to child academic achievement. This study illustrates the application of optimal matching with observational data.

1. *Conceptual Issues and Substantive Interests.* As described in Chapter 1, prior research has shown that both childhood poverty and childhood welfare dependency have an impact on child development. In general, growing up in poverty adversely affects a child's life prospects, and the consequences become more severe as poverty persists (Smith & Yeung, 1998). Duncan et al. (1998) found that family economic conditions in early childhood had the greatest impact on achievement, especially among children in families with low incomes. Foster and Furstenberg (1998, 1999) found that the most disadvantaged children tended to live in female-headed households with low incomes, receive public assistance, and/or have unemployed heads of household. In their study relating patterns of childhood poverty to children's IQ and behavior problems, Duncan, Brooks-Gunn, and Klebanov (1994) found that the duration of economic deprivation was a significant predictor of both outcomes. Focusing on the effects of the timing, depth, and length of poverty on children, Brooks-Gunn and Duncan's study (1997) supported the conclusion that family income has selective but significant effects on the well-being of children and adolescents, with greater impacts on ability and achievement than on emotional development. In addition, Brooks-Gunn and Duncan found that poverty had a far greater influence on child development if children experience poverty during early childhood.

The literature clearly indicates a link between intergenerational welfare dependence and child developmental outcomes. From the perspective of the resources model (Wolock & Horowitz, 1981), this link is repetitive and leads to a vicious cycle that traps generations in poverty. Children born to families with intergenerational dependence on welfare may lack sufficient resources to achieve academic goals, which will ultimately affect their own employability and increase their risk for using public assistance in adulthood.

Corcoran and Adams (1997) developed four models to explain poverty persistence across generations: (1) the lack of economic resources hinders human capital development; (2) parent's noneconomic resources, which are correlated with their level of poverty, determine children's poverty as adults; (3) the welfare system itself produces a culture of poverty shared by parents and children; and (4) structural-environmental factors associated with labor market conditions, demographic changes, and racial discrimination shape intergenerational poverty. Corcoran and Adams's findings support all these models to some extent, with the strongest supports established for the economic resources argument.

Prior research on poverty and its impact on child development has shed light on the risk mechanisms linking resources and child well-being. Some of these findings have shaped the formation of welfare reform policies; some have fueled the ongoing debate about the impacts of welfare reform; and still other findings remain controversial. There are two major methodological limitations in this literature. First, prior research did not analyze a broad range of child outcomes (i.e., physical health, cognitive and emotional development, and academic achievement). Second, and more central to this example, prior research implicitly assumed a causal effect of poverty on children's academic achievement. However, most such studies used covariance control methods such as regression or regression-type models without explicit control for sample selection and confounding covariates. As we have shown earlier, studies using covariance control may fail to draw valid causal inferences.

2. *Data and Results of Optimal Matching.* This study uses the 1997 Child Development Supplement (CDS) to the Panel Study of Income Dynamics (PSID) and the core PSID annual data from 1968 to 1997 (Hofferth et al., 2001). The core PSID is comprised of a nationally representative sample of families. In 1997, the Survey Research Center at the University of Michigan collected information on 3,586 children between the ages of birth and 12 years who resided in 2,394 PSID families. Information was collected from parents, from teachers, and from the children themselves. The objective was to provide researchers with comprehensive and nationally representative data about the effects of maternal employment patterns, changes in family structure, and poverty on child health and development. The CDS sample contained data on academic achievement for 2,228 children associated with 1,602 primary caregivers. To address the research question about intergenerational dependence on welfare, we analyze a subset of this sample. Children included in the study were those who had valid data on receipt of welfare programs in childhood (e.g., AFDC [Aid to Families With Dependent Children]) and whose caregivers were aged 36 years or younger in 1997. The study involved a careful examination of 30 years of data using the 1968 PSID ID number of primary caregivers as a key. Due to limited information, the study could not distinguish between the types of assistance programs. The study criteria defined a child as a recipient of public assistance (e.g., AFDC) in a particular year if his or her caregiver ever received the AFDC program in that year and defined a caregiver as a recipient of AFDC in a particular year if the caregiver's primary caregiver (or the study child's grandparent) ever received the program in that year. The definition of receipt of AFDC in a year cannot disentangle short-term use (e.g., receipt of AFDC for only a single month) from long-term use (e.g., all 12 months). One limitation of the study is posed by the discrete nature of AFDC data and the fact that the AFDC study variable (i.e., "caregiver's number of years using AFDC in childhood") was treated as a continuous variable in the analysis,

which may not accurately measure the true influence of AFDC. After screening the data, applying the inclusion criteria, and deleting missing data listwise, the study sample comprised 1,003 children associated with 708 caregivers.

For this illustration, we report findings that examine one domain of academic achievement: the age-normed "letter-word identification" score of the Woodcock-Johnson Revised Tests of Achievement (Hofferth et al., 2001). A high score on this measure indicates high achievement. The score is defined as the outcome variable for this study and has the dual virtues of being standardized and of representing a key concept in the welfare reform debate—the impact of welfare reform on children's educational attainment.

Table 5.8 shows the level of intergenerational dependence on welfare in this sample. Of 1,003 children whose welfare status was compared with that of their caregivers, 615 or 61.3% remained in the same status of not using welfare, 114 or 11.4% showed upward social mobility (i.e., their caregivers used welfare between ages 6 and 12, but the next generation did not use welfare from birth to their current age in 1997), 154 or 15.4% showed downward social mobility (i.e., their caregivers did not use welfare, but the next generation used welfare at some point in their lives), and 120 or 12.0% remained in the same status of using welfare as their caregivers. Thus, the overall level of intergenerational dependence on welfare, as defined within a two-generation period, was 12.0%.

Based on the research question, we classified the study participants into two groups: those who ever used AFDC from birth to current age in 1997 and those who never used AFDC during the same period. Thus, this dichotomous variable defines the treatment condition in the study: those who ever used AFDC versus controls who never used AFDC. To assess the treatment effect (i.e., child's use of AFDC) on academic achievement, the analyst must control for many covariates or confounding variables. For the purpose of illustration, we considered the following covariates: current income or poverty status, measured as the ratio of family income to poverty threshold in 1996; caregiver's

Table 5.8   Status of Child's Use of AFDC by Status of Caregiver's Use of AFDC in Childhood (Example 5.9.2)

Child's AFDC Use From Birth to Current Age in 1997	Caregiver's AFDC Use in Childhood (Ages 6 to 12)		Total
	Never Used	Used	
Never used	615 (61.3)	114 (11.4)	729 (72.7)
Used	154 (15.4)	120 (12.0)	274 (27.3)
Total	769 (76.7)	234 (23.3)	1,003 (100)

NOTE: p < .001, chi-square test; each percentage (in parentheses) is obtained by dividing the observed frequency by the sample total 1,003.

education in 1997, which was measured as years of schooling; caregiver's history of using welfare, which was measured as the number of years (i.e., a continuous variable) caregiver used AFDC between ages 6 and 12; child's race, which was measured as African American versus non–African American; child's age in 1997; and child's gender, which was measured as male versus female.

Table 5.9 shows sample descriptive statistics, an independent sample t test on the ATE, and an OLS regression evaluating the ATE. The p values associated with covariates were provided by the Wilcoxon rank-sum (Mann-Whitney) test. As the table shows, except for child's gender, the difference on each covariate between treated and control groups is statistically significant. The treated group tended to be those who were poorer in 1996 ($p < .000$), whose caregivers had a lower level of education in 1997 ($p < .000$), and whose caregivers used AFDC for a longer time in childhood ($p < .000$). In addition, the treated group had a larger percentage of African Americans ($p < .000$), and the treated group was on average older in 1997($p < .001$) than the control group. Without controlling for these covariates, the study's estimate of treatment effect would be biased. The table also presents two estimates of ATE. One estimate was derived from the independent sample t test, which shows that the treated group on average has a mean letter-word identification score that was 9.82 points lower than that of the control group ($p < .000$). The other estimate was obtained from the OLS regression with robust estimates of standard errors to control for the clustering of children within families. This second estimate is the model most studies would use, which shows that controlling for these covariates, the treated group on average has a letter-word identification score that is 4.73 points lower than that of the control group ($p < .01$). Based on Chapters 2 and 3, we could say with reasonable confidence that both estimates are likely to be biased and inconsistent.

Our next step was to conduct a propensity score analysis. At Step 1, we used GBM (i.e., Stata *boost*) to estimate the propensity scores of receiving treatment. The GBM showed that the ratio of family income to poverty in 1996 had the strongest influence on the likelihood function (86.74%), caregiver's use of AFDC in childhood has the next strongest influence (6.03%), and the remaining influences included caregiver's education in 1997 (3.70%), child's race (2.58%), and child's age in 1997 (0.95%). Gender was not a significant predictor shown by the Wilcoxon rank-sum test; therefore, the GBM did not use child's gender.

Figure 5.4 shows the box plots and histograms of the estimated propensity scores by treatment status. As the figure indicates, the two groups differ from each other on the distribution of estimated propensity scores. The common support region is especially problematic because the treated group's 25th percentile is equal to the control group's upper adjacent value. The two groups share a very narrow common support region. If we applied nearest neighbor

Table 5.9 Sample Description and Results of Regression Analysis (Example 5.9.2)

Variable	Child AFDC Use From Birth to Current Age in 1997 [% or Mean (SD)]			Estimated Regression Coefficient (Robust SE) Using Letter-Word Score in 1997 as Outcome
	Never	Used	p-value	
Outcome: letter-word identification score in 1997	103.98 (16.8)	94.16 (14.8)	.000[a]	
Covariate				
Ratio of family income to poverty line in 1996	3.16 (2.76)	1.05 (1.00)	.000[b]	1.13 (0.33)**
Caregiver's education in 1997 (years of schooling)	13.2 (1.88)	11.6 (1.55)	.000[b]	0.91 (0.34)**
Caregiver's number of years using AFDC in childhood	0.48 (1.36)	1.86 (2.59)	.000[b]	−0.76 (0.31)*
Child's race: African American (reference: other)	36.20%	78.10%	.000[b]	−1.88 (1.23)
Child's age in 1997	6.50 (2.78)	7.11 (2.81)	.001[b]	0.87 (0.17)***
Child's gender: male (reference: female)	53.50%	51.80%	.636[b]	−2.00 (0.99)*
Child's AFDC use status: used (reference: never)				−4.73 (1.39)**
Constant				84.85 (4.5)***
R^2				0.15
Number of children (number of families)	729 (506)	274 (202)		1,003 (708)

a. Independent-sample t test, two-tailed.
b. Wilcoxon rank-sum (Mann-Whitney) test.
*$p < .05$, **$p < .01$, ***$p < .001$.

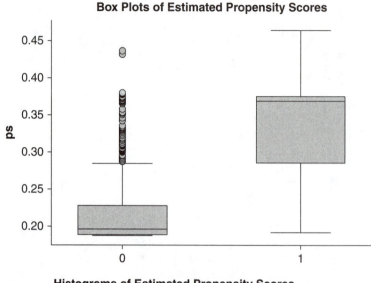

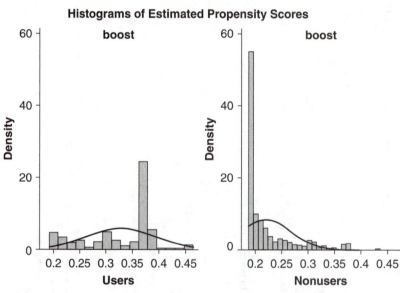

Figure 5.4   Distribution of Estimated Propensity Scores (Example 5.9.2)

matching within caliper or other types of greedy matching, it is likely that the narrow common support region would produce a nontrivial loss of matched participants. In addition, greedy matching identifies matches for each treated participant in a nonoptimal fashion. Based on these concerns, we decided to use optimal matching.

Before running optimal matching, we need to carefully examine the ratio of treated participants to control participants to determine matching schemes. In this sample, we have 274 treated participants and 729 controls or a ratio of treated participants to controls of approximately 0.38:1. With these data, a 1-to-1 or 1-to-2 pair matching scheme is feasible. However, optimal pair matching is not generally optimal, especially when compared with full matching based on the same data (see Section 5.4.2), and a 1-to-1 matching will make $729 - 274 = 455$ controls nonusable, and a 1-to-2 matching will make $729 - (274 \times 2) = 181$ controls nonusable. For purposes of illustration, we decided to run a 1-to-1 pair matching. We also found that with these data, a full matching or variable matching is permissible. Using principles and considerations described in Section 5.4.2, we used the following matching schemes: (a) full matching; (b) Variable Matching 1, which uses at least 1 and at most 4 controls for each treated participant; (c) Variable Matching 2, which uses at least 2 and at most 4 controls for each treated participant; (d) Variable Matching 3, which uses Hansen's equation (i.e., it specifies that the minimum number of matched controls is $.5(1 - \hat{P}) / \hat{P} = .5(1 - .273) / .273 = 1.33$, and the maximum number of matched controls is $2(1 - \hat{P}) / \hat{P} = 2(1 - .273) / .273 = 5.32$); (e) Variable Matching 4, which uses at least 2 and at most 7 controls for each treated participant; and (f) Pair Matching 1-to-1.

We ran *optmatch* in R to implement optimal matching with these schemes. Table 5.10 presents optimal matching results. Two useful statistics produced by *optmatch* are shown in the table: stratum structure and total distance. The stratum structure is a count of matched sets in terms of the ratio of the number of treated participants to the number of controls. For instance, the full matching produced a total of 135 matched sets; of these, 1 set had 22 treated participants and 1 control, 40 sets had 1 treated participant and 1 control, 20 sets had 1 treated participant and 2 controls, and so forth. Note that one set contains 1 treated participant and 245 controls. Total distance is the sum of differences on propensity scores between treated and control participants over all matched sets, and thus, total distance is an overall measure of the closeness of matching, with a small number indicating a close match. Using total distance, we found that among all six schemes, full matching offered the best approach, and pair matching was the second best. Of the four variable-matching schemes, Variable Matching 3 using Hansen's equation worked the best. It is worth noting that the 1-to-1 pair matching did not use 455 controls, which means there was a loss of 45.4% of study cases. Although pair matching showed closeness in matching, it eliminates study cases, which exerts an undesirable impact on study power. Our data confirmed previous findings that full matching is the best.

With matched samples, we always want to know how well matching has reduced bias. The level of bias reduction can be shown by a comparison

Table 5.10 Results of Optimal Matching (Example 5.9.2)

Matching Scheme	Stratum Structure Ratio of "Treatment:Control" (Number of Matched Sets)	Total Distance	Percentage of Cases Lost in Matching
Full matching	22:1 20:1 15:1 14:1 10:1 8:1 7:1 6:1 5:1 4:1 3:1 2:1 1:1 (1 1 1 1 2 2 1 1 1 3 4 8 40 1:2 1:3 1:4 1:5 1:6 1:7 1:8 1:9 1:10 1:12 1:13 1:14 1:18 20 12 6 6 4 3 2 2 2 2 3 1 1 1:19 1:27 1:29 1:245 2 1 1 1)	31831	0.0%
Variable matching 1 (at least 1, at most 4)	1:1 1:3 1:4 (122 1 151)	241707	0.0%
Variable matching 2 (at least 2, at most 4)	1:2 1:3 1:4 (182 1 90)	255603	0.0%
Variable matching 3 (Hansen's equation)	1:1 1:6 (183 91)	198664	0.0%
Variable matching 4 (at least 2, at most 7)	1:2 1:3 1:7 (237 1 36)	228723	0.0%
Pair matching	1:1 0:1 (274 455)	40405	45.40%

between the absolute standardized differences in covariate means before and after matching (i.e., a comparison between d_x and d_{xm}). Table 5.11 presents this information. Full matching greatly reduces covariate imbalance on all variables except gender. Taking the ratio of family income to poverty in 1996 as an example, before matching, the treated and control groups differ on this variable by more than 100% of a standard deviation, whereas after full matching, the standard bias is only 4% of a standard deviation. Nearly all matching schemes reduce bias for almost all variables to some extent, but some do more, and some do less. In terms of covariate balancing, Table 5.11 confirms that full matching worked the best and pair matching the second best.

5.9.3 POST–FULL MATCHING ANALYSIS USING THE HODGES-LEHMANN ALIGNED RANK TEST

The Hodges-Lehmann aligned rank test can be used on matched samples created by full matching or variable matching. Because none of the variable matching schemes showed satisfactory bias reduction, we ruled out these matched samples for further analysis. Table 5.12 presents results of the post–full matching analysis.

As the table shows, we used full matching and found that children who used AFDC had a letter-word identification score in 1997 that was, on average, 1.97 points lower than those who had never used AFDC; the difference was statistically significant at a .05 level. We used the Hodges-Lehmann test to gauge the statistical significance. The study also detected an effect size of .19, which is a small effect size in terms of Cohen's (1988) criteria.

5.9.4 POST–PAIR MATCHING ANALYSIS USING REGRESSION OF DIFFERENCE SCORES

A regression of difference scores may be performed on matched samples created by optimal pair matching. Based on the pair-matched sample, we first calculated difference scores between treated and control cases for each pair on all study variables (i.e., on the outcome variable and all covariates). We then regressed the difference score of the outcome on the difference scores of covariates. In addition, note that our model includes a correction for clustering effects (i.e., children are nested within caregivers) that we accomplished by using robust estimates of standard errors. Table 5.13 presents results of the post–pair matching analysis of regression adjustment.

As explained in Section 5.5.5, the intercept of a difference score regression indicates the ATE of the sample. The estimated intercept from this model is -3.17 ($p < .05$). Thus, using pair matching and regression adjustment, the

Table 5.11   Covariate Imbalance Before and After Matching by Matching Scheme
(Example 5.9.2)

Covariate and Matching Scheme	d_x	d_{xm}
Ratio of family income to poverty line in 1996	**1.02**	
Full matching		0.04
Variable Matching 1 (at least 1, at most 4)		0.79
Variable Matching 2 (at least 2, at most 4)		0.94
Variable Matching 3 (Hansen's equation)		0.74
Variable Matching 4 (at least 2, at most 7)		0.87
Pair matching		0.25
Caregiver's education in 1997 (years of schooling)	**0.91**	
Full matching		0.17
Variable Matching 1 (at least 1, at most 4)		0.72
Variable Matching 2 (at least 2, at most 4)		0.87
Variable Matching 3 (Hansen's equation)		0.75
Variable Matching 4 (at least 2, at most 7)		0.81
Pair matching		0.34
Caregiver's number of years using AFDC in childhood	**0.67**	
Full matching		0.01
Variable Matching 1 (at least 1, at most 4)		0.56
Variable Matching 2 (at least 2, at most 4)		0.65
Variable Matching 3 (Hansen's equation)		0.55
Variable Matching 4 (at least 2, at most 7)		0.63
Pair matching		0.40
Child's race: African American (reference: other)	**0.93**	
Full matching		0.01
Variable Matching 1 (at least 1, at most 4)		0.79
Variable Matching 2 (at least 2, at most 4)		0.92
Variable Matching 3 (Hansen's equation)		0.73
Variable Matching 4 (at least 2, at most 7)		0.85
Pair matching		0.44
Child's age in 1997	**0.22**	
Full matching		0.07
Variable Matching 1 (at least 1, at most 4)		0.21
Variable Matching 2 (at least 2, at most 4)		0.22

Covariate and Matching Scheme	d_x	d_{xm}
Variable Matching 3 (Hansen's equation)		0.19
Variable Matching 4 (at least 2, at most 7)		0.24
Pair matching		0.14
Child's gender: male (reference: female)	0.03	
Full matching		0.05
Variable Matching 1 (at least 1, at most 4)		0.02
Variable Matching 2 (at least 2, at most 4)		0.05
Variable Matching 3 (Hansen's equation)		0.01
Variable Matching 4 (at least 2, at most 7)		0.03
Pair matching		0.09

NOTE: Absolute standardized difference in covariate means, before (d_x) and after matching (d_{xm}).

study found that, on average, children who used AFDC had a letter-word identification score in 1997 that was 3.17 points lower than children who never used AFDC; this finding was statistically significant.

5.9.5 PROPENSITY SCORE WEIGHTING

In this example, we illustrate the analysis of propensity score weighting, a model described in Section 5.6. Denoting estimated propensity scores as P, we first created two weight variables: (1) to estimate ATE, we created weights for treated participants as $1/P$, and for controls as $[1/(1-P)]$; and (2) to estimate ATT, we created weights for treated participants as 1 and for controls as $[P/(1-P)]$.

Recall that propensity score weighting does not use matching, and therefore, the sample is the same as the original sample. Given this, the method of checking covariate imbalance described in Section 5.5.3 is not applicable. We need to employ a different approach to check imbalance, one that is suitable to weighted analysis. The approach we chose is a *weighted simple regression* or *weighted simple logistic regression*. Specifically, we ran a weighted regression using a continuous covariate as the dependent variable and using a dichotomous treatment variable as the single independent variable. If our covariate being tested is a dichotomous variable, then we would run a weighted logistic regression using the dichotomous covariate as the dependent variable, and again using the dichotomous treatment variable as the single independent variable. Table 5.14 presents results of imbalance checking based on this method. The table presents p values associated with the regression coefficients of the treatment variable (i.e., child's use of AFDC) for ATE and ATT weights, respectively.

Table 5.12 Estimated Average Treatment Effect on Letter-Word Identification Score in 1997 With Hodges-Lehmann Aligned Rank Test (Matching Scheme: Full Matching) (Example 5.9.3)

Average Treatment Effect (Effect Size: Cohen's d)	Point Estimate of the Hodges-Lehmann Test Statistic $(\hat{W}_s - E(\hat{W}_s))$	Standard Error of the Test Statistic $\sqrt{Var(\hat{W}_s)}$	Z $(\hat{W}_s - E(\hat{W}_s))/\sqrt{Var(\hat{W}_s)}$	p value
−1.97 (.190)	−7051.766	3247.278	−2.172	.015

Table 5.13   Regressing Difference Score of Letter-Word Identification on
              Difference Scores of Covariates After Pair Matching (Example 5.9.4)

Covariate Difference Score	Estimated Regression Coefficient (Robust SE)
Constant	−3.17 (1.73)*
Ratio of family income to poverty line in 1996	3.14 (1.67)*
Caregiver's education in 1997 (years of schooling)	.01 (.51)
Child's race: African American (reference: other)	−3.34 (2.33)
Child's age in 1997	.23 (.33)
Child's gender: male (reference: female)	−1.21 (1.84)
Number of children (number of families)	274 (202)

*$p < .05$, one-tailed test.

Results were not as good as we hoped. All covariates for both weights, except for gender, were imbalanced to a statistically significant degree between treated participants and controls. This finding suggests that for this data set propensity score weighting cannot remove covariate imbalance, and therefore, the results of weighted analysis may remain biased.

For our demonstration purposes, we ran a propensity score weighting analysis, and these results are presented in Table 5.15. The analysis showed that children who used AFDC had an average score of letter-word identification that was 5.16 points lower than children who never used AFDC ($p < .001$). From a perspective of treatment effects for the treated (i.e., what is the effect if the analyst considers only those individuals assigned to the treatment condition?), we find that children who used AFDC had an average score of letter-word identification that was 4.62 points lower than children who never used AFDC ($p < .01$). Comparing with ATE, ATT decreases both in size and in the level of significance.

5.9.6 MODELING DOSES OF TREATMENT

In this example, we illustrate the analysis of propensity score matching that investigates the impact of multiple doses of treatment on outcome. The illustration employs the same data and research questions as those for Section 5.9.2, with one exception. In the analysis of Section 5.9.2, the treatment "child's use of AFDC" is binary—whether or not a study child ever used AFDC from birth to current age in 1997. The authors of this study also examined each year's PSID data for all study children from their birth to 1997 and obtained, for each child, the percentage of time using AFDC from birth to their current age in 1997. This example, then, assesses the impact of

Table 5.14 Covariate Imbalance After Propensity Score Weighting (Example 5.9.5)

Covariate (Employed as Dependent Variable in Regression)	p Value of Regression Coefficient of Treatment (Child Use AFDC)	
	ATE	ATT
Ratio of family income to poverty line in 1996	.000***	.000***
Caregiver's education in 1997 (years of schooling)	.000***	.000***
Child's race: African American (reference: other)	.000***	.000***
Child's age in 1997	.005**	.011*
Child's gender: male (reference: female)	.688	.590
Number of children (number of families)	1,003 (708)	1,003 (708)

NOTES: The balance check used regression for a continuous dependent variable and logistic regression for a dichotomous dependent variable.

ATE = average treatment effect where weight for a treated case is $1/p$, for a control case is $1/(1-p)$; ATT = average treatment effect for the treated where weight for a treated case is 1 and for a control case is $p/(1-p)$.

*$p < .05$, **$p < .01$, ***$p < .001$, two-tailed test.

Table 5.15   Regression Analysis of Letter-Word Identification Score in 1997 With Propensity Score Weighting (Example 5.9.5)

Predictor Variable	Estimated Regression Coefficient (Robust SE)	
	ATE	ATT
Ratio of family income to poverty line in 1996	1.14 (0.32)***	1.26 (0.34)***
Caregiver's education in 1997 (years of schooling)	0.99 (0.36)**	0.93 (0.37)*
Child's race: African American (reference: other)	−2.50 (1.35)	−2.75 (1.42)
Child's age in 1997	0.74 (0.18)***	0.62 (0.20)**
Child's gender: male (reference: female)	−1.62 (1.09)	−1.59 (1.15)
Child's AFDC use status: used (reference: never)	−5.16 (1.42)***	−4.62 (1.41)**
Constant	84.20 (4.83)***	85.29 (5.05)***
R^2	.142	.132
Number of children (number of families)	1,003 (708)	1,003 (708)

NOTE: ATE = average treatment effect where the weight for a treated case is $1/p$, for a control case is $1/(1-p)$; ATT = average treatment effect for the treated where the weight for a treated case is 1 and for a control case is $p/(1-p)$.

*$p < .05$, **$p < .01$, ***$p < .001$, two-tailed test.

using different levels of AFDC participation on academic achievement. We hypothesize that an increase in levels of using AFDC decreases academic achievement in a linear fashion.

After examining the distribution of percentages of time using AFDC from birth to current age, we create three dose groups: never used, 1% to 33.9% of the time using AFDC from birth to current age in 1997, and 34% to 100% of the time using AFDC from birth to current age in 1997. The distribution of these three doses is shown in Table 5.16.

To investigate the impact of doses on academic achievement, we employ Imbens's (2000) method. A multinomial logit model was first estimated. Variables showing statistically significant differences from bivariate analyses were entered into the multinomial model as predictors. Results of the multinomial model are shown in Table 5.17.

In Stata, we used the **predict** command following the estimation of the multinomial model to predict the generalized propensity scores for all study children. Doing so, we obtain three scores for each child, and each score

Table 5.16   Distribution of Dose Categories (Example 5.9.6)

Percentage of Time Child Used AFDC From Birth to Current Age in 1997	n	%
Dose Category 1: Never used	729	72.68
Dose Category 2: 1% to 33.9% of Time	135	13.46
Dose Category 3: 34% to100% of Time	139	13.86
Total	1,003	100

Table 5.17   Multinomial Logit Model Predicting Generalized Propensity (Example 5.9.6)

	Dose 1: Never Used		Dose 2: 1%–33.9%	
Covariate	Coefficient	SE	Coefficient	SE
Ratio of family income to poverty line in 1996	1.77	0.186***	1.03	0.194***
Caregiver's education in 1997 (years of schooling)	0.30	0.077***	0.04	0.081
Caregiver's number of years using AFDC in childhood	−.28	0.055***	−.16	0.054**
Child's race: African American (reference: other)	−.99	0.303**	−.28	0.335
Child's age in 1997	−.12	0.043**	−.08	0.045
Constant	−2.69	0.979**	−.41	1.017

NOTE: LR $\chi^2(10) = 486.78$, $p < .000$, Pseudo $R^2 = .313$; $p < .05$, $**p < .01$, $***p < .001$, two-tailed test. The multinomial logit model employs Dose Category 3 "34% to 100% of time" as a reference (omitted) category.

indicates the generalized propensity of never using AFDC, the generalized propensity of using AFDC for 1% to 33.9% of the time, and the generalized propensity of using AFDC for 34% to 100% of the time. We then define the inverse of the propensity score predicting the study child's probability of using the *actual* dose as a sampling weight. We use the letter-word identification score in 1997 as the outcome variable to measure academic achievement and run a weighted OLS regression that also controls for clustering effect in the outcome analysis. Results of the outcome analysis are shown in Table 5.18.

   In the OLS regression, we create two dummy variables to measure dose categories, with the highest dose "34% to 100% of time using AFDC" as a

Table 5.18 Regression Analysis of the Impact of Dosage of Child AFDC Use on the Letter-Word Identification Score in 1997 With and Without Propensity-Score Adjustment (Example 5.9.6)

	Estimated Regression Coefficient (Robust SE)	
Covariate	Regression Without Propensity Score Adjustment	Regression With Propensity Score Adjustment
Ratio of family income to poverty line in 1996	1.13 (.332)**	1.03 (.281)***
Caregiver's education in 1997 (years of schooling)	.91 (.336)**	1.26 (.378)**
Caregiver's number of years using AFDC in childhood	−.73 (.318)*	−.48 (.536)
Child's race: African American (reference: other)	−1.86 (1.232)	−1.58 (1.95)
Child's age in 1997	.878 (.169)***	.66 (.208)**
Child's gender: male (reference: female)	−2.02 (.987)*	−1.17 (1.30)
Dose of treatment: % of time child used AFDC (reference: used 34%–100% of time)		
Never used	5.45 (1.798)**	7.48 (1.07)***
Used 1%–33.9% of time	1.30 (1.979)	3.29 (1.735)+
Constant	79.41 (4.444)***	73.32 (3.996)***
R^2	.154	.230
Number of children (number of families)	1,003 (708)	1,003 (708)

NOTE: $+p < .1$, $*p < .05$, $**p < .01$, $***p < .001$, two-tailed test.

reference group. The outcome analysis further controls for child's age, gender, and race, caregiver's education in 1997, number of years the caregiver used AFDC in childhood, and poverty (i.e., ratio of family income to poverty line in 1996). For the purpose of comparison, we also run a nonweighted OLS regression that does not control for propensity scores (i.e., a regression without using estimated propensity scores as sampling weights).

The results of the regression with propensity score adjustment show that, other things being equal, (a) "using AFDC 1% to 33.9% of the time" decreases the academic achievement from "never used" by 4.19 units (i.e., 7.48 − 3.29 = 4.19), (b) "using AFDC 34% to 100% of the time" decreases the academic achievement from "using AFDC 1% to 33.9% of the time" by 3.29 units, and (c) "using AFDC 34% to 100% of the time" decreases the academic achievement from "never used" by 7.48 units. All differences are statistically significant at a .05 level from a one-tailed test. Thus, the results confirm a linear impact of doses of using AFDC on academic achievement: The longer the time a study child was exposed to poverty, the lower his or her academic achievement. Such a conclusion cannot be drawn by analyses treating the child's use of AFDC as a binary variable.

In contrast, the regression model without controlling for observed covariates showed the impact of poverty in a lesser amount for all possible contrasts, and only one of the two treatment variables is statistically significant. This conclusion exemplifies, in a sense, the importance of controlling for covariates using propensity scores.

Together the above analysis shows that modeling multiple doses of treatment is informative. It allows us to test research hypotheses not only centering on treatment effectiveness but also concerning the impact of dose levels.

5.9.7 COMPARISON OF MODELS AND CONCLUSIONS OF THE STUDY OF THE IMPACT OF POVERTY ON CHILD ACADEMIC ACHIEVEMENT

We have demonstrated different methods to evaluate the impact of poverty (i.e., welfare dependence) on child academic achievement. Our examples represent a common practice in propensity score analysis: that is, instead of using a single method in a specific study, we use multiple methods, conduct follow-up comparisons and sensitivity analyses, and attempt to draw conclusions based on a thorough investigation of convergences and divergences across models.

So what do the results of different models tell us about the impact of poverty on academic achievement? Table 5.19 compares results from analyses defining a binary treatment condition. We exclude results from Section 5.9.6 to achieve the highest level of comparability across models. Based on this table, we can draw the following conclusions.

Table 5.19   Comparison of Findings Across Models Estimating the Impact of
Poverty on Children's Academic Achievement (Example 5.9.7)

| | Estimated Average | |
Model	Treatment Effect	Treatment Effect for the Treated
Independent-sample *t* test	−9.82***	
OLS regression	−4.73***	
Optimal matching (full) with Hodges-Lehmann aligned rank test	−1.97*	
Regressing difference score of outcome on difference scores of covariates after pair matching	−3.17*	
Regression analysis with propensity score weighting	−5.16***	−4.62***

*$p < .05$, ***$p < .001$, one-tailed test.

First, all models estimated a significant treatment effect. However, it is worth mentioning that this may not always be the case when performing propensity score analysis. We know that a simple covariance control, such as regression or regression-type analysis, ignores selection bias and, therefore, tends to produce biased and inconsistent estimates about treatment effects. After correcting for this violation by using propensity score models, estimates may be different. The phenomenon under investigation for this study (i.e., the effect of AFDC participation on academic achievement) is strong and, therefore, a significant estimate by a noncorrective approach (e.g., *t* test or OLS regression) remained significant even after we introduced sophisticated controls.

Second, among all estimates, which are more accurate or acceptable? Of the six models, *t* test and OLS regression by design did not control for selection bias; the propensity score weighting for ATE and ATT attempted to control it but failed. Therefore, the only acceptable estimates are those offered by optimal full matching using the Hodges-Lehmann aligned rank test and the difference score regression after pair matching. The estimate from optimal full matching is −1.97 ($p < .05$) and from pair matching is −3.17 ($p < .05$). Based on these findings and in the context of the observed covariates, we may conclude that poverty on average *causes*[3] a reduction in letter-word identification score by a range of 2 to 3 points. In contrast, both independent *t* test and OLS regression exaggerated the impact: *t* test exaggerated the impact by 398% (i.e., $\frac{9.82-1.97}{1.97} \times 100 = 398\%$)

or 210% (i.e., $\frac{9.82-3.17}{3.17} \times 100 = 210\%$), and OLS regression exaggerated the impact by 140% (i.e., $\frac{4.73-1.97}{1.97} \times 100 = 140\%$) or 49% (i.e., $\frac{4.73-3.17}{3.17} \times 100 = 49\%$). Note that not only were the estimated effects exaggerated by t test and OLS regression, but also the estimated significance level was exaggerated (i.e., both models show a p value less than .001).

Finally, the examples underscore the importance of conducting propensity score analyses. Researchers should explore the use of these types of models simply because they appear—when compared with traditional approaches—to provide more precise estimates of treatment effects in observational studies where measures of potential selection artifacts are collected.

5.9.8 COMPARISON OF RAND-*GBM* AND STATA'S *BOOST* ALGORITHMS

In this final example, we want to show results of a study serving solely a methodological purpose: comparison of the GBM procedure developed by McCaffrey et al. (2004) to the general GBM algorithm. As discussed in Section 5.3.4, GBM aims to minimize sample prediction error; that is, the algorithm stops iterations when the sample prediction error is minimized. This is the standard setup for both Stata's *boost* program and R's *gbm* program. McCaffrey et al. (2004) altered this procedure by stopping the algorithm at the number of iterations that minimizes the ASAM in covariates (thus we refer to this algorithm as Rand-*gbm*). In other words, the Rand-*gbm* is a procedure tailored more specifically to propensity score analysis. We are interested in the following question: To what extent do results from Rand-*gbm* differ from those produced by a regular GBM algorithm? To answer this question, we compared the Rand-*gbm* with Stata's *boost*. We have applied different data sets to conduct the comparisons and obtained almost same findings. Although we believe that our findings are sufficiently important to be shared with readers, we should emphasize that our comparison is not a Monte Carlo study and results may not hold in other settings or types of data.

We performed our comparison using data from an evaluation of school-based intervention in which group randomization failed. For more information about the study sample and data, see Section 4.5.2. We used boosted regression to estimate propensity scores. Figure 5.5 presents distributions of estimated propensity scores by Rand-*gbm* and Stata's *boost*.

Using exactly the same set of conditioning variables, Rand-*gbm* and Stata's *boost* produced quite different propensity scores. The Rand-*gbm* propensity scores, shown in the histograms in Figure 5.5, have a high level of dispersion with a range between .2 and .8. In contrast, propensity scores estimated by Stata's *boost* have a low level of dispersion and concentrate on the range

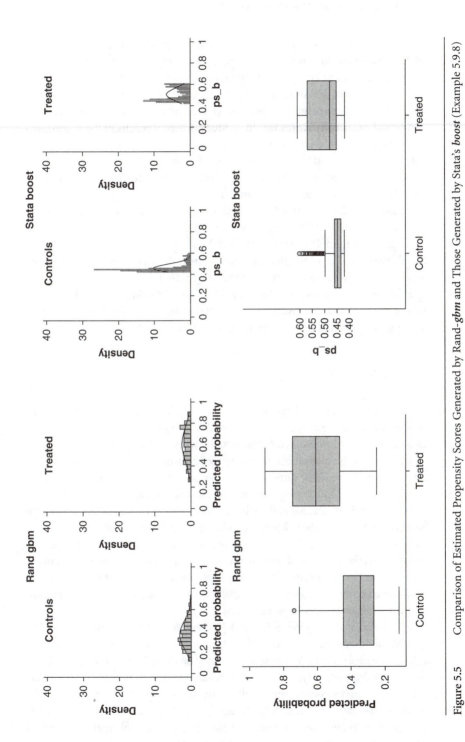

Figure 5.5   Comparison of Estimated Propensity Scores Generated by Rand-*gbm* and Those Generated by Stata's *boost* (Example 5.9.8)

between .4 and .6. As a result, the two methods produced very different box plots: Rand-*gbm* does not show much overlapping of the propensity scores between treated and control participants while Stata's *boost* does.

Using the two sets of propensity scores, we then conducted pair matching and postmatching analysis. We were interested in the question of whether the two sets of propensity scores would produce different results in the subsequent analyses. Table 5.20 shows results of covariate-imbalance check. As the last two columns of Table 5.20 (i.e., the statistics of d_{xm}) show, even though the two sets of propensity scores have different distributions, both corrected for (or failed to correct for) imbalance in very similar fashion.

Based on the pair-matched samples, we then analyzed difference score regression (Table 5.21). Results show that the treatment effect (i.e., constant of the regression model) estimated by Rand-*gbm* is 0.15 and that estimated by Stata's *boost* is 0.13. Both effects are not statistically significant. Thus, the two sets of data provide approximately the same findings.

Given the results, we conclude that Rand-*gbm* and Stata's *boost* do not lead to different results on covariate control and estimates of treatment effects, although this finding needs to be verified in future studies.

5.10 Conclusions

Before we conclude this chapter, we need to point out some limitations of propensity score matching. According to Rubin (1997), propensity scores (a) cannot adjust for unobserved covariates, (b) work better in larger samples, and (c) do not handle a covariate that is related to treatment assignment, but not related to outcome, in the same way as a covariate with the same relation to treatment assignment but strongly related to outcome. Rubin recommended performing sensitivity analysis and testing different sets of conditioning variables to address the first limitation. Michalopoulos et al. (2004) also noted that propensity scores correct less well for studies in which the treated and nontreated groups are not from the same social context/milieu and, therefore, are not exposed to the same ecological influences. This is a special case of being unable to adjust for unobserved covariates common in social service and other program evaluations that compare across service jurisdictions.

It is also worth saying again that propensity score matching is a rapidly growing field of study and many new developments are still in a testing stage. Additional problems as well as strategies may be identified as researchers move to new developments. We agree that even randomized clinical trials are imperfect ways of determining the result of treatment for every member of the population—treated or not. Nor can propensity score methods provide definitive answers to questions of treatment effectiveness. Multiple methods for estimating program effects are indicated for use within and across studies.

Table 5.20 Comparison of Covariate Imbalance Before and After Matching Between Rand-*gbm* and Stata's *boost* (Example 5.9.8)

Covariate	d_x	d_{xm}	
		Rand-*gbm*	Stata's *boost*
Age at baseline	0.07	0.03	0.03
Gender: female (reference: male)	0.11	0.08	0.08
Race (reference: other)			
African American	0.59	0.54	0.56
White	0.51	0.53	0.57
Hispanic	0.24	0.16	0.09
Caregiver's education	0.17	0.12	0.08
Ratio of income to poverty line	0.14	0.16	0.18
Caregiver employment: full time (reference: other)	0.14	0.13	0.10
Father: presence at home (reference: absence)	0.23	0.19	0.20

NOTE: Absolute standardized difference in covariate means, before (d_x) and after matching (d_{xm}).

Table 5.21 Regressing Difference Score of Outcome (i.e., Change of Academic Competence in Third Grade) on Difference Scores of Covariates After Pair Matching: Comparison of Results Between Rand-*gbm* and Stata's *boost* (Example 5.9.8)

Covariate Difference Score	Rand-gbm		Stata's boost	
	Coefficient	Robust S.E.	Coefficient	Robust S.E.
Age at baseline	0.10	0.09	−0.04	0.13
Gender: female (reference: male)	−0.21	0.23	−0.05	0.08
Race African American (reference: other)	−0.17	0.27	0.16	0.25
White	−0.28	0.33	0.04	0.22
Hispanic	−0.63	0.28+	−0.38	0.37
Caregiver's education	−0.03	0.02	−0.01	0.02
Ratio of income to poverty line	0.00	0.00	0.00	0.00
Caregiver employment: full time (reference: other)	−0.09	0.22	0.10	0.13
Father: presence at home (reference: absence)	0.35	0.21	−0.12	0.12
Constant	0.15	0.13	0.13	0.12

$p < .05$, $+p < .10$, two-tailed test.

Researchers using propensity score matching should be cautious about these limitations and make efforts to warrant that interpretation of study results does not go beyond the limits of data and analytical methods. Nonetheless, from this chapter we have seen that propensity score matching is a promising approach that offers a growing evidentiary base for observational studies facing violations of the unconfoundedness assumption and selection biases.

Notes

1. We will describe the bias-adjusted matching estimator in Chapter 6.
2. The calculation of ASAM will be described in Section 5.5.3.
3. Note that we are using the term *causes*. Indeed, it took a long journey before we could finally use the term! And our use here must still be conditioned. It is causal only in the context of observed heterogeneity, that is, selection effects for which we have adequate measurement.

6

Matching Estimators

This chapter focuses on a collection of matching estimators, including the simple matching estimator, the bias-corrected matching estimator, the variance estimator assuming a constant treatment effect and homoscedasticity, and the variance estimator allowing for heteroscedasticity (Abadie et al., 2004; Abadie & Imbens, 2002, 2006). Matching is a shared characteristic among these estimators and those described in Chapters 5 and 7. However, the matching estimators presented in this chapter do not use logistic regression to predict propensity scores. They require fewer decisions, are easy to implement, and do not involve nonparametric estimation of unknown functions. Because of these advantages, these matching estimators are an attractive approach for solving many problems encountered in program evaluation.

Section 6.1 provides an overview of the matching estimators. We compare these methods with those discussed in other chapters and give particular attention to similarities and differences in the methodological features. Our intent in this section is to provide contextual information sufficient to differentiate matching estimators from other methods of propensity score analysis. Section 6.2 is the core of the chapter and describes the methodology of matching estimators. Although the focus of this review is on the simple matching estimator and the bias-corrected matching estimator, we review Abadie and Imbens's (2006) study on the large sample properties of matching estimators. Section 6.3 summarizes key features of the Stata *nnmatch* program, which can be used to run matching estimators. Section 6.4 gives detailed examples. Because matching estimators estimate average treatment effects for the treated, Section 6.4 also shows how to use matching estimators to conduct efficacy subset analyses (ESAs) that test hypotheses pertaining to levels of treatment exposure (i.e., dose analyses). Section 6.5 concludes the chapter.

6.1 Overview

As discussed in Chapter 2, a seminal development in the conceptualization of program evaluation was the Neyman-Rubin counterfactual framework. The key assumption of the framework is that individuals selected into treatment and nontreatment groups have potential outcomes in both states: the one in which the outcomes are observed and the one in which the outcomes are not observed. Thus, for the treated group, in addition to an observed mean outcome under the condition of treatment $E(Y_1|W=1)$, the framework assumes that an unobserved mean outcome exists under the condition of nontreatment $E(Y_0|W=1)$. Similarly, participants in the control group have both an observed mean $E(Y_0|W=0)$ and an unobserved mean $E(Y_1|W=0)$. The unobserved potential outcomes under either condition are missing data. Based on the counterfactual framework, the matching estimators directly impute the missing data at the unit level by using a vector norm. That is, at the unit level, a matching estimator imputes potential outcomes for each study participant. Specifically, it estimates the value of $Y_i(0)|W_i=1$ (i.e., potential outcome under the condition of control for a treatment participant) and the value of $Y_i(1)|W_i=0$ (i.e., potential outcome under the condition of treatment for a control participant). After imputing the missing data, matching estimators can be used to estimate various average treatment effects, including the sample average treatment effect (SATE), the population average treatment effect (PATE), the sample average treatment effect for the treated (SATT), the population average treatment effect for the treated (PATT), the sample average treatment effect for the controls (SATC), and the population average treatment effect for the controls (PATC).

As previously mentioned in Section 5.2, a central challenge inherent to observational study is the dimensionality of covariates or matching variables. In essence, as the number of matching variables increases, so does the difficulty of using exact matching to find a match for a given treated participant. The methods described in Chapters 5 and 7 use logistic regression to predict propensity scores that, in turn, are used to reduce multiple matching variables to a single score. They solve the dimensionality problem. As such, treated and control participants who have the same propensity score values are deemed to have the same distributions on the observed covariates. In contrast, the matching estimators do not use logistic regression to predict propensity scores. Instead, these methods use a *vector norm* to calculate distances on the observed covariates between a treated case and each of its potential control cases. The vector norm is used to choose the outcome of a control case whose distance on covariates is the shortest vis-à-vis other control cases. This outcome serves as the counterfactual for the treated case. Similarly, the matching estimators can be used to choose the outcome of a treated case whose distance on covariates is the shortest vis-à-vis other treated cases. This outcome serves as the counterfactual for the control case.

To calculate a vector norm, the matching estimators choose one of two kinds of variance matrices: the inverse of the sample variance matrix or the inverse of the sample variance-covariance matrix. When choosing the inverse of the sample variance-covariance matrix to calculate the vector norm, the matching estimator calculates Mahalanobis metric distances. Thus, as described in Section 5.4.1, this method becomes Mahalanobis metric matching, which was developed prior to the invention of propensity score matching (Cochran & Rubin, 1973; Rubin, 1976, 1979, 1980a). Notably, even though these methods share features of conventional approaches, the matching estimators developed by Abadie and Imbens (2002, 2006) expand the Mahalanobis metric matching in several important ways. They estimate (a) average treatment effects for the controls, (b) treatment effects that include both sample and population estimates, (c) variances and standard errors for statistical significance tests, and (d) a bias correction for finite samples (when the Mahalanobis metric matching is not exact).

Two assumptions are embedded in the matching estimators: (1) the assignment to treatment is independent of the outcomes, conditional on the covariates, and (2) the probability of assignment is bounded away from 0 and 1, which is also known as an overlap assumption requiring sufficient overlap in the distributions of the observed covariates (Abadie et al., 2004).

The first assumption is the same as the *strongly ignorable treatment assignment* assumption (Rosenbaum & Rubin, 1983), the fundamental assumption that we have examined in several discussions in this book. This is a restrictive and strong assumption, and, in many cases, it may not be satisfied. However, the assumption is a conceptual link that connects observational studies with nonrandomized designs to theories and principles developed for randomized designs. At some point and in one way or another, most evaluation analyses are conditioned on the ignorable treatment assignment assumption (Abadie et al., 2004; Imbens, 2004).

The overlap assumption requires some overlapping of the estimated propensity scores. It implies that the treated and control groups share a common support region of propensity scores in the sample data. If this assumption is violated, it is not appropriate to use the matching estimators. Under such a condition, researchers might want to consider using optimal matching (see Section 5.4.2), which is more robust against violations of overlap.

As discussed earlier, matching estimators share a key characteristic with all other methods introduced in Chapters 5 and 7, that is, matching estimators match a treated case to a control (or vice versa) based on observed covariates. However, the matching estimators use a simpler mechanism for matching (i.e., a vector norm), which makes the approach much easier to implement than other methods. Unlike other methods in which matching is based on propensity scores (e.g., nearest neighbor [Section 5.4.1] or optimal matching using the network flow theory [i.e., Section 5.4.2]), the matching estimators do not require postmatching analysis such as survival analysis, hierarchical linear modeling, the Hodges-Lehmann aligned rank test, or regression adjustment based on difference scores. Omitting the postmatching analysis is an advantage because reducing the number of analytic procedures involved in matching also reduces the number of subjective decisions a researcher has to make. This is usually desirable.

The matching estimators allow evaluators to estimate effects for both the sample and the population. The difference between the SATE and the PATE is based on whether the effect can be inferred beyond the study sample. As such, SATE and PATE are useful for answering different questions. Abadie et al. (2004) used a job-training program to illustrate the differences. SATE is useful in evaluating whether the job-training program (i.e., the sample data at hand) was successful. In contrast, if the evaluator wants to know whether the same program would be successful in a second sample from the population, PATE is useful. Although SATE and PATE produce the same coefficients, they estimate standard errors differently. Indeed, the estimated standard error of the population effect is, in general, larger than that of the sample effect and, therefore, may lead to different conclusions about significance. This variation makes sense because a successful treatment may exist in one sample and not in another sample from the same population. The relationship between SATE and PATE is similar to the relationship between SATT and PATT and to the relationship between SATC and PATC.

The SATC and PATC are estimable only by the matching estimators. In the case of SATC, the average treatment effect for the control group is based on this question: What would the sample treatment effect for the treated group look like if the controls received the treatment condition and the treatment cases received the control condition? Similarly, for PATC, the average treatment effect for the control condition indicates what the population treatment effect for the treated group would look like if the controls received the treatment condition and the treatment cases received the control condition. If the variables used in matching accounted for all cases of selection bias and the evaluation data met all the assumptions of the matching estimators, SATT and SATC would appear to have values of similar magnitude. Thus, the difference between the two coefficients holds the potential to indicate both the level of hidden selection bias and the departure of data from model assumptions. However, although this sounds quite useful, it has a serious limitation. In practice, both problems may work together to exert joint and entangled effects. Nonetheless, a difference between SATT and SATC can be an indicator that alternative analyses with different assumptions should be tried.

In Chapter 2, we reviewed the *stable unit treatment value assumption*, or SUTVA (Rubin, 1986), which holds that the potential outcomes for any unit do not vary with the treatments assigned to any other units, and there are no different versions of the treatment. As described in Chapter 2, SUTVA is an assumption that facilitates the investigation or estimation of counterfactuals as well as a conceptual perspective that underscores the importance of using appropriate estimators when analyzing differential treatment effects. The SUTVA assumption imposes *exclusion* restrictions on outcome differences. Based on these restrictions, economists underscore the importance of analyzing average treatment effects for the subpopulation of treated units because that analysis is frequently more important than the effect on the population as a whole. Analysis for the subpopulation is especially a concern when evaluating the importance of a narrowly targeted intervention, such as a labor market program (Heckman, Ichimura, & Todd, 1997, 1998; Imbens, 2004). Matching estimators offer an easy tool for this task. What statisticians and econometricians have called *evaluating average treatment effects for the treated* (also treatment of the treated, TOT) is similar to the ESA framework that is found in the intervention research literature (Lochman, Boxmeyer, Powell, Roth, & Windle, 2006). Because the matching estimators evaluate a potential outcome for each study unit, it is not cumbersome to estimate the average treatment effects for user-defined subsets of units and then test research hypotheses related to differential treatment exposure.

Following the tradition of the econometric literature, but in contrast to some of the statistical literature, the matching estimators focus on matching with replacement (Abadie & Imbens, 2006). That is, the matching estimators allow individual observations to be used as a match more than once. Matching with replacement makes matching estimators different from all matching methods described in Chapter 5. Denoting $K_M(i)$ as the number of times unit i is used as a match—given that M matches per unit are used—the matching estimators described in this chapter allow $K_M(i) > 1$, whereas the matching methods described in Chapter 5 require $K_M(i) \leq 1$. According to Abadie and Imbens (2006),

> Matching with replacement produces matches of higher quality than matching without replacement by increasing the set of possible matches. In addition, matching with replacement has the advantage that it allows us to consider estimators that match all units, treated as well as controls, so that the estimand is identical to the population average treatment effect. (p. 240)

When matching with replacement, the distribution of $K_M(i)$ plays an important role in the calculation of the variance of the estimator and, therefore, $K_M(i)$ is a key factor in examining the large sample properties of matching estimators (Abadie & Imbens, 2006).

The development of variances for various matching estimators is especially attractive, because it makes possible significance testing of treatment effects estimated by matching estimators. Unlike matching with nonparametric regression that relies on bootstrapping to draw statistical inferences (i.e., described in Chapter 7), the matching estimators offer a consistent estimator for variance estimation, and thus allow for a more rigorous significance test than those methods that use bootstrapping. Research to date suggests that "bootstrapping methods for estimating the variance of matching estimators do not necessarily give correct results" (Abadie et al., 2004, p. 300).

Although the matching estimators offer several advantages, they also share a common limitation with other matching methods. That is, most matching estimators contain a conditional bias term whose stochastic order increases with the number of continuous variables. As a result, matching estimators are not $1/\sqrt{N}$ consistent. Abadie and Imbens (2006) have shown that the variance of matching estimators remains relatively high. Consequently, matching with a fixed number of matches does not lead to an efficient estimator; specifically, it does not achieve the semiparametric efficiency bound calculated by Hahn (1998). These problems are discussed in Section 6.2. From our own simulation studies, we have found that the matching estimators are more sensitive to the violation of the *strongly ignorable assumption* than other methods, which means that, under certain conditions (e.g., high selection), matching estimators will produce larger bias than other methods. We discuss this issue in Chapter 8 when comparing the sensitivity of different estimators to selection bias.

6.2 Methods of Matching Estimators

In this section, we review the basic methodological features of the matching estimators. Primarily, our review follows the work of Abadie et al. (2004) and Abadie and Imbens (2006). We focus on two issues: (1) the point estimates of various treatment effects (i.e., the coefficients of SATE, PATE, SATT, PATT, SATC, and PATC) using either a simple matching estimator or a bias-corrected matching estimator and (2) the estimates of the variances of various treatment effects (i.e., the standard errors of SATE, PATE, SATT, PATT, SATC, and PATC) assuming homoscedasticity or allowing for heteroscedasticity. In practice, researchers will use a combination of one method for estimating the coefficient and one method for estimating the variance (equivalently the standard error). A typical evaluation often uses bias-corrected matching with a variance estimation allowing for heteroscedasticity. All methods described in this section are considered jointly as the collection of matching estimators.

6.2.1 SIMPLE MATCHING ESTIMATOR

For each observation i, the unit-level treatment effect is $\tau_i = Y_i(1) - Y_i(0)$. As discussed earlier, one of the two outcomes is always missing (i.e., either $Y_i(0)$

or $Y_i(1)$ is missing, depending on whether the unit's treatment condition is $W_i = 1$ or $W_i = 0$. Under the exogeneity and overlap assumptions, the simple matching estimator imputes the missing potential outcome by using the average outcome for individuals with "similar" values on observed covariates.

We first consider matching with *one observed covariate*. This is the most basic case of simple matching, under which condition the estimator simply takes the outcome value of the controlled case that is the closest match on the observed covariate. Under the condition of tie, the estimator takes the mean value of outcomes of the tied cases. Similarly, the estimator takes the outcome value of the treated case that is the closest match on the observed covariate for a control case.

Let $J_M(i)$ denote the set of indices for the matches for unit i that are at least as close as the Mth match and $\# J_M(i)$ denote the number of elements of $J_M(i)$; the simple matching estimator can be expressed by the following equations:

$$\hat{Y}_i(0) = \begin{cases} Y_i & \text{if } W_i = 0 \\ \frac{1}{\#J_M(i)} \sum_{l \in J_M(i)} Y_l & \text{if } W_i = 1 \end{cases} \qquad (6.1)$$

and

$$\hat{Y}_i(1) = \begin{cases} \frac{1}{\#J_M(i)} \sum_{l \in J_M(i)} Y_l & \text{if } W_i = 0 \\ Y_i & \text{if } W_i = 1. \end{cases}$$

Table 6.1 illustrates how the simple matching estimator works when there is a single covariate x_1. In this example, three units are controls ($W = 0$), four units are treated ($W = 1$), and one covariate x_1 is observed. The imputed values of potential outcomes are highlighted. For the first unit (a control case with $x_1 = 2$), the exact match is the fifth unit (because x_1 for $i = 5$ is also equal to 2); so the imputed value $y(1)$ (i.e., the potential outcome under the condition of treatment for $i = 1$) is 8 by taking the value of y for $i = 5$. For the second unit (a control case with $x_1 = 4$), treated units 4 and 6 are equally close (both with $x_1 = 3$, a closest value to 4); so the imputed potential outcome $y(1)$ is the average of the outcomes for units 4 and 6, namely $(9 + 6)/2 = 7.5$. Note that for $i = 2, \#J_M(i) = 2$, because two matches ($i = 4$ and $i = 6$) are found for this participant.

We next consider matching with *more than one observed covariate*. Under this condition, the simple matching estimator uses the vector norm (i.e., $\|x\|v = x'Vx$ with positive definite matrix V, see Note 1 at the end of the chapter) to calculate distances between one treated case and each of its multiple possible controls and chooses the outcome of the control case whose distance is the shortest among all as the potential outcome for the treated case. In a similar fashion, the simple matching estimator imputes the missing outcome under the

Table 6.1   An Example of Simple Matching With One Observed Covariate for Seven Observations

i	w_i	x_{1i}	y_i	$\hat{y}_i(0)$	$\hat{y}_i(1)$
1	0	2	7	7	8
2	0	4	8	8	7.5
3	0	5	6	6	7.5
4	1	3	9	7.5	9
5	1	2	8	7	8
6	1	3	6	7.5	6
7	1	1	5	7	5

condition of treatment for a control case. Specifically, we define $\|z - x\|v$ as the distance between vectors x and z, where x represents the vector of the observed covariate values for observation i, and z represents the vector of the covariate values of a potential match for observation i. There are two choices for v: the inverse of the sample variance matrix (i.e., a diagonal matrix with all off-diagonal elements constrained to be zero) or the inverse of the sample variance-covariance matrix (i.e., a nondiagonal matrix with all off-diagonal elements to be nonzero covariances). When the inverse of the sample variance-covariance matrix is used, $\|z - x\|v$ becomes a Mahalanobis metric distance. As noted in Chapter 5, some researchers (e.g., D'Agostino, 1998) defined the variance-covariance matrix differently than the matching estimators described in this chapter for the Mahalanobis metric matching, where they use the inverse of the variance-covariance matrix of the *control* observations, rather than that of all *sample* observations (i.e., both the treated and control observations) to define v.

Let $d_M(i)$ denote the distance from the covariates for unit i, X_p, to the Mth nearest match with the opposite treatment condition. Allowing for the possibility of ties, at this distance fewer than M units are closer to unit i than $d_M(i)$ and at least M units are as close as $d_M(i)$. With multiple covariates, $J_M(i)$ still represents the set of indices of the matches for unit i that are at least as close as the Mth match, but M matches are chosen using a vector norm that meets the condition of nearest distances as follows:

$$J_M(i) = \{l = 1, \ldots, N \mid W_l = 1 - W_i, \ \|X_l - X_i\|v \le d_M(i)\}.$$

If there are no ties, the number of elements in $J_M(i)$ is M, but may be larger. Previously we mentioned that $K_M(i)$ is the number of times unit i is used as a match given that M matches per unit are used. With the new notation introduced, we can now define $K_M(i)$ more precisely as the number of times i is used as a match for all observations l of the opposite treatment condition, each time weighted by the total number of matches for observation *l*:

$$K_M(i) = \sum_{l=1}^{N} 1\{i \in J_M(l)\} \frac{1}{\#J_M(l)},$$

where $1\{\cdot\}$ is the indicator function, which is equal to 1 if the expression in brackets is true but equal to 0 otherwise. If we denote the total sample size as N, the number of treated cases as N_1, and the number of controls as N_0, then we can express these numbers by using the notation we just introduced:

$$N = \sum_i K_M(i), \ N_1 = \sum_{i:W_i=0} K_M(i), \ \text{and} \ N_0 = \sum_{i:W_i=1} K_M(i).$$

Under the condition of more than one observed covariate, the missing potential outcome for each unit is imputed by using the same equations as Equation 6.1. Unlike the case of one observed covariate, the imputation of potential outcomes now uses the vector norm. Using hypothetical data, we provide an illustration of simple matching using the vector norm when matching uses multiple covariates. Table 6.2 shows an example of simple matching using three observed covariates for seven observations. The table illustrates minimum distance for each unit that was calculated by the vector norm using the inverse of a sample variance matrix. Details of calculating the minimum distances are explained below.

First, an x vector is formed by the values of x_1, x_2, and x_3 for the unit being considered for imputation of a potential outcome. Thus, the x vector for $i = 1$ is (2 4 3). This unit is a control case ($W = 0$). Because there are four treated units in the sample data (i.e., $i = 4, 5, 6,$ and 7), there are four distances for unit $i = 1$.

Table 6.2   An Example of Simple Matching With Three Observed Covariates for Seven Observations With Minimum Distance Determined by Vector Norm Using the Inverse of a Sample Variance Matrix

i	w_i	x_1	x_2	x_3	y	Minimum Distance	i Whose Distance Is Minimum	$\hat{y}_i(0)$	$\hat{y}_i(1)$
1	0	2	4	3	7	1.0648	5	7	8
2	0	4	2	2	8	1.1053	6	8	6
3	0	5	5	4	6	4.8119	4	6	9
4	1	3	4	6	9	4.8119	3	6	9
5	1	2	3	4	8	1.0648	1	7	8
6	1	3	3	2	6	1.1053	2	8	6
7	1	1	1	3	5	5.5263	1	7	5

For each of the treated units, a z vector is formed by the values of x_1, x_2, and x_3; that is, z for $i = 4$ is (3 4 6), z for $i = 5$ is (2 3 4), z for $i = 6$ is (3 3 2), and z for $i = 7$ is (1 1 3).

With seven observations, the sample variances are $Var(x_1) = 1.810$, $Var(x_2) = 1.810$, and $Var(x_3) = 1.952$. Hence, the inverse of the sample variance matrix is

$$\begin{bmatrix} 1.810 & 0 & 0 \\ 0 & 1.810 & 0 \\ 0 & 0 & 1.952 \end{bmatrix}^{-1} = \begin{bmatrix} .553 & 0 & 0 \\ 0 & .553 & 0 \\ 0 & 0 & .512 \end{bmatrix}.$$

Using vector norm $\|z - x\|v$, where v is the inverse of the sample variance matrix, we calculate the four distances for $i = 1$ as follows:

Distance between the pair of $i = 1$ and $i = 4$:

$$[(2 \quad 4 \quad 3) - (3 \quad 4 \quad 6)] \begin{bmatrix} .553 & 0 & 0 \\ 0 & .553 & 0 \\ 0 & 0 & .512 \end{bmatrix} \left[\begin{pmatrix} 2 \\ 4 \\ 3 \end{pmatrix} - \begin{pmatrix} 3 \\ 4 \\ 6 \end{pmatrix} \right]$$

$$= (-1 \quad 0 \quad -3) \begin{bmatrix} .553 & 0 & 0 \\ 0 & .553 & 0 \\ 0 & 0 & .512 \end{bmatrix} \begin{pmatrix} -1 \\ 0 \\ -3 \end{pmatrix} = 5.1624.$$

Distance between the pair of $i = 1$ and $i = 5$:

$$[(2 \quad 4 \quad 3) - (2 \quad 3 \quad 4)] \begin{bmatrix} .553 & 0 & 0 \\ 0 & .553 & 0 \\ 0 & 0 & .512 \end{bmatrix} \left[\begin{pmatrix} 2 \\ 4 \\ 3 \end{pmatrix} - \begin{pmatrix} 2 \\ 3 \\ 4 \end{pmatrix} \right]$$

$$= (0 \quad 1 \quad -1) \begin{bmatrix} .553 & 0 & 0 \\ 0 & .553 & 0 \\ 0 & 0 & .512 \end{bmatrix} \begin{pmatrix} 0 \\ 1 \\ -1 \end{pmatrix} = 1.0648.$$

Distance between the pair of $i = 1$ and $i = 6$:

$$[(2 \quad 4 \quad 3) - (3 \quad 3 \quad 2)] \begin{bmatrix} .553 & 0 & 0 \\ 0 & .553 & 0 \\ 0 & 0 & .512 \end{bmatrix} \left[\begin{pmatrix} 2 \\ 4 \\ 3 \end{pmatrix} - \begin{pmatrix} 3 \\ 3 \\ 2 \end{pmatrix} \right]$$

$$= (-1 \quad 1 \quad 1) \begin{bmatrix} .553 & 0 & 0 \\ 0 & .553 & 0 \\ 0 & 0 & .512 \end{bmatrix} \begin{pmatrix} -1 \\ 1 \\ 1 \end{pmatrix} = 1.6175.$$

Distance between the pair of $i = 1$ and $i = 7$:

$$
\left[(2 \quad 4 \quad 3) - (1 \quad 1 \quad 3) \right]
\begin{bmatrix} .553 & 0 & 0 \\ 0 & .553 & 0 \\ 0 & 0 & .512 \end{bmatrix}
\left[\begin{pmatrix} 2 \\ 4 \\ 3 \end{pmatrix} - \begin{pmatrix} 1 \\ 1 \\ 3 \end{pmatrix} \right]
$$

$$
= (1 \quad 3 \quad 0)
\begin{bmatrix} .553 & 0 & 0 \\ 0 & .553 & 0 \\ 0 & 0 & .512 \end{bmatrix}
\begin{pmatrix} 1 \\ 3 \\ 0 \end{pmatrix} = 5.5263.
$$

The above calculation verifies two values shown in Table 6.2: for $i = 1$ (i.e., the first row), the minimum distance equals 1.0648, which is the distance between $i = 1$ and $i = 5$ because 1.0648 is the smallest of the four distances we just calculated; and the imputed value $\hat{y}_{i=1}(1) = 8$, which is the observed outcome for $i = 5$. We chose the outcome of $i = 5$ as $\hat{y}_{i=1}(1)$ because $i = 5$ has the minimum distance on observed covariates from $i = 1$. Potential outcomes for all other units can be obtained by replicating the above process.

Minimum distance can also be determined by the Mahalanobis metric distance that defines v as the inverse of the sample variance-covariance matrix. Table 6.3 illustrates the calculation of minimum distance for each unit that is based on the inverse of a sample variance-covariance matrix. The distances so calculated are exactly the same as Mahalanobis metric distances. We explain the main features of this calculation on the following pages.

Table 6.3   An Example of Simple Matching With Three Observed Covariates for Seven Observations With Minimum Distance Determined by Vector Norm Using the Inverse of a Sample Variance-Covariance Matrix

i	w_i	x_1	x_2	x_3	y	Minimum Distance	i Whose Distance Is Minimum	$\hat{y}_i(0)$	$\hat{y}_i(1)$
1	0	2	4	3	7	2.5518	6	7	6
2	0	4	2	2	8	3.4550	6	8	6
3	0	5	5	4	6	4.1035	6	6	6
4	1	3	4	6	9	5.2330	3	6	9
5	1	2	3	4	8	3.0790	1	7	8
6	1	3	3	2	6	2.5518	1	7	6
7	1	1	1	3	5	5.6580	2	8	5

First, the $\mathbf{x}$ vector for $i = 1$ is (2 4 3), and the $\mathbf{z}$ vector for $i = 4$ is (3 4 6), for $i = 5$ is (2 3 4), for $i = 6$ is (3 3 2), and for $i = 7$ is (1 1 3). The inverse of the sample variance-covariance matrix for this data set of seven observations is

$$
\begin{bmatrix} 1.810 & 1.024 & .071 \\ 1.024 & 1.810 & .929 \\ .071 & .929 & 1.952 \end{bmatrix}^{-1} = \begin{bmatrix} .917 & -.663 & .282 \\ -.663 & 1.211 & -.552 \\ .282 & -.552 & .764 \end{bmatrix}.
$$

Using vector norm $\|\mathbf{z} - \mathbf{x}\|\mathbf{v}$, we can calculate the four distances for $i = 1$ as follows:

Distance between the pair of $i = 1$ and $i = 4$:

$$
[(2\ \ 4\ \ 3) - (3\ \ 4\ \ 6)]\begin{bmatrix} .917 & -.663 & .282 \\ -.633 & 1.211 & -.552 \\ .282 & -.552 & .764 \end{bmatrix}\left[\begin{pmatrix} 2 \\ 4 \\ 3 \end{pmatrix} - \begin{pmatrix} 3 \\ 4 \\ 6 \end{pmatrix}\right]
$$

$$
= (-1\ \ 0\ \ -3)\begin{bmatrix} .917 & -.663 & .282 \\ -.633 & 1.211 & -.552 \\ .282 & -.552 & .764 \end{bmatrix}\begin{pmatrix} -1 \\ 0 \\ -3 \end{pmatrix} = 9.4877.
$$

Distance between the pair of $i = 1$ and $i = 5$:

$$
[(2\ \ 4\ \ 3) - (2\ \ 3\ \ 4)]\begin{bmatrix} .917 & -.663 & .282 \\ -.633 & 1.211 & -.552 \\ .282 & -.552 & .764 \end{bmatrix}\left[\begin{pmatrix} 2 \\ 4 \\ 3 \end{pmatrix} - \begin{pmatrix} 2 \\ 3 \\ 4 \end{pmatrix}\right]
$$

$$
= (0\ \ 1\ \ -1)\begin{bmatrix} .917 & -.663 & .282 \\ -.633 & 1.211 & -.552 \\ .282 & -.552 & .764 \end{bmatrix}\begin{pmatrix} 0 \\ 1 \\ -1 \end{pmatrix} = 3.0790.
$$

Distance between the pair of $i = 1$ and $i = 6$:

$$
[(2\ \ 4\ \ 3) - (3\ \ 3\ \ 2)]\begin{bmatrix} .917 & -.663 & .282 \\ -.633 & 1.211 & -.552 \\ .282 & -.552 & .764 \end{bmatrix}\left[\begin{pmatrix} 2 \\ 4 \\ 3 \end{pmatrix} - \begin{pmatrix} 3 \\ 3 \\ 2 \end{pmatrix}\right]
$$

$$
= (-1\ \ 1\ \ 1)\begin{bmatrix} .917 & -.663 & .282 \\ -.633 & 1.211 & -.552 \\ .282 & -.552 & .764 \end{bmatrix}\begin{pmatrix} -1 \\ 1 \\ 1 \end{pmatrix} = 2.5518.
$$

Distance between the pair of $i = 1$ and $i = 7$:

$$[(2 \quad 4 \quad 3) - (1 \quad 1 \quad 3)] \begin{bmatrix} .917 & -.663 & .282 \\ -.633 & 1.211 & -.552 \\ .282 & -.552 & .764 \end{bmatrix} \left[\begin{pmatrix} 2 \\ 4 \\ 3 \end{pmatrix} - \begin{pmatrix} 1 \\ 1 \\ 3 \end{pmatrix} \right]$$

$$= (1 \quad 3 \quad 0) \begin{bmatrix} .917 & -.663 & .282 \\ -.633 & 1.211 & -.552 \\ .282 & -.552 & .764 \end{bmatrix} \begin{pmatrix} 1 \\ 3 \\ 0 \end{pmatrix} = 7.8365.$$

The above calculation verifies two values shown in Table 6.3: for $i = 1$ (i.e., the first row), the minimum distance = 2.5518, which is the distance between $i = 1$ and $i = 6$, because 2.5518 is the smallest of the four distances; and the imputed value = 6, which is the observed outcome for $i = 6$. Potential outcomes for all other units can be obtained by replicating the above process.

Finally, we consider the calculation of *various treatment effects*. After imputing the missing potential outcomes, we now have two outcomes for each study unit: one is an observed outcome, and the other is an imputed potential outcome (or the counterfactual). Taking average values in a varying fashion, we obtain point estimates of various treatment effects as follows:

Sample average treatment effect (SATE):

$$\hat{\tau}^{\text{average}} = \frac{1}{N} \sum_{i=1}^{N} \{\hat{Y}_i(1) - \hat{Y}_i(0)\} = \frac{1}{N} \sum_{i=1}^{N} (2W_i - 1)\{1 + K_M(i)\} Y_i. \quad (6.2)$$

Sample average treatment effect for the treated (SATT):

$$\hat{\tau}^{t} = \frac{1}{N_1} \sum_{i:w_i=1} \{Y_i - \hat{Y}_i(0)\} = \frac{1}{N_1} \sum_{i=1}^{N} \{W_i - (1 - W_i)K_M(i)\} Y_i. \quad (6.3)$$

Sample average treatment effect for the controls (SATC):

$$\hat{\tau}^{c} = \frac{1}{N_0} \sum_{i:w_i=0} \{\hat{Y}_i(1) - Y_i\} = \frac{1}{N_0} \sum_{i=1}^{N} \{W_i K_M(i) - (1 - W_i)\} Y_i. \quad (6.4)$$

As noted earlier, a population effect is exactly the same as the point estimate of its corresponding sample effect. Thus, PATE = SATE, which can be obtained by applying Equation 6.2; PATT = SATT, which can be obtained by applying Equation 6.3; and PATC = SATC, which can be obtained by applying Equation 6.4.

When running matching estimators, the number of matches for each unit must be considered. In the above examples, we used all units in the opposite treatment condition as potential matches and selected a single unit based on the minimum distance. Because matching estimators use matching with replacement, in theory, all controls can be selected as matches for a treated unit, and all treated units can be selected as matches for a control unit. Going to the other extreme, the researcher can choose only one match for each unit using the nearest neighbor. The drawback of using one match is that the process uses too little information in matching. Like all smoothing parameters, the final inference of matching estimators can depend on the choice of the number of matches. To deal with this issue, Abadie et al. (2004) recommend using *four matches* for each unit, "because it offers the benefit of not relying on too little information without incorporating observations that are not sufficiently similar" (p. 298).

6.2.2 BIAS-CORRECTED MATCHING ESTIMATOR

Abadie and Imbens (2002) found that when the matching is not exact, the simple estimator will be biased in finite samples. Specifically, with k continuous covariates, the estimator will have a bias term that corresponds to the matching discrepancies (i.e., the differences in covariates between matched units and their matches). To remove some of the bias that remains after matching, Abadie and Imbens (2002) developed a *bias-corrected matching estimator*. The adjustment uses a least squares regression to adjust the difference within the matches for the differences in their covariate values.

The regression adjustment is made in four steps:

1. Suppose that we are estimating the SATE. In this case, we run regressions using only the data in the matched sample. Define $\mu_w(x) = E\{Y(w)|X = x\}$ for $w = 0$ (control condition) or $w = 1$ (treatment condition). Using the data of the matched sample, we run two separate regression models: one uses the data of $w = 0$ and the other uses the data of $w = 1$. Each regression model uses $Y(w)$ as a dependent variable, and all covariates are used as independent variables.

2. At this stage, we have obtained two sets of regression coefficients, one for $w = 0$ and one for $w = 1$. Let the intercept of the regression function be $\hat{\beta}_{w0}$ and the slope vector be $\hat{\beta}'_{w0}$. We choose $\hat{\beta}_{w0}$ and $\hat{\beta}'_{w1}$ that minimize the weighted sum of squared residuals using $K_M(i)$ as a weight. Precisely, the adjustment term $\hat{\mu}_w$, for $w = 0, 1$, is a predicted value based on the following equation: $\hat{\mu}_w = \hat{\beta}_{w0} + \hat{\beta}'_{w1}x$, where

$$(\hat{\beta}_{w0}, \hat{\beta}'_{w1}) = \arg\ \min_{\{\beta_{w0}, \beta_{w1}\}} \sum_{i: W_i = w} K_M(i)(Y_i - \hat{\beta}_{w0} - \hat{\beta}'_{w1}X_i)^2.$$

In other words, we choose $\hat{\beta}_{w0}$ and $\hat{\beta}_{w1}^y$ that minimize the weighted sum of squared residuals $\displaystyle\sum_{i:W_i=w} K_M(i)(Y_i - \hat{\beta}_{w0} - \hat{\beta}_{w1}'X_i)^2$.

3. After obtaining the adjustment term $\hat{\mu}_w$ for $w = 0$ and $w = 1$, we can use the term to correct the bias embedded in the simple matching estimator. The bias-corrected estimator then uses the following equations to impute the missing potential outcomes:

$$\tilde{Y}_i(0) = \begin{cases} Y_i & \text{if } W_i = 0 \\ \frac{1}{\#J_M(i)} \displaystyle\sum_{l \in J_M(i)} Y_l\{Y_l + \hat{\mu}_0(X_i) - \hat{\mu}_0(X_l)\} & \text{if } W_i = 1. \end{cases} \quad (6.5)$$

and

$$\tilde{Y}_i(1) = \begin{cases} \frac{1}{\#J_M(i)} \displaystyle\sum_{l \in J_M(i)} Y_l\{Y_l + \hat{\mu}_1(X_i) - \hat{\mu}_1(X_l)\} & \text{if } W_i = 0 \\ Y_i & \text{if } W_i = 1. \end{cases}$$

4. The above steps illustrating the bias correction process use the point estimate SATE as an example (equivalently PATE, because the two coefficients are exactly the same). If we are interested in estimating the SATT or PATT, we then need only estimate the regression function for the controls, $\hat{\mu}_0$. If we are interested in estimating the SATC or PATC, we then need only estimate the regression function for the treated, $\hat{\mu}_1$.

The bias-corrected matching estimator can always be used to replace the simple matching estimator. It is especially useful when matching on several covariates of which at least one is a continuous variable. This correction is needed because exact matching is seldom exact when matching uses continuous covariates. For an application of the bias-corrected matching estimator, see Hirano and Imbens (2001). On balance, the most important function of bias-corrected matching is to adjust for poor matches for the covariates at hand, and the method does not really correct for bias except in some special and unrealistic instances. For instance, the method cannot (and is not designed to) correct bias generated by the omission of important covariates or bias due to hidden selection.

6.2.3 VARIANCE ESTIMATOR ASSUMING HOMOSCEDASTICITY

Abadie and Imbens (2002) developed a variance estimator for various treatment effects. They first considered a variance estimator under two assumptions: (1) the unit-level treatment effect $\tau_i = Y_i(1) - Y_i(0)$ is constant and (2) the conditional variance of $Y_i(W)$ given X_i does not vary with either the covariates x or the treatment w, which is known as an assumption of homoscedasticity. Under these assumptions, we can obtain variances for various effects by applying the following formulas.

Variance of SATE:

$$\hat{V}^{SATE} = \frac{1}{N^2} \sum_{i=1}^{N} \{1 + K_M(i)\}^2 \frac{1}{2N} \sum_{i=1}^{N} \left[\frac{1}{J_M(i)} \sum_{l \in J_M(i)} \{W_i(Y_i - Y_l - \hat{\tau}) + (1 - W_i)(Y_l - Y_i - \hat{\tau})\}^2 \right].$$ (6.6)

Variance of SATT:

$$\hat{V}^{SATT} = \frac{1}{N_1^2} \sum_{i=1}^{N} \{W_i - (1 - W_i)K_M(i)\}^2 \frac{1}{2N_1} \sum_{i:W_i=1}^{N} \left[\frac{1}{J_M(i)} \sum_{l \in J_M(i)} \{(Y_i - Y_l - \hat{\tau})^2\} \right].$$ (6.7)

Variance of SATC:

$$\hat{V}^{SATC} = \frac{1}{N_0^2} \sum_{i=1}^{N} \{W_i K_M(i) - (1 - W_i)\}^2 \frac{1}{2N_0} \sum_{i:W_i=0}^{N} \left[\frac{1}{J_M(i)} \sum_{l \in J_M(i)} \{(Y_l - Y_i - \hat{\tau})^2\} \right].$$ (6.8)

Variance of PATE:

$$\hat{V}^{PATE} = \frac{1}{N^2} \sum_{i=1}^{N} \left[\{\hat{Y}_i(1) - \hat{Y}_i(0) - \hat{\tau}\}^2 + \{K_M^2(i) + 2K_M(i) - K_M'(i)\} \frac{1}{2N} \sum_{i=1}^{N} \frac{1}{J_M(i)} \sum_{l \in J_M(i)} \{W_i(Y_i - Y_l - \hat{\tau}) + (1 - W_i)(Y_l - Y_i - \hat{\tau})\}^2 \right]$$ (6.9)

where

$$K_M'(i) = \sum_{l=1}^{N} 1\{i \in J_M(l)\} \left\{ \frac{1}{\#J_M(l)} \right\}^2.$$

Variance of PATT:

$$\hat{V}^{PATT} = \frac{1}{N_1^2} \sum_{i=1}^{N} \left[W_i\{Y_i(1) - \hat{Y}_i(0) - \hat{\tau}^t\}^2 + (1 - W_i)\{K_M^2(i) - K_M'(i)\} \right.$$

$$\left. \frac{1}{2N_1} \sum_{i:W_1=0} \left\{ \frac{1}{J_M(i)} \sum_{l \in J_M(i)} \{(Y_l - Y_i - \hat{\tau})^2\} \right\} \right]. \quad (6.10)$$

Variance of PATC:

$$\hat{V}^{PATC} = \frac{1}{N_0^2} \sum_{i=1}^{N} \left[(1 - W_i)\{\hat{Y}_i(1) - Y_i(0) - \hat{\tau}^t\}^2 + W_i\{K_M^2(i) - K_M'(i)\} \right.$$

$$\left. \frac{1}{2N_0} \sum_{i:W_i=0} \left\{ \frac{1}{J_M(i)} \sum_{l \in J_M(i)} \{(Y_l - Y_i - \hat{\tau})^2\} \right\} \right]. \quad (6.11)$$

Taking the square root of each variance term, we obtain a standard error of the point estimate. We can then use the standard error either to calculate a 95% confidence interval of the point estimate or to perform a significance test at a given level of statistical significance. The ratio of the point estimate over the standard error follows a standard normal distribution, which allows users to perform a z test.

6.2.4 VARIANCE ESTIMATOR
ALLOWING FOR HETEROSCEDASTICITY

The assumptions about a constant treatment effect and homoscedasticity may not be valid for certain types of evaluation data. In practice, evaluators may have data in which a term of conditional error variance in Equations 6.6 to 6.11 varies by treatment condition w and covariate vector x. To deal with this problem, Abadie and Imbens (2002) developed a robust variance estimator that allows for heteroscedasticity. The crucial feature of this robust estimator is its utility for estimating a conditional error variance for all sample points. The algorithm developed by Abadie and Imbens includes a second matching procedure such that it matches treated units to treated units and control units to controls. When running a computing software package, such as **nnmatch** in Stata, users must specify the number of matches used in the second matching stage across observations of the same treatment condition. This number need not be the same as the number of matches used in estimating the treatment effect itself. For details of the robust estimator, see Abadie et al. (2004, p. 303).

6.2.5 LARGE SAMPLE PROPERTIES AND CORRECTION

Although matching estimators, including those developed by other authors like Cochran and Rubin (1973), Rosenbaum and Rubin (1983), and Heckman and Robb (1985) have great intuitive appeal and are widely used in practice, their formal large sample properties were not established until recently. This delay was due in part to the fact that matching with a fixed number of matches is a highly nonsmooth function of the distribution of the data, which is not amenable to standard asymptotic methods for smooth functionals. Recently, however, Abadie and Imbens (2006) developed an analytic approach to show the large sample properties of matching estimators. Below, we briefly highlight their findings.

Abadie and Imbens's (2006) results indicated that some of the formal large sample properties of matching estimators are not very attractive. First, they demonstrated that matching estimators include a conditional bias term whose stochastic order increases with the number of continuous matching variables. The order of this bias term may be greater than $1/\sqrt{N}$, and therefore, matching estimators are not $1/\sqrt{N}$ consistent. Second, in general, the simple matching estimator is $1/\sqrt{N}$ consistent. However, the simple matching estimator does not achieve the semiparametric efficiency bound as calculated by Hahn (1998). For cases when only a single continuous covariate is used to match, Abadie and Imbens have shown that the efficiency loss can be made arbitrarily close to zero by allowing a sufficiently large number of matches. Third, despite these poor formal properties, matching estimators are extremely easy to implement and do not require consistent nonparametric estimation of unknown functions. As such, matching estimators have several attractive features that may account for their popularity. Fourth, Abadie and Imbens proposed an estimator of the conditional variance of the simple matching estimator that does not require consistent nonparametric estimation of unknown functions. This conditional variance is essentially estimated by the variance estimator that allows for heteroscedasticity and involves a second matching procedure (see Section 6.2.4). The crucial idea is that instead of matching treated units to controls, the estimator of the conditional variance matches treated units to treated units and control units to controls in the second stage. Finally, based on results of the large sample properties of matching estimators, Abadie and Imbens concluded that bootstrapping is not valid for matching estimators.

6.3 Overview of the Stata Program *nnmatch*

Software for implementing the matching estimators is available in Stata, Matlab, and R. In this section, we review a user-developed program in Stata called ***nnmatch.*** It processes all the estimators described in this chapter. As a user-developed program, *nnmatch* is not included in the regular Stata package. To search the Internet for this software, users can use the *findit* command, followed by *nnmatch* (i.e., *findit nnmatch*), and then follow the online instructions to download and install the program. After installation, users should check the help file to obtain basic instructions for running the program. The work of Abadie et al. (2004), published by *The Stata Journal*, was written specifically to address how to use *nnmatch* for evaluating various treatment effects discussed in this chapter; users may find this reference helpful.

The *nnmatch* program can be initiated using the following basic syntax:

nnmatch depvar treatvar varlist, tc(att) m(#) metric(maha) biasadj(bias) ///
robusth(#) population keep(filename)

In this command, *depvar* is the outcome variable on which users want to assess the difference between treated and control groups, that is, the outcome variable showing treatment effect; *treatvar* is the binary treatment membership variable that indicates the intervention condition; and *varlist* specifies the covariates to be used in the matching. These statements are required and must be specified. The rest of the statements are optional, and omission of their specifications calls for default specifications. The term *tc* specifies the type of treatment effects to be evaluated, and three values may be specified in the parentheses: *ate*, *att*, and *atc*, which stand for *average treatment effect*, *average treatment effect for the treated*, and *average treatment effect for the controls*, respectively. By default, *nnmatch* estimates *ate*. The *m* specifies the number of matches that are made per observation. Users replace # in the syntax with a specific number and include that number in the parentheses. The term *metric* specifies the metric for measuring the distance between two vectors of covariates, or the type of variance matrix users selected to use in the vector norm. By default, *nnmatch* uses the inverse of sample variance matrix; *metric(maha)* causes *nnmatch* to use the inverse of sample variance-covariance matrix in the vector norm and to evaluate Mahalanobis metric distances. The term *biasadj* specifies that the bias-corrected matching estimator be used. By default, *nnmatch* uses the simple matching estimator. If the user specifies *biasadj(bias)*, *nnmatch* uses the same set of matching covariates, *varlist*, to estimate the linear regression function in the bias adjustment. Alternatively, the user can specify a new list of variables in the parentheses, which causes *nnmatch* to enter a different set of covariates in the regression adjustment. The term *robusth(#)* specifies that *nnmatch* estimate heteroscedasticity-consistent standard errors using # matches in the second matching stage. A specific number is used to replace # in the above syntax, and users should include that number in the parentheses. The number need not be the same as that specified in *m(#)*. By default, *nnmatch* uses the homoscedastic/constant variance estimator. The term *population* causes *nnmatch* to estimate a population treatment effect (i.e., *ate*, *att*, or *atc*). By default, *nnmatch* estimates a sample treatment effect. Last, *keep(filename)* saves a temporary matching data set in the file *filename.dta*. A set of new variables are created and saved in the new Stata data file that may be used for follow-up analysis.

Table 6.4 exhibits the syntax and output of six *nnmatch* examples: estimation of SATE, PATE, SATT, PATT, SATC, and PATC using four matches per observation, and using bias-corrected and robust variance estimators that allow for heteroscedasticity. Substantive findings of these examples are discussed in the next section.

6.4 Examples

In this section, we use two examples to illustrate the application of matching estimators in program evaluation. The first example shows the evaluation of six treatment effects concerning causality using a bias-corrected estimator with variance estimator allowing for heteroscedasticity. The second example illustrates the application of matching estimators to an ESA for which doses of treatment becomes a central concern. The second example also runs all analyses following missing data imputation based on 50 imputed files. As such, the second example mimics study conditions that are likely to be found in real program evaluation.

6.4.1 MATCHING WITH BIAS-CORRECTED AND ROBUST VARIANCE ESTIMATORS

This section presents an example showing the application of matching with bias-corrected and robust variance estimators. Using a sample and matching variables similar to those in the example given in Section 5.9.2, we analyze a subsample of 606 children from the panel data of the 1997 Child Development Supplement (CDS) to the Panel Study of Income Dynamics (PSID) and the core PSID annual data from 1968 to 1997 (Hofferth et al., 2001). The primary study objective was to test a research hypothesis regarding the causal effect of childhood poverty on developmental outcomes, specifically academic achievement. The study tested the effect of participation in a welfare program—an indicator of poverty—on test performance. We refer readers to Section 5.9.2 for a review of the conceptual framework and substantive details of the study.

For this application, we report findings from the examination of a single domain of academic achievement: the age-normed *passage comprehension* score of the Woodcock-Johnson Revised Tests of Achievement (Hofferth et al., 2001). Higher scores on this measure are considered an indication of higher academic achievement. The passage comprehension score is used as the outcome variable in this study. Readers should note that this study analyzed data for 606 children rather than the 1,003 children included in the analysis

Table 6.4   Exhibit of Stata *nnmatch* Syntax and Output Running Bias-Corrected
Matching Estimators With Robust Standard Errors

```
//Chapter 6 Example 1
//use the PSID-CDS pcss97 (passage comprehension) to illustrate
nnmatch
cd "D:\Sage\ch6"
use cds_pcss97, replace

//(SATE) Sample average treatment effect
nnmatch pcss97 kuse male black age97 pcged97 mratio96 pcg_adc, ///
m(4) tc(ate) bias(bias) robusth(4)

(Output)
Matching estimator:  Average Treatment Effect ate

Weighting matrix: inverse variance        Number of obs     =     606
                                    Number of matches   (m) =       4
                                    Number of matches,
                          robust std. err. (h) =                     4


pcss97 |    Coef.    Std. Err.     z     P>|z|      [ 95% Conf. Interval]
-------+------------------------------------------------------------------
SATE  |  -4.703774    1.76969   -2.66   0.008    -8.172303   -1.235244

Matching variables:  male black age97 pcged97 mratio96 pcg_adc
Bias-adj variables:  male black age97 pcged97 mratio96 pcg_adc

//(PATE) Population average treatment effect
nnmatch pcss97 kuse male black age97 pcged97 mratio96 pcg_adc, ///
m(4) tc(ate) population bias(bias) robusth(4)

(Output)
Matching estimator: Population Average Treatment Effect ate

Weighting matrix: inverse variance        Number of obs     =     606
                            Number of matches   (m) =         4
                            Number of matches,
                          robust std. err. (h) =           4


pcss97 |    Coef.    Std. Err.     z     P>|z|      [ 95% Conf. Interval]
-------+------------------------------------------------------------------
PATE  |  -4.703774    1.765187  -2.66   0.008    -8.163476   -1.244072

Matching variables:  male black age97 pcged97 mratio96 pcg_adc
Bias-adj variables:  male black age97 pcged97 mratio96 pcg_adc

//(SATT) S average treatment effect for the treated
nnmatch pcss97 kuse male black age97 pcged97 mratio96 pcg_adc, ///
m(4) tc(att) bias(bias) robusth(4)
```

(Continued)

Table 6.4 (Continued)

```
(Output)
Matching estimator:  Average Treatment Effect for the Treated

Weighting matrix: inverse variance          Number of obs        =      606
                          Number of matches  (m) =           4
                          Number of matches,
                             robust std. err. (h) =          4


pcss97 |   Coef.    Std. Err.     z     P>|z|     [ 95% Conf. Interval]
-------+-------------------------------------------------------------------
SATT |  -5.229651   1.781161    -2.94   0.003    -8.720663    -1.738639

Matching variables:   male black age97 pcged97 mratio96 pcg_adc
Bias-adj variables:   male black age97 pcged97 mratio96 pcg_adc
```

//(PATT) P average treatment effect for the treated
nnmatch pcss97 kuse male black age97 pcged97 mratio96 pcg_adc, ///
m(4) tc(att) population bias(bias) robusth(4)

```
(Output)
Matching estimator: Population Average Treatment Effect for the Treated
Weighting matrix: inverse variance          Number of obs        =      606
                          Number of matches  (m) =           4
                          Number of matches,
                             robust std. err. (h) =          4


pcss97 |    Coef.    Std. Err.     z     P>|z|     [ 95% Conf. Interval]
-------+-------------------------------------------------------------------
PATT |  -5.229651    1.72059     -3.04   0.002    -8.601946    -1.857356

Matching variables:   male black age97 pcged97 mratio96 pcg_adc
Bias-adj variables:   male black age97 pcged97 mratio96 pcg_adc
```

//(SATC)S average treatment effect for the controls
nnmatch pcss97 kuse male black age97 pcged97 mratio96 pcg_adc, ///
m(4) tc(atc) bias(bias) robusth(4)

```
(Output)
Matching estimator:  Average Treatment Effect for the Controls

Weighting matrix: inverse variance          Number of obs        =      606
                          Number of matches  (m) =           4
                          Number of matches,
                             robust std. err. (h) =          4


pcss97 |    Coef.    Std. Err.     z     P>|z|     [ 95% Conf. Interval]
-------+-------------------------------------------------------------------
SATC |  -4.467255   2.133536    -2.09   0.036    -8.648908    -.2856015
```

```
Matching variables:   male black age97 pcged97 mratio96 pcg_adc
Bias-adj variables:   male black age97 pcged97 mratio96 pcg_adc


//(PATC)P average treatment effect for the controls
nnmatch pcss97 kuse male black age97 pcged97 mratio96 pcg_adc, ///
m(4) tc(atc) population bias(bias) robusth(4)

(Output)
Matching estimator: Population Average Treatment Effect for the Controls

Weighting matrix: inverse variance          Number of obs       =      606
                               Number of matches  (m) =       4
                               Number of matches,
                                  robust std. err. (h) =       4
```

pcss97	Coef.	Std. Err.	z	P>|z|	[95% Conf. Interval]
PATC	-4.467255	2.135647	-2.09	0.036	-8.653045 -.2814641

```
Matching variables:   male black age97 pcged97 mratio96 pcg_adc
Bias-adj variables:   male black age97 pcged97 mratio96 pcg_adc
```

presented in Section 5.9.2. Because 397 cases were missing passage comprehension scores, those cases were removed from the analytic sample.

Based on the research question, we classified the study participants into two groups: children who ever used AFDC (Aid to Families With Dependent Children) from birth to current age in 1997 and those who never used AFDC during the same period. Thus, this dichotomous variable indicated the treatment condition in the study: those who ever used AFDC versus controls who never used AFDC. Of the 606 study children, 188 had used AFDC and were considered as the treated group, and 418 participants had never used AFDC and were considered the control group.

To assess the treatment effect (i.e., participation in the AFDC program as an indicator of poverty) on academic achievement, we considered the following covariates or matching variables: (a) current income or poverty status, which was measured as the ratio of family income to poverty threshold in 1996; (b) caregiver's education in 1997, which was measured as years of schooling; (c) caregiver's history of using welfare, which was measured as the number of years (i.e., a continuous variable) the caregiver used AFDC during the caregiver's childhood (i.e., ages 6 to 12 years); (d) child's race, which was measured as African American versus non–African American; (e) child's age in 1997; and (f) child's gender, which was measured as male versus female.

It is worth noting that of the six matching variables, four were continuous variables and only child's race and gender were categorical variables. Given this condition, it is impossible to conduct exact matching for this data set, and it is therefore important to use the bias-corrected matching estimator to correct for bias corresponding to the matching discrepancies between matched units and their matches on the four continuous covariates. In our example, we used the same set of matching variables as the independent variables for the regression adjustment in the bias correction process. Following the recommendation of Abadie et al. (2004), we chose four matches per observation in the analysis.

The choice of a variance estimator deserves some explanation. Note that the homoscedastic variance estimator assumes that the unit-level treatment effect is constant and that the conditional variance of $Y_i(w)$ given X_i does not vary with either covariates or the treatment. To test whether our data met the homoscedastic assumption, we first ran a regression of the passage comprehension scores on the six matching variables plus the binary treatment variable. We then performed the Breusch-Pagan and Cook-Weisberg tests of heteroscedasticity for each of the seven independent variables. Our results from the Breusch-Pagan and Cook-Weisberg tests showed that child's age was statistically significant ($p < .000$) and indicated that the conditional variance of the outcome variable was not constant across levels of child's age. Based on this finding, we decided to use the robust variance estimator that allows for heteroscedasticity. We used the same number of matches (i.e., four matches) in the second matching stage to run the robust variance estimator.

Table 6.5 presents results of our analysis of the study data. The interpretation and findings of the study may be summarized as follows. First, as noted earlier, a specific sample effect is the same as its corresponding population effect in magnitude (e.g., both SATE and PATE are equal to –4.70). The two effects differ from each other only on the standard error (e.g., the standard error for SATE was 1.76969, whereas the standard error for PATE was 1.765187). Second, our results suggested that childhood poverty strongly affected the children's academic achievement. On average, children who used AFDC in childhood had a passage comprehension score 4.7 units lower than that of children who had never used AFDC in childhood. This finding held true after we took selection bias into consideration for six observed covariates. With regard to the subpopulation of treated participants, the treatment effect was even larger: –5.23, or 0.63 units larger than the sample (or population) average treatment effect. Third, had all controls (i.e., children who never used AFDC) used AFDC and all treated children not used AFDC, then on average, the control children would have a passage comprehension score 4.47 units lower than their counterparts. Note that in this study, SATT equaled –5.23 and SATC equaled –4.47, or a difference of 0.76 units. This difference is attributable either to additional selection bias that was not accounted for in the study or to study data that violated assumptions of matching estimators, which suggests the need for further scrutiny. Fourth, a population effect indicates whether the tested intervention will be effective in a second sample taken from the same population. Taking SATT ($p = .003$) and

Table 6.5 Estimated Treatment Effects (Effects of Child's Use of AFDC) on Passage Comprehension Standard Score in 1997 Using Bias-Corrected Matching With Robust Variance Estimators (Example 6.4.1)

Treatment Effect	Coefficient	Standard Error	z	p Value	95% CI
Sample average treatment effect (SATE)	−4.70	1.76969	−2.66	.008	[−8.17, −1.24]
Population average treatment effect (PATE)	−4.70	1.765187	−2.66	.008	[−8.16, −1.24]
Sample average treatment effect for the treated (SATT)	−5.23	1.781161	−2.94	.003	[−8.72, −1.74]
Population average treatment effect for the treated (PATT)	−5.23	1.72059	−3.04	.002	[−8.60, −1.86]
Sample average treatment effect for the controls (SATC)	−4.47	2.133536	−2.09	.036	[−8.65, −0.29]
Population average treatment effect for the controls (PATC)	−4.47	2.135647	−2.09	.036	[−8.65, −0.28]

NOTE: 95% CI = 95% confidence interval.

PATT ($p = .002$) as examples, the study indicated that the treatment effect for the treated group was statistically significant in the sample at a level of .01. If we take a second sample from the population, we are likely to observe the same level of treatment effect for the treated, and the effect should remain statistically significant at a level of .01. Finally, our results showed that all six treatment effects were statistically significant ($p < .05$), and all 95% confidence intervals did not contain a zero. Thus, we concluded that the study data could not reject a null hypothesis of a zero treatment effect, and represented by participation in the AFDC program and conditioned on the available data, childhood poverty appears to be an important factor causing children's poor achievement in passage comprehension.

6.4.2 EFFICACY SUBSET ANALYSIS WITH MATCHING ESTIMATORS

In this example, we illustrate how to use matching estimators to conduct an ESA, that is, to estimate treatment effects by dosage or exposure level. In Section 5.7, we described the method of dose analysis that uses either ordered logistic regression or multinomial logistic regression to predict propensity scores, and then performs either a nonbipartite matching or propensity score weighting analysis. Because the matching estimators directly evaluate sample and population average treatment effects for the treated, they permit a direct ESA to test hypotheses regarding treatment dosage. Therefore, we selected the matching estimators to conduct the dosage analysis.

This example uses a sample and matching variables similar to those of the example described in Section 4.5.2. The findings reported here represent preliminary findings from an evaluation of the "Social and Character Development" program that was implemented in North Carolina. The study sample comprised more than 400 study participants.

The North Carolina intervention included a skills-training curriculum, *Making Choices*, which was designed for elementary school students. The primary goals of the Making Choices curriculum were to increase students' social competence and to reduce students' aggressive behavior. The treatment group received a multi-element intervention, which included 29 Making Choices classroom lessons delivered over the course of the third-grade year, and eight follow-up or "booster shot" classroom lessons delivered in each of the fourth- and fifth-grade years. Students assigned to the control group received regular character development and health education instruction and did not receive any lessons from the Making Choices curriculum.

Several valid and reliable instruments were used to measure the students' social competence and aggressive behavior, and outcome data were collected for both the treated students and controls at the beginning and end points of each academic year for the third-, fourth-, and fifth-grade years. Therefore, six waves of panel data were available for the evaluation. The example used only the outcome data collected during the fourth and fifth grades, that is, the data measuring behavior change 1 or 2 years after the intervention. For most students, outcome data were collected by different teachers at different grades, and these data were likely to reflect some raters' effect. To remove the raters' effect, we analyzed change scores within a grade (i.e., using an outcome variable at the end point of a grade minus the outcome at the beginning point of the same grade). Typically scores within a grade were made by the same teachers.

As mentioned in Section 4.5.2, the Making Choices intervention used group randomization with 14 schools (Cohort 1 = 10 schools; Cohort 2 = 4 schools). However, in spite of the cluster randomized design, a preliminary analysis showed that the sample data were not balanced on many observed covariates. This indicated that the group randomization had not worked as planned. In some school districts, as few as four schools met the study criteria and were eligible for participation. As a result of the smaller than anticipated sample, the two intervention schools differed systematically on covariates from the two control schools. When the investigators compared data from these schools, they found that the intervention schools differed from the control schools in several significant ways: the intervention schools had (a) lower academic achievement scores on statewide tests (Adequate Yearly Progress), (b) a higher percentage of students of color, (c) a higher percentage of students receiving free or reduced-price lunches, and (d) lower mean scores on behavioral composite scales at baseline. Using bivariate tests and logistic regression models, the researchers found that these differences were statistically significant at the .05 level. The researchers were confronted with the failure of randomization. Had these selection effects not been taken into consideration, the evaluation of the program effectiveness would be biased. Hence, it is important to use propensity score approaches, including matching estimators, to correct for the selection bias in evaluation.

Like most evaluations, the data set contained missing values for many study variables. Before performing the evaluation analysis, we conducted a missing data imputation using the multiple imputation method (Little & Rubin, 2002; Schafer, 1997). With this method, we generated 50 imputed data files for each outcome variable. Results from our analysis showed that with 50 data sets, the imputation achieved a relative efficiency of 99%. Following the convention of practice in imputing missing data, we imputed missing values for all cases; but cases that had missing data on the outcome variable were deleted. As such, the sample size for final analysis varies by outcome variable.

With multiply imputed files (i.e., 50 distinct data files in this example), we first ran *nnmatch* for each file and then used Rubin's rule (Little & Rubin, 2002; Schafer, 1997) to aggregate the point estimates and standard errors to generate one set of statistics for the significance test for each outcome variable. We refer readers to this text's companion Web page, where we provide the syntax for running *nnmatch* 50 times for each outcome and for the aggregation using Rubin's rule.

To analyze the outcome changes that occurred in the fourth- and fifth-grade years, we first analyzed the change scores for the sample as a whole using three methods: (1) optimal pair matching, which was followed by regression adjustment; (2) optimal full matching, which was followed by the Hodges-Lehmann aligned rank test; and (3) matching estimators. We provide results of these analyses in Table 6.6.

Results from the first two methods (i.e., optimal pair matching and optimal full matching) were not promising. Based on the design of the intervention program, we expected positive findings (i.e., the intervention would be effective in changing behavioral outcomes); however, none of the results from the optimal pair matching with regression adjustment was statistically significant. The situation improved slightly with the results of the optimal full matching with the Hodges-Lehmann test, in which some outcomes showed a statistical trend ($p < .10$), and two variables (i.e., social competence and prosocial) showed statistical significance ($p < .05$). When faced with such situations, researchers need to seek a plausible explanation. We thought that there were at least two plausible explanations for the nonsignificant findings: one is that the intervention was not effective, and the other is that our evaluation data violated assumptions embedded in the evaluation methods we had used and, therefore, the results reflect methodological artifacts. In practice, there is no definitive way to find out which explanation is true. However, the results of the third analytic method, matching estimators, showed more significant findings than the previous two analyses.

It is worth noting that even when using matching estimators, we are still likely to find that the intervention was not effective. Moreover, we might also find that results from the matching estimators are methodologically erroneous, but in this case, the error goes in the other direction: the violation of model assumptions might have produced "overly optimistic" findings. The discussion of which of the three methods is most suitable to our study and which set of results is more trustworthy is beyond the scope of this chapter. However, we emphasize that using multiple approaches in evaluation research is important, particularly for programs whose effects may be marginal, or whose effect sizes fall into the "small" category defined by Cohen (1988). As underscored by Cohen, detecting a small effect size is often an important objective for a new inquiry in social behavioral research. This point was also emphasized by Sosin (2002), whose study used varying methods, including sample selection, conventional control variable, instrumental variable, and propensity score matching, to examine a common data set. Sosin found that the various methods provided widely divergent estimates. In light of this finding, he suggested that researchers regularly compare estimates across multiple methods.

Returning to our example, we first considered findings for the whole sample and then divided the sample into subsets based on treatment dosage. We

Table 6.6 Estimated Treatment Effects Measured as Change Scores in the Fourth and Fifth Grades by Three Estimators (Example 6.4.2)

Outcome Variable	Hypothetical Sign	Change Score Using Optimal Pair Matching and Regression Adjustment		Change Score Using Optimal Full Matching and Hodges-Lehmann Aligned Rank Test		Change Score Using Matching Estimator (SATT)	
		Grade 4	Grade 5	Grade 4	Grade 5	Grade 4	Grade 5
ICSTAGG—Aggression	−	−.28	−.18	−.33+	−.02	−.47***	−.15
ICSTACA—Academic competence	+	−.09	.23	−.18	.23	−.09	.15
ICSTINT—Internalizing	−	−.03	−.21	.16	−.30+	.05	−.30*
CCCSCOM—Social competence	+	.19	.27	.12	.26*	.21*	.27**
CCCPROS—Prosocial	+	.17	.28	.11+	.27*	.20*	.29**
CCCEREG—Emotion regulation	+	.21	.26	.20+	.26+	.22*	.26**
CCCRAGG—Relational aggression	−	−.16	−.21	−.13+	−.21+	−.27**	−.18+

***$p < .001$, **$p < .01$, *$p < .05$, +$p < .1$, two-tailed.

then conducted ESA on each of the subsets. In this study, we defined *dosage* as the number of minutes a student received Making Choices classroom lessons. Based on this definition (which relied on teacher reports of implementation and fidelity), the dosage for all controls was zero. Furthermore, we defined three subsets on the basis of grade-level booster shot dosage: (1) adequate (and recommended) exposure to intervention, that is, students who received Making Choices for more than 240 minutes and less than 379 minutes; (2) high exposure to intervention, that is, students who received unusually high program exposure of more than 380 minutes of Making Choices lessons; and (3) the control group, who received zero minutes of Making Choices. Table 6.7 shows the sample size and distribution of the three subsets by grade.

With regard to ESA, the subsets defined above allowed us to make two comparisons for each grade: the adequate exposure group versus the comparison group, and the high exposure group versus the comparison group. Our general hypothesis was that the Making Choices intervention produces desirable behavioral changes for the treatment group, and students who have adequate or more exposure to the intervention will show higher levels of changed behavior.

The matching we conducted for our analysis had the following characteristics: we chose bias-corrected matching that used the same set of matching variables in the regression model as for the bias correction, we used four matches per observation, our matching strategy used the robust variance estimator to allow for heteroscedasticity and used four matches per observation in the second matching stage, and our matching estimated SATT. We present the results of the ESA in Table 6.8.

On balance, the ESA confirmed the research hypotheses for two core outcomes for the fourth-grade year (i.e., aggression and relational aggression). That is, the treatment effects for the treated on these outcomes were not only statistically significant but also in the same direction of the hypothetical sign. Furthermore, the high exposure group exhibited a larger effect than the adequate exposure group. Second, for the social competence measures, including prosocial behavior and emotional regulation, the direction and size of the findings are relatively consistent over the fourth and fifth grades. This pattern is consistent also with the hypotheses. Greater variation is observed in the high-exposure groups where sample sizes are smaller. Finally, the sample treatment effects for the treated group were statistically significant for most outcomes in the analyses of adequate exposure group in both the fourth- and fifth-grade years. In sum, the ESA suggested that the recommended exposure level of 240 minutes of program content produces positive effects. At that dose level, the Making Choices program appears to promote prosocial behavior and reduce aggressive behavior. Program exposure at a higher level may have some gain for fourth-grade children, but it appears to have a negligible effect for fifth-grade children. On balance, this conclusion is consistent with the theory, objective, and design of the Making Choices intervention.

Table 6.7 Sample Size and Distribution of Exposure Time to Program Intervention ("Dosage") by Grade (Example 6.4.2)

| | Grade 4 | | | Grade 5 | | |
Outcome Variable	Total	Adequate	High	Comp.	Total	Adequate	High	Comp.
ICSTAGG—Aggression	413	130	60	223	434	148	33	253
ICSTACA—Academic competence	413	130	60	223	434	148	33	253
ICSTINT—Internalizing	413	130	60	223	434	148	33	253
CCSCOM—Social competence	414	130	61	223	433	149	33	251
CCCPROS—Prosocial	414	130	61	223	433	149	33	251
CCCEREG—Emotion regulation	414	130	61	223	433	149	33	251
CCCRAGG—Relational aggression	414	130	61	223	433	149	33	251

NOTE: Adequate = adequate exposure to intervention (240 to 379 minutes); High = high exposure to intervention (380 or more minutes); Comp. = comparison group (0 minutes).

Table 6.8   Efficacy Subset Analysis Using Matching Estimators: Estimated Average Treatment Effects for the Treated (SATT) by Dosage (Example 6.4.2)

Outcome Variable	Hypothetical Sign	Adequate Exposure (240–379) Versus Comp. (0) Grade 4	High Exposure (380+) Versus Comp. (0) Grade 4	Adequate Exposure (240–379) Versus Comp. (0) Grade 5	High Exposure (380+) Versus Comp. (0) Grade 5
ICSTAGG—Aggression	−	−.34*	−.78***	−.15	−.14
ICSTACA—Academic competence	+	−.21	.18	.11	.33
ICSTINT—Internalizing	−	−.05	.31+	−.33*	.09
CCCSCOM—Social competence	+	.21*	.18	.27**	.27
CCCPROS—Prosocial	+	.22*	.15	.30**	.25
CCCEREG—Emotion regulation	+	.21*	.22	.25*	.30+
CCCRAGG—Relational aggression	−	−.20*	−.45***	−.18+	−.25

NOTE: ***$p < .001$, **$p < .01$, *$p < .05$, +$p < .1$, two-tailed.

6.5 Conclusions

This chapter discussed a method that is widely used in observational studies: matching. The discussion focuses on the collection of matching estimators developed by Abadie and Imbens (2002). These include the vector norm, bias correction using a linear regression function, and robust variance estimation involving second-stage matching. Recently, formal study has examined the large sample properties of various matching estimators, including those developed by researchers other than Abadie and Imbens. In general, matching estimators are not $1/\sqrt{N}$ consistent, and therefore, the results of large sample properties are not so attractive. However, using a correction that matches treated units to treated units and control units to controls, the robust estimator of asymptotic variance has proven to be promising.

Matching estimators are intuitive and appealing, but caution is warranted in at least two areas of application. First, matching with continuous covariates poses ongoing challenges. When matching variables are continuous, users need to be cautious and make adjustments, such as using a bias-corrected matching estimator, for bias. Note that the presence of continuous matching variables does not appear to be a severe problem in propensity score matching, where continuous as well as categorical variables are treated as independent variables in the logistic regression predicting the propensity scores. Second, bootstrapping to estimate variances is problematic, and the direct estimation procedure developed by Abadie and Imbens (2002) appears preferable. Note that the bootstrapping method is used in statistical inference for matching with nonparametric regression, which is a topic we tackle in the next chapter.

The matching estimators discussed in this chapter handle heteroscedasticity, but they do not correct for inefficiency induced by clustering. In social behavioral research, intraclass correlation, or clustering, is frequently encountered in program evaluation (Guo, 2005). However, the current version of matching estimators does not take the issue of clustering into consideration. The developers of matching estimators are aware of this limitation and are working toward making improvements to matching estimators that will provide adjustments for clustering (G. Imbens, personal communication, October 10, 2007).

Despite limitations, the collection of matching estimators described in this chapter offers a clear and promising approach to balancing data when treatment assignment is nonignorable. The method is easy to implement and requires of the researcher few subjective decisions.

Note

1. The vector norm originally defined by Abadie et al. (2004) was $\|x\|v = (x'Vx)^{1/2}$, that is, a square root of the quantity $x'Vx$. We verified results from the Stata program *nnmatch* developed by Abadie et al. We found that the vector norm used by *nnmatch* did not take the square root. Therefore, in our presentation of the vector norm, we removed the square root and defined the vector norm as $\|x\|v = x'Vx$. This modification did not change the nature of minimum distance because, if $\sqrt{a_1}$ is a minimum value among $\sqrt{a_1}, \sqrt{a_2}, \ldots, \sqrt{a_n}$, a_1 remains a minimum value among $a_1, a_2, \ldots, a_n$.

7

Propensity Score Analysis With Nonparametric Regression

I n this chapter, we review a final propensity score method—*propensity score analysis with nonparametric regression.* This method was developed by Heckman, Ichimura, and Todd (1997, 1998). The 1997 paper from Heckman et al. shows the application of propensity score analysis with nonparametric regression to evaluations of job-training programs, whereas the 1998 paper presents a rigorous distribution theory for the method. A central feature of this method is the application of nonparametric regression (i.e., local linear regression with a tricube kernel, also known as *lowess*) to smooth unknown and possibly complicated functions. The method allows estimation of treatment effects for the treated by using information from all possible controls within a predetermined span. Because of this feature, the method is sometimes called *kernel-based matching* (Heckman et al., 1998).[1] The model is sometimes called a *difference-in-differences* approach (Heckman et al., 1997), when it is applied to two-time-point data (i.e., analyzing pre- and posttreatment data) to show change triggered by an intervention in a dynamic fashion. The three terms— propensity score analysis with nonparametric regression, kernel-based matching, and the difference-in-differences method—are used interchangeably in this chapter. Although the asymptotic properties of *lowess* have been established, it is technically complicated to program and calculate standard errors based on these properties. Therefore, to implement the estimator, bootstrapping is used to draw statistical inferences. In general, this method uses propensity scores derived from multiple matches to calculate a weighted mean that is used as a counterfactual. As such, kernel-based matching is a robust estimator.

Section 7.1 provides an overview of propensity score matching with nonparametric regression. Section 7.2 describes the approach by focusing on three topics: the kernel-based matching estimators and their applications to two-time-point data, a heuristic review of *lowess*, and a review of issues pertaining to the asymptotic and finite-sample properties of kernel-based matching. Section 7.3 summarizes key features of two computing programs in Stata (i.e., *psmatch2* and *bootstrap*) that can be used to run all the models described in this chapter. Section 7.4 presents an application with two-point data. Because local linear regression can be applied to postintervention outcomes that do not constitute a difference-in-differences, we also show how to use it in evaluations with one-point data. Section 7.5 concludes the chapter.

7.1 Overview

In contrast to kernel-based matching, most matching algorithms described in previous chapters are 1-to-1 or 1-to-n (where n is a fixed number) matches. That is, 1-to-1 and 1-to-n methods are designed to find one control or a fixed number of controls that best match a treated case on a propensity score or on observed covariates X. In practice, this type of matching is not very efficient because we may find controls that sum to more than n for each treated case within a predetermined caliper. Often, the number of controls close to a treated case varies within a caliper, but information on the relative closeness of controls is ignored. Kernel-based matching constructs matches using all individuals in the potential control sample in such a way that it takes more information from those who are closer matches and downweights more distal observations. By doing so, kernel-based matching uses comparatively more information than other matching algorithms.

Both kernel-based matching and optimal matching use a varying number of matches for each treated case; however, the two methods employ different approaches. Kernel-based matching uses nonparametric regression, whereas optimal matching uses network flow theory from operations research. Optimal matching aims to minimize the total distance and uses differential weights to take information from control cases. But it does so by optimizing a "cost" defined by a network flow system. As described in Section 5.4.2 and Equation 5.6, the total distance an optimal matching algorithm aims to minimize is a weighted average, which is similar to kernel-based matching. However, optimal matching uses three methods to choose a weight function, all of which depend on the proportion of the number of treated cases (or the proportion of the number of controls) in a matched set to the total number of treated cases (or controls), or the proportion of both treated and control cases falling in set s among the sample total. It is the choice of the weighting function that makes kernel-based matching fundamentally different from optimal matching. In choosing a weighting function, kernel-based matching uses *lowess*, a nonparametric method for smoothing unknown and complicated functions.

As discussed in Chapter 2, Heckman sharply contrasted the econometric model of causality to the statistical model of counterfactuals. This is reflected in the development of kernel-based matching. Heckman and his colleagues argued that two assumptions embedded in Rosenbaum and Rubin's (1983) framework for propensity score matching (i.e., the *strongly ignorable treatment assignment* and *overlap assumptions*) were too strong and restrictive. Under these conditions, conceptually different parameters (e.g., the mean effect of treatment on the treated, the mean effect of treatment on the controls, and the mean effect of randomly assigning persons to treatment) become the same (Heckman et al., 1998). Heckman and his colleagues thought that these three effects should be explicitly distinguished, and for the evaluation of narrowly targeted programs such as a job-training program, they argued that the mean effect of treatment on the treated is most important. Kernel-based matching serves this purpose, and it is a method for estimating the average effect of treatment on the treated (ATT). For an elaboration of the ATT perspective, see Heckman (1997) and Heckman and Smith (1998).

To overcome what they saw as the limitations of Rosenbaum and Rubin's framework and to identify the treatment effect on the treated, Heckman and colleagues developed a framework that contained the following key elements. First, instead of assuming strongly ignorable treatment assignment, or $(Y_0, Y_1) \perp W | X$, they imposed a weaker condition assuming $Y_0 \perp W | X$. Under this assumption, only the outcome under the control condition is required to be independent of the treatment assignment, conditional on observed covariates. Second, instead of assuming full independence, the Heckman team imposed mean independence, or $E(Y_0 | W = 1, X) = E(Y_1 | W = 0, X)$. That is, conditional on covariates, only the mean outcome under the control condition for the treated cases is required to be equal to the mean outcome under the treated condition for the controls. Third, their framework included two crucial elements: *separability* and *exclusion restrictions*. Separability divides the variables that determine outcomes into two categories: observables and unobservables. This separation permits the definition of parameters that do not depend on unobservables. Exclusion restrictions isolate different variables that determine outcomes and program participation. Specifically, the exclusion restriction partitions the covariate X into two sets of variables (T, Z), where the T variables determine outcomes, and the Z variables determine program participation.[2] Putting these elements together, Heckman and his colleagues developed a framework suitable for the evaluation of the treatment effect for the treated, or $E(Y_1 - Y_0 | W = 1, X)$, rather the average treatment effect $E(Y_1 | W = 1) - E(Y_0 | W = 0)$. This framework extended Rosenbaum and Rubin's framework to a more general and feasible case by considering $U_0 \perp W | X$, where U_0 is an unobservable determining outcome Y_0.

Under the exclusion restrictions, Heckman and colleagues further assumed that the propensity score based on X (i.e., $P(X)$) equals the propensity score based on program participation (i.e., $P(Z)$), or $P(X) = P(Z)$, which leads to $U_0 \perp W|P(Z)$. By these definitions and assumptions, Heckman et al. argued that their framework no longer implied that the average unobservables under the control condition, conditional on the propensity score of program participation, equaled zero. Similarly, the Heckman group argued that their framework no longer implied that the average unobservables under the treatment condition, conditional on the propensity score of program participation, equaled zero. More formally, there is no need under this framework—they contended—to assume $E(U_0|P(Z)) = 0$ or $E(U_1|P(Z)) = 0$, where U_1 is an unobservable determining outcome Y_1. The only required assumption under the framework is that the distributions of the unobservables are the same in the populations of $W = 1$ and $W = 0$, once data are conditioned on $P(Z)$. Thus, Heckman and his colleagues devised an innovative general framework for the evaluation of program effects, and they addressed constraints in Rosenbaum and Rubin's framework, which rely on stronger and more restrictive assumptions.

With a fixed set of observed covariates X, should an analyst develop propensity score $P(X)$ first using X and then match on $P(X)$, or should an analyst match directly on X? Heckman and colleagues (1998) compared the efficiency of the two estimators (i.e., an estimator that matches on propensity score $P(X)$ and an estimator that matches on X directly), and concluded, "There is no unambiguous answer to this question" (Heckman et al., 1998, p. 264). They found that neither estimator was necessarily more efficient than the other and that neither estimator was practical because both assume that the conditional mean function and $P(X)$ are known values, whereas in most evaluations these values must be estimated. When the treatment effect is constant, as in the conventional econometric evaluation models, they reported an advantage for matching on X rather than on $P(X)$. However, when the outcome Y_1 depends on X only through $P(X)$, there is no advantage to matching on X over matching on $P(X)$. Finally, when it is necessary to estimate $P(X)$ or $E(Y_0|W = 0, X)$, the dimensionality of the X is indeed a major drawback to the practical application of the matching method. Both methods (i.e., matching on X or matching on $P(X)$) are "data-hungry" statistical procedures.

Based on their study, Heckman and his colleagues recognized that matching on $P(X)$ avoids the dimensionality problem by estimating the mean function conditional on a one-dimensional score. The advantage of using the propensity score is simplicity in estimation. Therefore, in the kernel-based matching, Heckman et al. developed a two-stage procedure to first estimate $P(X)$, and then use nonparametric regression to match on $P(X)$.

It is important to note that there exists some skepticism about the weaker assumptions developed by Heckman and his colleagues. The question is whether or not the so-called weaker assumptions are so much weaker that in practice they would be any easier to evaluate than the stronger assumptions associated with Rosenbaum and Rubin's framework. For instance, Imbens (2004) notes that although the mean-independence assumption

> is unquestionably weaker, in practice it is rare that a convincing case is made for the weaker assumption without the case being equally strong for the stronger version. The reason is that the weaker assumption is intrinsically tied to functional-form assumptions, and as a result one cannot identify average effects on transformations of the original outcome (such as logarithms) without the stronger assumption. (p. 8)

7.2 Methods of Propensity Score Analysis With Nonparametric Regression

We will start the discussion of the kernel-based matching by highlighting a few key features of the method. Specifically, we note how kernel-based matching constructs a weighted average of counterfactuals for each treated case, and then how this method calculates the sample average treatment effect for all treated cases. In this case, we assume that the weighting procedure is given; that is, we ignore the details of nonparametric regression. Indeed, applying nonparametric regression to propensity score analysis was a creative approach that was a significant contribution made by Heckman and his colleagues. However, the development of nonparametric regression stems from empirical problems in smoothing complicated mathematical functions, and these have general objectives beyond the field of observational studies. To more fully explain the logic of nonparametric regression, we provide a heuristic review of *lowess*. Specifically, we focus on the main features of local linear regression and the determination of weights by using various functions. We conclude this section by summarizing the main findings of studies that have examined the asymptotic and finite-sample properties of *lowess*.

7.2.1 THE KERNEL-BASED MATCHING ESTIMATORS

The kernel and local linear matching algorithms were developed from the nonparametric regression method for curve smoothing (Heckman et al., 1997, 1998; Smith & Todd, 2005). These approaches enable the user to perform *one-to-many matching* by calculating the weighted average of the outcome variable for all nontreated cases, and then comparing that weighted average with the outcome of the treated case. The difference between the two terms yields an estimate of the treatment effect for the treated. A sample average for all treated cases (denoted as ATT in the equation below) is an estimation of the sample average treatment effect for the treated group. Hence, the one-to-many matching method combines matching and analysis (i.e., comparison of mean difference on outcome measure) into one procedure.

Denote I_0 and I_1 as the set of indices for controls and program participants, respectively; and Y_0 and Y_1 as the outcomes of control cases and treated cases, respectively. To estimate a treatment effect for each treated case $i \in I_1$, outcome Y_{1i} is compared with an average of the outcomes Y_{0j} for matched case $j \in I_0$ in the untreated sample. Matches are constructed on the basis of propensity scores $P(X)$ that are estimated using the logistic regression on covariates X. Precisely, when the estimated propensity score of an untreated control is closer to the treated case $i \in I_1$, the untreated case gets a higher weight when constructing the weighted average of the outcome. Denoting the average treatment effect for the treated as ATT, the method can be expressed by the following equation:

$$\text{ATT} = \frac{1}{n_1} \sum_{i \in I_1 \cap S_p} \left[Y_{1i} - \sum_{j \in I_0 \cap S_p} W(i,j) Y_{0j} \right], \tag{7.1}$$

where n_1 is the number of treated cases, and the term $\sum_{j \in I_0 \cap S_p} W(i,j) Y_{0j}$ measures the weighted average of the outcome for all nontreated cases that match to participant i on the propensity score differentially. It is worth noting that in Equation 7.1 $\sum_{j \in I_0 \cap S_p} W(i,j) Y_{0j}$ sums over all controls $j \in I_0 \cap S_p$. This feature is a crucial element of kernel-based matching because it implies that each treated case matches on all controls falling into the common-support region rather than 1-to-1 or 1-to-n. Furthermore, this element implies that the estimator forms a weighted average by weighting the propensity scores differentially or using different weights of $W(i,j)$. In Equation (7.1), $W(i,j)$ is the weight or distance on propensity score between i and j. We explain how to determine $W(i,j)$ in the next subsection.

Heckman et al. (1997, 1998) used a difference-in-differences method, which is a special version of the estimated ATT effect. In this situation, Heckman and his colleagues used longitudinal data (i.e., the outcomes before and after intervention) to calculate the differences between the outcome of the treated cases and the weighted average differences in outcomes for the non-treated cases. Replacing Y_{1i} with $(Y_{1ti} - Y_{1t'i})$, and Y_{0j} with $(Y_{0tj} - Y_{0t'j})$, where t denotes a time point after treatment and t' a time point before treatment, we obtain the difference-in-differences estimator:

$$\text{DID} = \frac{1}{n_1} \sum_{i \in I_1 \cap S_p} \left\{ (Y_{1ti} - Y_{0t'i}) - \sum_{j \in I_0 \cap S_p} W(i,j)(Y_{0tj} - Y_{0t'j}) \right\}. \tag{7.2}$$

Note that each treated participant has a difference $(Y_{1ti} - Y_{0t'i})$ and his or her multiple matches have average differences $\Sigma_{i \in I_1 \cap Sp} W(i,j)(Y_{0tj} - Y_{0t'j})$. The difference between the two values yields the *difference-in-differences* that measures the average change in outcome that is the result of treatment for a treated case $i \in I_1$. Taking the average over all treated cases of the sample, that is, taking the average of $(1/n_1) \Sigma_{i \in I_1 \cap Sp} \{\bullet\}$, the analyst obtains the difference-in-differences estimate of sample treatment effects for the treated cases.

In Chapter 5, we discussed the common-support-region problem that is frequently encountered in propensity score matching. As illustrated in Figure 5.2, propensity score matching typically excludes participants from the study when they have no matches. Cases cannot be matched because treated cases fall outside the lower end of the common-support region (i.e., cases with low logit) and non-treated cases fall outside the upper end of the common-support region (i.e., cases with high logit). Even for matched cases, the potential for matches at the two ends of the region may be sparse, which means the estimation of treatment effects for the treated is not efficient. To deal with this problem, Heckman et al. (1997) recommended a trimming strategy: The analyst should discard the non-parametric regression results in regions where the propensity scores for the non-treated cases are sparse, and typically use different trimming specifications to discard 2%, 5%, or 10% of study participants at the two ends.[3] Researchers may treat these trimming specifications as sensitivity analyses. Under different trimming specifications, findings indicate the sensitivity of treatment effects for the treated to the distributional properties of propensity scores.

7.2.2 REVIEW OF THE BASIC CONCEPTS OF LOCAL LINEAR REGRESSION (*lowess*)

Nonparametric regression methods are used to determine $W(i, j)$ of Equations 7.1 and 7.2. In this subsection, we describe the general ideas of the kernel estimator and the local linear regression estimator that use a *tricube* kernel function, where the second estimator is known as *lowess*. Readers are referred to Hardle (1990) as a comprehensive reference that describes the general nonparametric regression approach. In addition, Fox (2000) is a helpful reference for kernel-based matching. This subsection is primarily based on Fox (2000), and we include an example originally developed by Fox to illustrate the basic concepts of nonparametric regression.[4]

As previously mentioned, nonparametric regression is a curve-smoothing approach (often called *scatterplot smoothing*), but what is curve smoothing? In typical applications, the nonparametric regression method passes a smooth curve through the points in a scatterplot of y against x. Consider the scatterplot in Figure 7.1 that shows the relationship between the life expectancy of females (the y variable) and gross domestic product (GDP) per capita (the x variable) for 190 countries (J. Fox, personal communication via e-mail, October 1,

2004). The dashed line was produced by a linear least square regression. The linear regression line is also known as a parametric regression line because it was produced by the parametric regression $\hat{y} = \hat{b}_0 + \hat{b}_1 x$, where $\hat{b}_0$ and $\hat{b}_1$ are the two parameters of interest.

In Figure 7.1, the linear regression line does not very accurately reflect the relationship between life expectancy (equivalently the mortality rate) and economic development. To improve the fit of the line, a researcher could analyze the relationship between y and x by seeking a mathematical function, and then using that function to draw a smooth curve that better conforms to the data. Unfortunately, the relationship between these two variables may be too complicated to be developed analytically. Now look at the solid line in Figure 7.1. Although it is imperfect, it is a more accurate representation of the relationship between life expectancy and economic development. This line is produced by nonparametric regression. We now turn to the central question "How do we draw a smooth curve using nonparametric regression?"

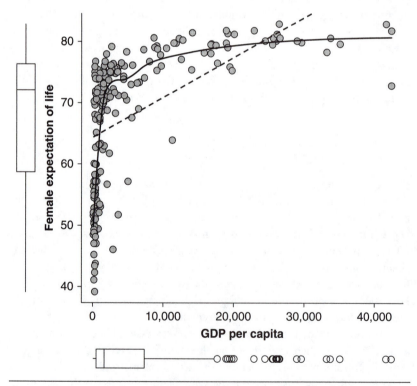

Figure 7.1   Illustration of the Need for a Better Curve Smoothing Using Nonparametric Regression

Before we can draw a smooth curve using nonparametric regression, we first need to do local averaging. Local averaging means that for any focal point x_0, the magnitude of its y value is a local weighted average determined by all y values neighboring x_0. Consider Figure 7.2, where we defined the 120th-ordered x value as our focal point or x_{120}. (That is, you sort the data by x in an ascending order first and then choose any one value of x as your focal point. By doing so, the neighboring observations close to the focal point, x_{120} in the current instance, are neighbors in terms of the x values.) The chosen country x_{120} in this example is the Caribbean nation of Saint Lucia, whose observed GDP per capita was \$3,183 and life expectancy for females was 74.8 years. We then define a window, called a span, that contains $0.5N = 95$ observations of the data using the focal point as the center. The span is bounded in the figure by the dashed lines. As the scatter shows, within the span, some countries had female life expectancies greater than that of Saint Lucia (i.e., > 74.8), whereas some countries had female life expectancies less than that of Saint Lucia (i.e., < 74.8). Intuitively, the local average of the y value for x_{120}, denoted as $\hat{f}(x_{120})$, should be close to 74.8, but it should not be exactly 74.8. We take the y values of all neighboring points in the span into consideration, so that the smoothed curve represents the relationship between the x and y variables. Thus, the local average is a weighted mean of all y values falling into the span such that it gives greater weight to the focal point x_{120} and its closest neighbors and less weight to distant points when constructing the weighted mean. The method of constructing the weighted mean for a focal point (i.e., $\hat{f}(x_{120})$) using various kernel functions is called the *kernel estimator*.

Let $z_i = (x_i - x_0)/h$ denote the scaled, signed distance between the x value for the ith observation and the focal x_0, where the scale factor h is determined by the kernel function. The fraction that is used to determine the number of observations that fall into a span is called *bandwidth*. In our example, we defined a span containing 50% of the total observations centering on the focal point; thus, the value of bandwidth is 0.5. The kernel function, denoted as $G(z_i)$, is a function of z_i and is the actual weight to form the fitted value of $\hat{f}(x_0)$. Having calculated all $G(z_i)$s within a bandwidth for a focal point, the researcher can obtain a fitted value at x_0 by computing a weighted local average of the ys as

$$\hat{f}(x_0) = \hat{y}|x_0 = \frac{\sum\limits_{i=1}^{n} G(z_i) y_i}{\sum\limits_{i=1}^{n} G(z_i)}. \tag{7.3}$$

Several methods have been developed to estimate a kernel function. Common kernel functions include the following:

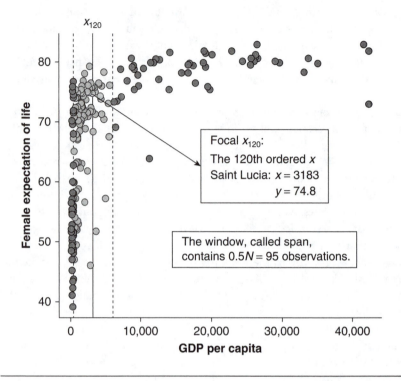

Figure 7.2   The Task: Determining the y Value for a Focal Point x_{120}

1. The *tricube kernel*:

$$G_T(z_i) = \begin{cases} (1 - |z_i|^3)^3 & \text{for } |z_i| < 1. \\ 0 & \text{for } |z_i| \geq 1. \end{cases} \qquad (7.4)$$

For this kernel function, h is the number of observations falling into a span centered at the focal x_0 when calculating $z_i = (x_i - x_0)/h$.

2. The *normal kernel* (also known as the *Gaussian kernel*) is simply the standard normal density function:

$$G_N(z_i) = \frac{1}{\sqrt{2\pi}} e^{-z_i^2/2}. \qquad (7.5)$$

For this kernel function, h is the standard deviation of a normal distribution centered at x_0 when calculating $z_i = (x_i - x_0)/h$.

Other kernel functions include (a) the *rectangular kernel* (also known as the *uniform kernel*), which gives equal weight to each observation in a span centered at x_0, and therefore, produces an unweighted local average, and (b) the *Epanechnikov kernel,* which has a parabolic shape with support $[-1, 1]$ and the kernel is not differentiable at $z = \pm 1$.

The tricube kernel is a common choice for the kernel-based matching. As shown in Figure 7.3, the weights determined by a tricube kernel function follow a normal distribution within the span.

Below, we use the data from the example to show the calculation of the weighted mean $\hat{f}(x_{120})$ for the focal point x_{120}. Figure 7.4 shows results of the calculation. In our calculation, the focal point x_{120} is Saint Lucia, whose GDP per capita is \$3,183 (i.e., $x_{120} = 3,183$). For South Africa (the nearest neighbor country below Saint Lucia on the data sheet in Figure 7.4), the z and $G_T(z)$ can be determined as follows:

$$z = \frac{x_i - x_0}{h} = \frac{3230 - 3183}{95} = .4947;$$

because $|.4947| < 1$, we obtain $G_T(z)$ by taking a tricube function using Equation 7.4: $G_T(z) = (1 - |z|^3)^3 = (1 - |.4947|^3)^3 = .68$. Taking another

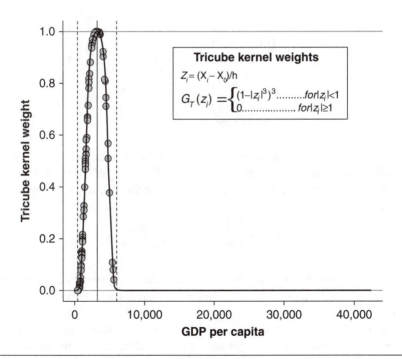

Figure 7.3   Weights Within the Span Can Be Determined by the Tricube Kernel Function

neighboring country (see Figure 7.4), Slovakia, as an example, the z and $G_T(z)$ can be determined as

$$z = \frac{x_i - x_0}{h} = \frac{3266 - 3183}{95} = .8737;$$

because $|.8737| < 1$, we obtain $G_T(z)$ by taking a tricube function using Equation 7.4: $G_T(z) = (1 - |z|^3)^3 = (1 - |.8737|^3)^3 = .04$. Taking Venezuela as an example, the z and $G_T(z)$ can be determined as

$$z = \frac{x_i - x_0}{h} = \frac{3496 - 3183}{95} = 3.2947;$$

because $|3.2947| > 1$, we obtain $G_T(z) = 0$ using Equation 7.4. Note that with respect to the closeness of x_i to x_0, Venezuela is viewed as a distal country relative to Saint Lucia, because Venezuela's $|z|$ is greater than 1. As such, Venezuela's weight is $G_T(z) = 0$, which means that Venezuela contributes nothing to the calculation of the weighted mean $\hat{f}(x_{120})$. Using Equation 7.3, we obtain the weighted mean for the focal point x_{120} as

$$\hat{f}(x_{120}) = \hat{y}|x_{120} = \frac{\sum\limits_{i=1}^{n} G(z_i) y_i}{\sum\limits_{i=1}^{n} G(z_i)}$$

$$= \frac{(0 \times 75.7) + (.23 \times 71.1) + (1 \times 74.8) + (.68 \times 68.3) + (.04 \times 75.8) + (0 \times 75.7)}{0 + .23 + 1 + .68 + .04 + 0} = 72.183.$$

In the illustration, some 94 countries fell into the span, but we used only five proximally neighboring countries to calculate $\hat{f}(x_{120})$. Indeed, of the five, only three qualify as having an absolute value of z that is less than 1 under

$$G_T(z_i) = \begin{cases} (1 - |z_i|^3)^3 & \text{for } |z_i| < 1 \\ 0 & \text{for } |z_i| \geq 1. \end{cases}$$

All other countries within the span would have a zero weight $G_T(z)$ like that for Poland and Venezuela. They would contribute nothing to the calculation of $\hat{f}(x_{120})$.

The above calculation is for one focal country, Saint Lucia. We now replicate the above procedure for the remaining 189 countries to obtain 190 weighted means. We obtain a smooth curve by connecting all 190 weighted means, such as that shown in Figure 7.5. The procedure to produce this smooth curve is called *kernel smoothing*, and the method is known as a *kernel estimator*.

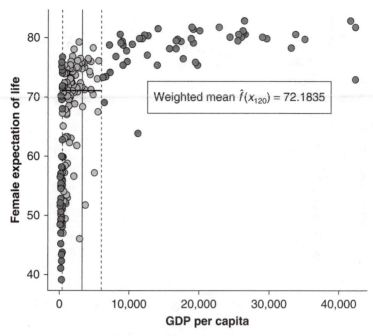

Calculation of the Weighted Mean $\hat{f}(x_{120})$ for x_{120}:

Country	Life Exp.	GDP	$|z|$	$G_T(z)$
Poland	75.7	3058	1.3158	0
Lebanon	71.7	3114	0.7263	0.23
Saint Lucia	74.8	3183	0	1.00
South Africa	68.3	3230	0.4947	0.68
Slovakia	75.8	3266	0.8737	0.04
Venezuela	75.7	3496	3.2947	0

Figure 7.4   The y Value at the Focal Point x_{120} Is a Weighted Mean

In contrast to the kernel estimator, *local linear regression* (also called *local polynomial regression* or *lowess*) uses a more sophisticated method to calculate the fitted y values. Instead of constructing a weighted average, local linear regression aims to construct a smooth local linear regression with estimated β_0 and β_1 that minimizes

$$\sum_{1}^{n} [Y_i - \beta_0 - \beta_1(x_i - x_0)]^2 G\left(\frac{x_i - x_0}{h}\right),$$

where $G((x_i - x_0)/h)$ is a tricube kernel. Note that

$$G\left(\frac{x_i - x_0}{h}\right) = G_T(z_i),$$

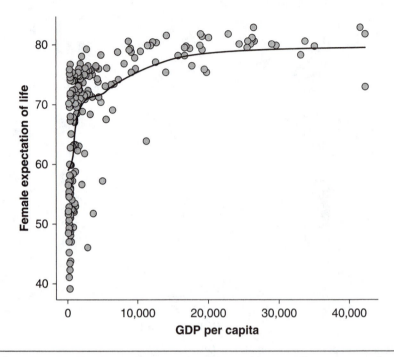

Figure 7.5   The Nonparametric Regression Line Connects All 190 Average Values

which can be determined by the same kernel estimator previously described. With a local linear regression, the fitted y value or $\hat{y}(x_0)$ is a predicted value falling onto a regression line, where the regression line is typically not parallel to the x-axis. Figure 7.6 shows how *lowess* predicts a fitted y value locally. Connecting all 190 fitted y values produced by the local linear regression, the researcher obtains a smoothed curve of *lowess* that should look similar to Figure 7.5.

As described previously, bandwidth is the fraction that is used to define the span (in the above example it was equal to 0.5). Therefore, bandwidth determines the value of h, which is the number of observations falling into the span (in the above example $h = 95$). The choice of bandwidth affects the level of smoothness of the fitted curve, and it is an important specification that affects the results of kernel-based matching. We reexamine this issue in our discussion of the finite-sample properties of *lowess*.

We have reviewed two methods of nonparametric regression: the kernel estimator and local linear regression. The primary purpose for this review is to describe the determination of $W(i, j)$ that is used in Equations 7.1 and 7.2. Now, thinking of the x-axis of a scatterplot as a propensity score, and the y-axis as an outcome variable on which we want to estimate average treatment effect for the treated, $W(i, j)$ becomes the weight derived from the distance of propensity score between a treated case $i \in I_1$ and each nontreated case $j \in I_0$. For each treated case i, there are n_0 weights or $n_0[W(i, j)]$s, where n_0 stands for the number of

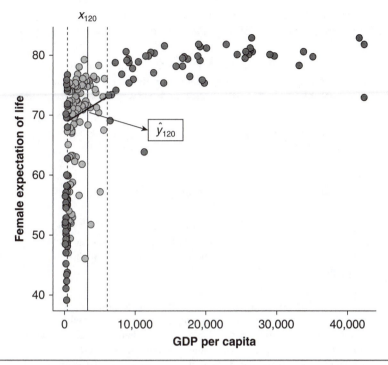

Figure 7.6   The Local Average Now Is Predicted by a Regression Line, Instead of a Line Parallel to the x-Axis

nontreated cases. This condition exists because each treated case i would have distances on propensity scores from all nontreated cases. By using either the kernel estimator or *lowess*, kernel-based matching calculates $W(i, j)$s for each i in such a way that it gives a large value of $W(i, j)$ to a j that has a short distance (is more proximal) on the propensity score from i, and a small value of $W(i, j)$ to a j that has a greater distance (is more distal) on the propensity score from i. Precisely, kernel matching employs the kernel estimator to determine $W(i, j)$ by using the following equation:

$$W(i, j) = \frac{G\left(\frac{P_j - P_i}{h}\right)}{\sum\limits_{k \in I_0} G\left(\frac{P_k - P_i}{h}\right)}, \tag{7.6}$$

where $G(\cdot)$ is a kernel function, h is the number of observations falling into the bandwidth, P_i, P_j, and P_k are estimated propensity scores, and P_i is a focal point within the bandwidth. At present, the tricube kernel is the most common and is the recommended choice for $G(\cdot)$. Unlike the notation we used for the previous review, P_i is the focal point or a propensity score for a treated case for which we want to establish the weighted average of counterfactuals, and P_j and P_k are the propensity scores of the jth and kth nontreated cases falling into the span, that is, $j \in I_0$ and $k \in I_0$.

Local linear matching employs local linear regression or *lowess* with a tricube function to determine $W(i, j)$ by using the following equation:

$$
W(i,j) = \frac{G_{ij} \sum\limits_{k \in I_0} G_{ik}(P_k - P_i)^2 - \left[G_{ij}(P_j - P_i)\right]\left[\sum\limits_{k \in I_0} G_{ik}(P_k - P_i)\right]}{\sum\limits_{j \in I_0} G_{ij} \sum\limits_{k \in I_0} G_{ij}(P_k - P_i)^2 - \left(\sum\limits_{k \in I_0} G_{ik}(P_k - P_i)\right)^2}, \quad (7.7)
$$

where $G(\cdot)$ is a tricube kernel function and $G_{ij} = ((P_j - P_i)/h)$. In evaluations, local linear matching based on Equation 7.7 appears somewhat more common than kernel matching based on Equation 7.6. In choosing one or the other, Smith and Todd (2005) advise:

> Kernel matching can be thought of as a weighted regression of Y_{0j} on an intercept with weights given by the kernel weights, $W(i, j)$, that vary with the point of evaluation. The weights depend on the distance between each comparison group observation and the participant observation for which the counterfactual is being constructed. The estimated intercept provides the estimate of the counterfactual mean. Local linear matching differs from kernel matching in that it includes in addition to the intercept a linear term in P_i. Inclusion of the linear term is helpful whenever comparison group observations are distributed asymmetrically around the participant observations, as would be the case at a boundary point of P or at any point where there are gaps in the distribution of P. (pp. 316–317)

7.2.3 ASYMPTOTIC AND FINITE-SAMPLE PROPERTIES OF KERNEL AND LOCAL LINEAR MATCHING

Several studies have examined the asymptotic properties of kernel and local linear matching methods (Fan, 1992, 1993; Hahn, 1998; Heckman et al., 1998). Between the two methods, local linear regression appears to have more promising sampling properties and a higher minimax efficiency (Fan, 1993). This may explain, in part, the predominance of local linear matching in practice applications.

Heckman et al. (1998) presented an asymptotic distribution theory for kernel-based matching estimators. Beyond the scope of this book, this theory involves proofs of the asymptotic properties of the kernel-based estimators. Heckman et al. (1998) argued that the proofs justify the use of estimated propensity scores (i.e., conducting kernel-based matching using estimated propensity score $P(X)$ rather than matching on X directly) under general conditions about the distribution of X.

Notwithstanding, in practice the implications of the asymptotic properties of kernel-based matching remain largely unknown. When applying nonparametric regression analysis of propensity scores to finite samples, especially small samples, the extent to which asymptotic properties apply or make sense in practice is far from clear. Under such a context, researchers should probably exercise caution in making statistical inferences, particularly when sample sizes are small.

Only recently has an investigation examined the finite-sample properties for the kernel and local linear matching. From a study by Frölich (2004), two implications are perhaps especially noteworthy: (1) it is important to seek the best bandwidth value through cross-validation of the nonparametric regression estimator and (2) trimming (i.e., discarding nonparametric regression results in regions where the propensity scores for the nontreated cases are sparse) seems not to be the best response to the variance problems associated with local linear matching.

Based on these implications, we tested various specifications of bandwidth and trimming strategies for simulated data and empirical data. We found that for empirical applications, methods need to be developed to seek the best bandwidth and to handle the variance problem induced by the common-support region. It appears important to test various bandwidth values and trimming schedules. As we mentioned earlier, it is prudent at this point to treat different specifications of bandwidth values and trimming schedules as sensitivity analyses and be cautious about final results if the estimated treatment effects for the treated vary by these specifications (i.e., when the study findings are sensitive to bandwidth values and trimming schedules).

The results of kernel and local linear estimations involve weighted average outcomes of the nontreated cases. Because the asymptotic distribution of weighted averages is relatively complicated to program, no software packages are currently available that offer parametric tests to discern whether a group difference is statistically significant. As a common practice, researchers use bootstrapping to estimate the standard error of the sample mean difference between treated and nontreated cases. Recently, however, Abadie et al. (2004) and Abadie and Imbens (2006) have warned that bootstrapping methods for estimating the variance of matching estimators do not necessarily give correct results. Thus, in practice, conducting a significance test of a treatment effect for the treated derived from kernel-based matching may be problematic, and researchers should be cautious when interpreting findings.

7.3 Overview of the Stata Programs *psmatch2* and *bootstrap*

To implement propensity score analysis with nonparametric regression, users can run a user-developed program *psmatch2* (Leuven & Sianesi, 2003) and the regular program ***bootstrap*** in Stata. Section 5.8 of this book provides the information for downloading and installing *psmatch2*.

The basic syntax to run ***psmatch2*** for estimating the treatment effect for the treated with a kernel matching is as follows:

> *psmatch2 depvar, kernel outcome(varlist) kerneltype(kernel_type) ///*
>
> *pscore(varname) bwidth(real) common trim(integer)*

In this command, *depvar* is the variable indicating treatment status, with *depvar* = 1 for the treated and *depvar* = 0 for the control observations. The keyword *kernel* is used to request a kernel matching. The term *outcome(varlist)* specifies on which outcome variable(s) users want to assess the treatment effect for the treated; users specify the name of the outcome variable(s) in the parentheses. To run difference-in-differences or two-time-point data, users may create a change-score variable (i.e., a difference in the outcome variable between time 2 and time 1) before running *psmatch2*, and then specify the change-score variable in the parentheses. The term *kerneltype(kernel_type)* specifies the type of kernel, and one of the following five values may be specified in the parentheses as a kernel type: *epan*— the Epanechnikov kernel, which is the default with the kernel matching; *tricube*— the tricube kernel, which is the default with the local linear regression matching; *normal*—the normal (Gaussian) kernel; *uniform*—the rectangular kernel; and *biweight*—the biweight kernel. The term *pscore(varname)* specifies the variable to be used as propensity score; typically, this is the saved variable of propensity score users created by using ***logistic*** before running *psmatch2*. The term *bwidth(real)* specifies a real number indicating the bandwidth for kernel matching or local linear regression matching; the default *bwidth* value is 0.06 for the kernel matching. The term *common* imposes a common-support region that causes the program to drop treatment observations with a propensity score higher than a maximum or less than a minimum propensity score of the controls. The term *trim(integer)* causes the program to drop "*integer* %" of the treatment observations at which the propensity score density of the control observations is the lowest. That is, suppose the user attempts to trim 5% of the treated cases; then the specification is *trim(5)*.

The basic syntax to run ***psmatch2*** for estimating a treatment effect for the treated with a local linear regression matching looks similar to the above syntax specifying kernel matching. The only change the user must make is to replace the keyword *kernel* with the keyword *llr*. Thus, the syntax for the local linear regression matching is as follows:

> *psmatch2 depvar, llr outcome(varlist) kerneltype(kernel_type) ///*
>
> *pscore(varname) bwidth(real) common trim(integer)*

Note that for local linear regression matching or *llr*, the default kernel type is *tricube*.

The Stata program **bootstrap** can be used to run bootstrap sampling to obtain an estimation of the standard error and a 95% confidence interval for the estimated average treatment effect for the treated. The **bootstrap** program can also be invoked by an abbreviation **bs**. The only required syntax of **bootstrap** consists of the following two elements: the previous command, which runs **psmatch2** for the kernel-based matching, and *r(att)*, which indicates that we want to do a bootstrap sampling on the estimated average treatment effect for the treated *att*. Each of the above two elements should be enclosed in double quotes. The following syntax shows the running of **psmatch2** to obtain an estimated treatment effect for the treated using local linear regression matching, and then the running of **bs** to obtain the bootstrap estimation of standard error:

> *psmatch2 aodserv, outcome(extern) pscore(logit) llr*
>
> *bs "psmatch2 aodserv, outcome(extern) pscore(logit) llr" "r(att)"*

Similar to running **psmatch2** for nearest neighbor matching or Mahalanobis matching (see Section 5.8), it is important to create a random variable and then sort data by this variable before invoking **psmatch2**. To guarantee that the same results are obtained from session to session, users control for seed number by using a *set seed* command.

Table 7.1 exhibits the syntax and output for a typical analysis estimating a treatment effect for the treated with a local linear regression matching. The analysis consists of the following steps: (a) we first run a logistic regression to obtain the predicted probabilities for all observations by using **logistic**; (b) we then create a logit score and define the logit as the propensity score; we also create a difference score that is a difference of the outcome variable between two time points—the difference score will be specified as the outcome variable in the subsequent analysis, and by doing so we are conducting a difference-in-differences analysis; we sort the sample data in a random order and set up a seed number to ensure that we will obtain the same results from session to session; (c) we run **psmatch2** using the keyword *llr* to request a local linear regression matching; note that without specifying *kerneltype* and *bwidth*, we use the default kernel of "tricube kernel" and a default bandwidth value of 0.8; we then run **bs** to obtain bootstrap estimation of the standard error and a 95% confidence interval; and (d) we run a similar analysis using a different bandwidth by specifying *bw(.01)*, and a similar analysis to trim 5% of treated cases by specifying *trim(5)*.

Table 7.1   Exhibit of Stata *psmatch2* and *bs* Syntax and Output Running Matching
With Nonparametric Regression

```
// Illustration of syntax for Chapter 7
cd "D:\sage\ch7"
clear
use ch7data1.dta, replace

//logistic regression to obtain predicted probabilities
logistic aodserv married high bahigh poverty2 poverty3 ///
         poverty4 poverty5 employ open black hispanic natam chdage2 ///
         chdage3 cgrage1 cgrage2 cgrage3 cra47a mental arrest psh17a ///
         sexual provide supervis other ra cidi cgneed

(Output)
Logistic regression                          Number of obs   =       1407
                                             LR chi2(28)     =     304.19
                                             Prob > chi2     =     0.0000
Log likelihood = -238.76283                  Pseudo R2       =     0.3891
```

aodserv	Odds Ratio	Std. Err.	z	P>|z|	[95% Conf. Interval]	
married	.8722329	.2539633	-0.47	0.639	.4929385	1.543377
high	.6797779	.1984999	-1.32	0.186	.3835396	1.204825
bahigh	1.025153	.374521	0.07	0.946	.5009768	2.097779
poverty2	1.040178	.3430974	0.12	0.905	.5449329	1.98551
poverty3	.6640589	.3026247	-0.90	0.369	.27183	1.622243
poverty4	1.662694	.7816965	1.08	0.279	.6616523	4.178255
poverty5	1.843642	.8439033	1.34	0.181	.7517098	4.521714
employ	.7969929	.2160853	-0.84	0.403	.4684592	1.35593
open	2.25983	.6765611	2.72	0.006	1.256719	4.063623
black	1.141407	.3522907	0.43	0.668	.6233334	2.090068
hispanic	.9598291	.4224611	-0.09	0.926	.4050833	2.274277
natam	3.76478	1.609303	3.10	0.002	1.628856	8.701548
chdage2	.8200867	.3220923	-0.51	0.614	.379793	1.770813
chdage3	1.098924	.3141514	0.33	0.741	.627528	1.92443
cgrage1	.2387381	.1755824	-1.95	0.051	.0564803	1.009129
cgrage2	.2919873	.2154312	-1.67	0.095	.0687605	1.239907
cgrage3	.3335309	.272129	-1.35	0.178	.0673966	1.650571
cra47a	1.000749	.2691986	0.00	0.998	.5906824	1.695493
mental	1.363044	.3589353	1.18	0.240	.8135041	2.283809
arrest	2.164873	.5697876	2.93	0.003	1.292406	3.626319
psh17a	2.874274	.8494826	3.57	0.000	1.610491	5.129772
sexual	.6673541	.3169791	-0.85	0.395	.2630593	1.693008
provide	1.068848	.441482	0.16	0.872	.4756984	2.401598
supervis	1.0883	.4047152	0.23	0.820	.5250555	2.255756
other	1.208436	.5186203	0.44	0.659	.5210947	2.802405
ra	10.44879	2.796397	8.77	0.000	6.183865	17.65517
cidi	2.924023	.763034	4.11	0.000	1.753302	4.876461
cgneed	2.574197	1.040365	2.34	0.019	1.16581	5.684025

```
//syntax to create logit and difference scores
predict p
g logit=log((1-p)/p) /* create logit using predicted probability */
g extern=bc3_ept-pbc_ept /* create difference score */
g x=uniform() /* create a random variable to sort sample data */
sort x
set seed 1000 /* use constant seed number to ensure same results */

//run psmatch2 and bootstrap
psmatch2 aodserv, outcome(extern) pscore(logit) llr
 /*use defualt bandwidth.*/

(Output)
There are observations with identical propensity score values.
The sort order of the data could affect your results.
Make sure that the sort order is random before calling psmatch2.
```

Variable	Sample	Treated	Controls	Difference	S.E.	T-stat
extern Unmatched	.151785714	-1.81621622	1.96800193	.990888957	1.99
ATT	.151785714	-3.21498278	3.36676849	.	.

psmatch2: Treatment assignment	psmatch2: Common support On suppor	Total
Untreated	1,295	1,295
Treated	112	112
Total	1,407	1,407

```
bs "psmatch2 aodserv, outcome(extern) pscore(logit) llr" "r(att)"
 /* run bootstrap */

(Output)
command:       psmatch2 aodserv , outcome(extern) pscore(logit) llr
statistic:     _bs_1     = r(att)
```

```
Bootstrap statistics                          Number of obs   =    1407
                                              Replications    =      50
```

Variable	Reps	Observed	Bias	Std. Err.	[95% Conf. Interval]	
_bs_1	50	3.366769	-.5101031	1.547192	.2575694 6.475968	(N)
					.2953247 5.525669	(P)
					.6939262 6.8301	(BC)

```
Note:  N   = normal
       P   = percentile
       BC  = bias-corrected
```

(Continued)

Table 7.1 (Continued)

```
//specify a different bandwidth=.01
sort x
set seed 1000
psmatch2 aodserv, outcome(extern) pscore(logit) llr bw(.01)
bs "psmatch2 aodserv, outcome(extern) pscore(logit) llr bw(.01)" ///
    "r(att)"

(Output)
. psmatch2 aodserv, outcome(extern) pscore(logit) llr bw(.01) /* specify
bandwidth=.01 */
There are observations with identical propensity score values.
The sort order of the data could affect your results.
Make sure that the sort order is random before calling psmatch2.
```

	Variable	Sample	Treated	Controls	Difference	S.E.	T-stat
	extern	Unmatched	.151785714	-1.81621622	1.96800193	.990888957	1.99
		ATT	.151785714	-3.81204898	3.9638347	.	.

psmatch2: Treatment assignment	psmatch2: Common support On suppor	Total
Untreated	1,295	1,295
Treated	112	112
Total	1,407	1,407

```
. bs "psmatch2 aodserv, outcome(extern) pscore(logit) llr bw(.01)" "r(att)"

command:      psmatch2 aodserv , outcome(extern) pscore(logit) llr bw(.01)
statistic:    _bs_1     = r(att)

Bootstrap statistics                          Number of obs   =     1407
                                              Replications    =       50
```

Variable	Reps	Observed	Bias	Std. Err.	[95% Conf. Interval]	
_bs_1	50	3.963835	-.434758	1.756217	.4345837	7.493086	(N)
					.4854501	7.276724	(P)
					1.241536	7.76199	(BC)

```
Note:  N  = normal
       P  = percentile
       BC = bias-corrected

// trim 5%
sort x
set seed 1000
psmatch2 aodserv, outcome(extern) pscore(logit) llr trim(5)
bs "psmatch2 aodserv, outcome(extern) pscore(logit) llr trim(5)" ///
    "r(att)"
```

(Output)

```
. psmatch2 aodserv, outcome(extern) pscore(logit) llr trim(5) /* trim 5% */
There are observations with identical propensity score values.
The sort order of the data could affect your results.
Make sure that the sort order is random before calling psmatch2.
```

Variable	Sample	Treated	Controls	Difference	S.E.	T-stat
extern	Unmatched	.151785714	-1.81621622	1.96800193	.990888957	1.
	ATT	.560747664	-2.81583993	3.37658759	.	.

psmatch2: Treatment assignment	psmatch2: Common support Off suppo On suppor		Total
Untreated	0	1,295	1,295
Treated	5	107	112
Total	5	1,402	1,407

```
. bs "psmatch2 aodserv, outcome(extern) pscore(logit) llr trim(5)" "r(att)"

command:     psmatch2 aodserv , outcome(extern) pscore(logit) llr trim(5)
statistic:   _bs_1    = r(att)

Bootstrap statistics                      Number of obs    =       1407
                                          Replications     =         50
```

Variable	Reps	Observed	Bias	Std. Err.	[95% Conf. Interval]	
_bs_1	50	3.376588	-.3977445	1.455069	.4525168	6.300658	(N)
					.4125744	5.858139	(P)
					1.074213	7.27467	(BC)

```
Note:  N  = normal
       P  = percentile
       BC = bias-corrected
```

7.4 Examples

We include two examples in this section to illustrate the application of kernel-based matching in program evaluation. The first example shows the application of local linear regression matching to the analysis of two-time-point data that uses a difference-in-differences estimator. In this example, we also illustrate how to specify different values of bandwidth and how to trim by using a varying schedule of trimming; then, combining these specifications together, we show how to conduct a sensitivity analysis. The second example illustrates the application of local linear regression matching to one-time-point data. We also compare the results of kernel-based matching with those produced by Abadie et al.'s (2004) matching estimators.

7.4.1 ANALYSIS OF DIFFERENCE-IN-DIFFERENCES

Here we use again the sample and matching variables from Section 4.5.1. This example shows the analysis of a subsample of data for 1,407 children obtained from the longitudinal data set of the National Survey of Child and Adolescent Well-Being (NSCAW). Of these study participants, 112 were children of caregivers who had received substance abuse treatment services (i.e., the treatment group), and the remaining 1,295 participants were children of caregivers who did not receive substance abuse treatment services (i.e., the comparison group). The research examined two study questions: At 18 months after their involvement with child protective services, how were the children of caregivers who received substance abuse treatment services faring? Did these children have more severe behavioral problems than their counterparts whose caregivers did not receive substance abuse treatment services?

In Section 4.5.1, we used the Heckit treatment effect model to analyze one-time-point data: The psychological status of children at the 18th month after caregivers became involved with child protective services. In the current analysis, we use two-time-point data: The psychological status of children measured at the NSCAW baseline and the same variables measured at 18 months after the baseline. As such, this analysis permits a longitudinal inquiry of a difference-in-differences, that is, a difference in the psychological status of children given a difference in the participation of caregivers in substance abuse treatment services. In addition to externalizing and internalizing function scores (measures of psychosocial status), the current analysis included an additional outcome variable: The total score of the Achenbach Children's Behavioral Checklist (CBCL/4–18; Achenbach, 1991). Based on caregiver ratings, the study uses three measures of child behavior as outcomes: externalizing scores, internalizing scores, and total scores. A high score on each of these measures indicates more severe behavioral problems. Based on the design of kernel-based matching, the current analysis is a one-to-many matching, and it is a comprehensive investigation of the treatment effect for the treated. In this case, treatment comprises child protective supervision and the participation of caregivers in a drug abuse intervention program. No specific services are provided to remediate child behavioral problems.

The difference-in-differences estimator in the current analysis uses local linear regression to calculate the weighted average difference for the nontreatment group using a tricube kernel and a default bandwidth. Based on the literature on the finite-sample properties of local linear matching, we tested the sensitivity of findings to different specifications on bandwidth and trimming. While holding trimming constant, three bandwidth values were used: 0.01, 0.05, and 0.8. We also tested the sensitivity of findings to variations in the trimming level. We applied the following three trimming schedules (i.e., imposed a common support by dropping treated observations whose propensity scores are higher than the maximum or less than the minimum propensity score of the nontreated observations) while fixing the bandwidth at the default level: 2%, 5%, and 10%. Standard errors for the difference-in-differences estimates were obtained through bootstrapping. The standard error of estimated difference-in-differences was used further to estimate a 95% bootstrap confidence interval for the average treatment effects for the treated. We report the 95% confidence interval using the bias-correction method, so that a meaningful effect is reasonably unlikely to occur by chance as indicated by a 95% confidence interval that does not include a zero.

Table 7.2 shows the estimated average treatment effects for the treated group. Taking the externalizing score as an example, the data indicate that the mean externalizing score for the treatment group increased from baseline to the 18th month by 0.15 units, and the mean score for the nontreatment group decreased from baseline to the 18th month by 1.82 units. The unadjusted mean difference between groups is 1.97, meaning that the average change for the externalizing score for the treatment group is 1.97 units higher (or worse) than that for the nontreatment group. The difference-in-differences estimation further adjusts for the heterogeneity of service participation by taking into consideration the distance on propensity scores between a treated case and its nontreated matches in the calculation of the treatment effects for the treated. The point estimate of the difference-in-differences on externalizing is 3.37, which falls into a 95% bootstrap confidence interval bounded by 0.27 and 5.43. That is, we are 95% confident that a nonzero difference on externalizing between treated and nontreated groups falls into this interval. The next significant difference on the adjusted mean is the total score: The point estimate of the difference-in-differences is 2.76, which falls into a 95% bootstrap confidence interval bounded by 0.96 and 5.12. The 95% bootstrap confidence interval of the difference-in-differences for the internalizing score contains a zero, and therefore we are uncertain whether such difference is statistically significant.

Sensitivity analyses of different bandwidth specifications and different trimming strategies tend to confirm the results. That is, for the externalizing and total scores, all analyses (except the total CBCL score associated with the large bandwidth) show a 95% bootstrap confidence interval bounded by nonzero difference-in-differences estimates. Similarly, for the internalizing score, all analyses show a 95% bootstrap confidence interval that includes a zero difference-in-differences estimate. The study underscores the importance of analyzing change of behavioral problems between service and nonservice groups using a corrective procedure such as propensity score analysis with

Table 7.2   Estimated Average Treatment Effects for the Treated on CBCL
              Change: Difference-in-Differences Estimation by Local Linear
              Regression (Example 7.4.2)

Group and Comparison	Outcome Measures: CBCL Scores		
	Externalizing	Internalizing	Total
Mean difference between 18 months and baseline			
Children whose caregivers received services ($n = 112$)	0.15	−2.09	−0.89
Children whose caregivers did not receive services ($n = 1,295$)	−1.82	−1.44	−1.92
Unadjusted mean difference[a]	1.97*	−0.65	1.03
Adjusted mean difference			
DID[b] point estimate (bias-corrected 95% confidence interval)	3.37*	0.84	2.76*
Sensitivity analyses			
DID point estimate (bias-corrected 95% confidence interval)			
Changing bandwidth			
Small bandwidth = 0.01	3.97*	1.3	3.35*
Small bandwidth = 0.05	3.52*	0.84	2.83*
Large bandwidth = 0.8	2.77*	0.08	2.10
Trimming			
2% (2 cases excluded)	3.31*	0.77	2.82*
5% (5 cases excluded)	3.38*	0.89	2.99*
10% (11 cases excluded)	3.50*	0.58	3.01*

NOTE: a. *t*-tests show that two unadjusted mean differences are not statistically different; b. Difference-in-differences; * The 95% confidence interval does not include a zero, or $p < .05$ for a two-tailed test.

nonparametric regression. The unadjusted mean differences based on the observational data underestimate the differences of behavior problems between the two groups, and the underestimation is $3.37 - 1.97 = 1.4$ units for the externalizing score, and $2.76 - 1.03 = 1.73$ for the total score.

It is worth noting that the analysis used bootstrapping for significance testing—which may be problematic—and thus is a limitation of the study. The results of the study should be interpreted with caution. Conditionally, the kernel-based matching analysis of NSCAW observational data suggest that the combination of protective supervision and substance abuse treatment for caretakers involved in the child welfare system should not alone be expected to produce developmental benefits for children. Additional confirmatory analyses and discussion of these findings are available elsewhere (Barth, Gibbons, & Guo, 2006).

7.4.2 APPLICATION OF KERNEL-BASED MATCHING TO ONE-POINT DATA

In this application, we use the same sample and variables presented in Section 6.4.1. The only difference is that Section 6.4.1 used the matching estimators, whereas the current analysis uses kernel-based matching. We have included the current illustration for two reasons: (1) We want to show that kernel-based matching can also be applied to one-point data for which the treatment effect for the treated is not a difference-in-differences, and (2) we want to compare kernel-based matching with the matching estimators.

With regard to the first purpose, we show in our syntax (available on the companion Web page of this book) that the analysis of one-point data is more straightforward than the analysis of difference-in-differences. That is, with one-point data, you can specify the outcome variable directly in the *psmatch2* statement, whereas with two-point data, you must first create a change-score variable using data from two time points, and then specify the change-score variable as the outcome in the *psmatch2* statement.

Recall from the example presented in Section 6.4.1 that the research objective for this study was to test a hypothesis pertaining to the causal effect of childhood poverty (i.e., children's use of the AFDC welfare program) on children's academic achievement. The study examined one domain of children's academic achievement: the age-normed "passage comprehension" score in 1997 (i.e., one time point) of the Woodcock-Johnson Revised Tests of Achievement (Hofferth et al., 2001). A higher score on this academic achievement measure indicated higher achievement. The study sample consisted of 606 children, of whom 188 had used AFDC at some time in their lives (i.e., ever used group) and who were considered as the treated cases. The remaining 418 children had never received AFDC (i.e., never used group) and were considered the comparison cases.

This study used the following six covariates: (1) current income or poverty status, measured as the ratio of family income to poverty threshold in 1996; (2) caregiver's education in 1997, which was measured as years of schooling; (3) caregiver's history of using welfare, which was measured as the number of years (i.e., a continuous variable) a caregiver participated in the AFDC program during his or her childhood ages of 6 to 12 years old; (4) child's race, which was measured as African American versus non–African American; (5) child's age in 1997; and (6) child's gender, which was measured as male versus female. Note that in Section 6.4.1, these covariates were used as matching variables in a vector-norm approach, but in the current analysis they are used in the estimation of a propensity score for each study child. In other words, the six covariates were included in a logistic regression, and then the propensity score was matched by the local linear regression.

Table 7.3 presents results of the study. Kernel-based matching estimated an average treatment effect for the treated of −4.85, meaning that children who used the AFDC welfare program during childhood score 4.85 units lower on the passage comprehension measure than those who did not use AFDC during childhood, after controlling for observed covariates. This effect was statistically significant at a .05 level. Note that the matching estimator produces a similar finding: The estimated treatment effect for the treated group is −5.23, and the effect is statistically significant at a .05 level. Although the matching estimator's estimate of the effect is slightly larger, both estimators find the effect statistically significant, and thus, lead to a consistent conclusion with regard to testing the research hypothesis. This study implies that (a) at least for this data set, both matching and kernel-matching estimators produce the same substantive findings and appear to be just about equally useful and (b) because the same issue was examined with different statistical methods and the findings converge, the substantive conclusion of the study is more convincing than that produced by either method alone. The application underscores an important point for observational studies: Researchers should compare estimates across multiple methods.

Table 7.3   Estimated Treatment Effect for the Treated (Child's Use of AFDC) on Passage Comprehension Standard Score in 1997: Comparing the Propensity Score Analysis With Nonparametric Regression With Bias-Corrected Matching and Robust Variance Estimator (Example 7.4.2)

Treatment Effect for the Treated	Coefficient	Standard Error	z	p Value	95% CI
Propensity score analysis with nonparametric regression and bootstrap	−4.85	NC	NC	NC	[−8.71, −0.26]
Bias-corrected matching and robust variance estimator (sample estimate or SATT)	−5.23	1.781	−2.94	.003	[−8.72, −1.74]

NOTE: 95% CI = 95% confidence interval; NC = not comparable due to bootstrap.

7.5 Conclusions

This chapter described the application of local linear regression matching with a tricube kernel to the evaluation of average treatment effects for the treated. Kernel-based matching was developed to overcome perceived limitations within the Rosenbaum and Rubin (1983) counterfactual framework. Heckman and his colleagues (1997, 1998) made important contributions to the field: (a) unlike traditional matching, kernel-based matching uses propensity scores differentially to calculate a weighted mean of counterfactuals, which is a creative way to use information from all controls; (b) applying kernel-based matching to two-time-point data, the difference-in-differences estimator permits analysis of treatment effects for the treated in a pre-/posttest trial fashion; and (c) by doing so, the estimator is more robust in terms of handling measurement errors: it eliminates temporarily invariant sources of bias that may arise, when program participants and nonparticipants are geographically mismatched or respond in systematically biased ways to survey questionnaires.

Kernel-based matching was developed also on the basis of a rigorously proven distribution theory. However, as far as we know, the distributional theory developed by Heckman and his colleagues is not incorporated into computing programs for estimating the standard errors of the estimated treatment effect for the treated. The use of bootstraping in statistical inference is a limitation, and results based on bootstrapping should be interpreted with caution.

Notes

1. In this chapter Heckman et al. (1998) refers to Heckman, Ichimura, and Todd (1998), and Heckman et al. (1997) refers to Heckman, Ichimura, and Todd (1997).

2. In practice, T and Z may or may not be comprised of the same variables.

3. Smith and Todd (2005) describe a more sophisticated procedure to determine the proportion of participants to be trimmed. They suggest a method for determining the density cutoff trimming level.

4. We are grateful to John Fox for providing this example and the R code to produce the figures for the example.

8

Selection Bias and Sensitivity Analysis

W e have introduced four methods or models designed to address selection bias, a key issue in observational studies and in many evaluations with control or comparison group designs. In this chapter, we address a series of questions that confront all evaluations in which propensity score analytic methods are used to estimate treatment effects. These questions include the following:

- What are the challenges that researchers face when doing observational studies in which selection bias, particularly hidden selection bias, is present?
- How well do corrective methods handle selection bias and what assumptions are embedded in each method?
- How might a researcher assess the sensitivity of findings to hidden selection bias?

Section 8.1 provides an overview of selection bias. It reviews the sources, types, and consequences of selection effects. In addition, it discusses strategies for correcting selection bias. Section 8.2 is the core of this chapter, and it presents a Monte Carlo study that compares four models under two settings of data generation: selection on the observables and selection on the unobservables. The implications of this study shed light on strategies for correcting selection bias. Section 8.3 describes Rosenbaum's (2002b, 2005) methods for evaluating the sensitivity of study findings. Section 8.4 reviews the Stata program *rbounds*, which can be used to run some of the Rosenbaum models. Section 8.5 presents examples of sensitivity analysis. Section 8.6 concludes the chapter.

8.1 Selection Bias: An Overview

Selection bias is sometimes called *selectivity bias* (Madalla, 1983), *overt* and/or *hidden bias* (Rosenbaum, 2002b), and the *selection problem* (Manski, 2007).

However, the labels *selection bias, selection effects,* and simply *selection* are most common, especially in observational and quasi-experimental studies. The selection problem does not exist in classic randomized experiments, as long as the kind of experiment envisioned by Fisher is successfully implemented. Therefore, a natural starting point for examining selection bias is to consider the conditions under which the classical argument and assumptions for randomized experiments do not hold. In the following passage, Manski (2007) describes three practical circumstances in which randomized experiments are likely to fail:

1. The classical argument supposes that subjects are drawn at random from the population of interest. Yet participation in experiments ordinarily cannot be mandated in democracies. Hence experiments in practice usually draw subjects at random from a pool of persons who volunteer to participate. So one learns about treatment response within the population of volunteers rather than within the population of interest.

2. The classical argument supposes that all participants in the experiment comply with their assigned treatments. In practice, subjects often do not comply.

3. The classical argument supposes that one observes the realized treatments, outcomes, and covariates of all participants in the experiment. In practice, experiments may have missing data. A particularly common problem is missing outcome data when researchers lose contact with participants before their outcomes can be recorded. (pp. 138–139)

The problems outlined by Manski are similar to the critiques of *social experiments* made by Heckman and Smith (1995) (see Section 1.3.3). Thus, as discussed in earlier chapters, selection bias becomes a problem pertaining to observational (including program evaluation) studies, where treatment assignment is not ignorable. The primary issues of selection bias are summarized below.

8.1.1 SOURCES OF SELECTION BIAS

Selection effects are often categorized on the basis of the source of bias, such as self-selection, researcher selection, administrative selection, geographic selection, measurement selection, and attrition selection. Self-selection is the most frequently occurring and widely studied source of bias in observational studies. In an early report, Roy (1951) discussed the problem using the example of individuals choosing between the professions of hunting and fishing based on their productivity in each pursuit. He argued that the two activities attract different kinds of people who self-select into hunting or fishing; and further that fundamental differences in hunters and fishermen, not the activities of hunting and fishing per se, account for any observed income differentials.

Self-selection is often involved in processes related to program recruitment and the assignment of study participants into treatment conditions. Researchers in the field of child mental health and child welfare have identified self-selection as a common problem in program evaluations. For instance, Littell (2001) found that in evaluations of family preservation services (FPS—a kind of family intervention designed to strengthen families and prevent out-of-home placements in child welfare) outcomes varied by the level of service participation, which in turn varied by program participant characteristics. Because FPS participation levels are largely a matter of self-selection—even when treatment is mandated—the outcomes of active participants were likely to differ from those of passive or resistant recipients. Studies of self-selection suggest that it is important to model the heterogeneity of service participation in evaluations.

When collecting and analyzing program data to determine study populations and samples, researchers often make decisions that influence selection. As far as we know, no researcher has ever intentionally introduced selection biases, but it is usually the case that early decisions affect the ignorable treatment assignment and other assumptions. Kennedy (2003) described this source of bias as a selection mechanism that determines which observations enter the study sample and which observations are excluded. It's worth noting that conceptually entering the study sample and entering treatment are different: the former affects external validity, while the latter affects internal validity. Although the two concepts should not be conflated, they both can be sources of selection bias in inferring causation from samples to populations. The classic example of selection bias introduced by researchers was the dramatic failure of political science researchers to predict Harry S. Truman's 1948 presidential victory:

> Surveys were taken via phones, which at that time were more likely to be owned by wealthy people. Wealthy people were also more likely to vote for Dewey. The unmeasured variable [of] wealth affected both the survey answer and the probability of being in the sample, creating the misleading result. (Kennedy, 2003, p. 286)

Today's use of the Internet as a primary means of data collection is analogous to the reliance on telephones for the presidential prediction in 1948. Although Internet surveys are popular and widely used in contemporary social behavioral research, the consequences and implications of this approach for statistical inferences remain unknown. Because access to the Internet is not yet universal, selection may influence the generalization of findings based on Internet surveys.

Guo and Hussey (2004) conducted a study evaluating the dilemmas, consequences, and strategies of using nonprobability sampling in social work research. By definition, nonprobability sampling refers to the selection of a sample using procedures that are not based on a predetermined probability but based instead on research purpose, subject availability, subjective judgment, or a variety of other nonstatistical criteria. In other words, nonprobability sampling is not based on random selection, whereas probability sampling is. Guo and Hussey used a Monte Carlo study to demonstrate the consequences of using nonprobability sampling procedures. They identified a set of research-related decisions that introduce *researcher selection effects* and should prompt caution in the design of studies.

Yet another form of selection bias is introduced when administrators make decisions to recruit participants and assign them to study conditions (i.e., treatment or control) based on *the administrators' evaluation* of participants' qualifications or eligibilities. Although this is an important and frequent source of bias, administrative selection is inadequately studied. Administrative selection occurs, for example, in the selection of students entering into either Catholic schools or public schools. In the school selection process, administrators in Catholic schools use minimum financial and academic qualifications as admission criteria. Comparison of the educational outcomes of students in Catholic versus public schools is confounded by these differences, which constitute a selection effect.

The problem of bias introduced by administrator selection is pronounced in evaluation studies that use nonexperimental designs. In the absence of randomization, evaluators often must compare groups of service recipients who are assigned to services on the basis of need, risk, or other criteria. Under such conditions, service participation will be related to the needs or risk factors of service participants, and service outcomes will be confounded with a selection effect; that is, service participants will differ systematically as a result of administrative decisions in assigning services. The problem of selection bias is pervasive across service systems. In child welfare, for example, Courtney (2000) found that comparisons of permanency outcomes between kinship care and foster care were affected by variation across jurisdictions in the selection process of entry to and exit from both types of care. Administrative rules, regulations, and decisions, he found, have a substantial impact on which children are eligible for kinship versus foster care and, more distally, on outcomes such as safety (i.e., having a safe place to live), permanency (i.e., having a nurturing and permanent family), and child well-being.

Often occurring in conjunction with administrative selection or self-selection, *geographic selection* refers to bias produced by geographic mismatch. Heckman et al. (1997) observed that geographic selection occurs when administrators and participants choose (or do not choose) a particular program because alternatives are not available in the area or labor market. Outcomes are then confounded with geographic site or other local effects.

From this more inclusive perspective on the term *selection*, yet another source of selection bias is measurement error. Psychometricians conventionally distinguish between two types of measurement errors: random errors and systematic errors (Nunnally & Bernstein, 1994). It is systematic error that causes problems and warrants an explicit control in data analysis. Systematic error introduces a selection effect when an instrument or item produces differential reactivity. That is, it produces bias when participants in intervention or control conditions respond differently because of the way a measure is written, its order of presentation, or some other feature of measurement. Systematic measurement error usually requires recalibration of instruments. In analysis, Heckman et al. (1997) considered measurement errors (e.g., differences in survey questionnaires) as temporarily invariant sources of bias, and in part, they developed the difference-in-differences estimator to correct for measurement errors.

Another important source of measurement error is the rater effect. As often is the case, data are collected from raters (e.g., teachers, caregivers, service managers) who score program participants using a protocol such as the Achenbach (1991) Child Behavior Check List (CBCL). Over the course of a long study, different raters may produce systematic measurement error that should be explicitly controlled in any evaluation project. Controlling for measurement error is critical because treatment effects in social behavioral studies tend to be small, and small effects can be easily "washed away" by rater effects. Using generalizability theory (Cronbach, Gleser, Nanda, & Rajaratnam, 1972; Shavelson & Webb, 1991), Guo and Hussey (1999) demonstrated seven sources of outcome variability in a crossed, two-facet random-effects design, which is equivalent to a longitudinal study using multiple raters. In this measurement procedure, inaccuracies between the sample data and the universe may come from several sources: participants (σ_s^2), raters (σ_r^2), occasions (σ_o^2), interaction of participant and rater (σ_{sr}^2), interaction of participant and occasion (σ_{so}^2), interaction of rater and occasion (σ_{ro}^2), and residual ($\sigma_{sro,e}^2$). The relative contribution of each source to the total measurement error varies by the intervention under study, the instrument used, and the study design. However, Guo and Hussey found that it was common for this kind of data to contain nonnegligible sources of variation associated with raters. For example, suppose an intervention under study was designed to change children's problematic behavior. When evaluating this kind of intervention, some raters may differ in their understanding of the rating rules, with some raters tending to give children more stringent ratings—a scenario likely to increase σ_r^2. Similarly, other raters may interpret the rating rules more leniently or allow the ratings to be influenced by personal favoritism toward a child—a scenario likely to increase σ_{sr}^2. Finally, a rater may give a more stringent rating one day than on other days simply because of unhappy events that may have occurred to the rater that day—a scenario likely to increase σ_{so}^2. When some or all of these conditions are present, rater effects can be systematic, and the investigator cannot ignore the rater influence. In their original study, Guo and Hussey developed a three-level random effects model that specified rater effects as random.

Parenthetically, in program evaluation using the corrective models described in this book, it may not be feasible to apply a three-level random effects model in conjunction with one of the four methods. Under such conditions, it remains important to control for nonnegligible systematic variation of rater effects, and this can sometimes be accomplished by changing the analysis plan. For instance, suppose an evaluator has four waves of student data: The first two waves of data were collected at the beginning and end of the fourth-grade term and were comprised of ratings by the same teachers (i.e., raters). The next two waves of data were collected at the beginning and end of students' fifth-grade term and were comprised of ratings by a second group of teachers. In this study, raters are the same only within a grade. Rather than following an analytic plan that estimates overall change from Wave 1 to Wave 4, the investigator can evaluate the intervention effect while controlling for rater effects by analyzing piecewise change (i.e., the change score within the fourth grade and the change score within the fifth grade). Whereas the drawback of this piecewise analysis is that the evaluator cannot estimate a model that describes the growth curve of overall change, the gain of such analysis is that the evaluator can control for rater effects. This analytic procedure is shown in Example 6.4.2. Our analyses suggest that, in the presence of rater effects, capturing a small treatment effect within grade may be more important than describing an overall change across grades. Indeed, the results of our study indicate that had we pooled all four waves of data for analysis, the small treatment effect would have been wiped out by rater effects.

Last but not least, attrition of study participants is an important source of selection bias, and it is a common problem in longitudinal studies. Attrition is one of the main sources of selection bias in Maddala's (1983) decision tree for evaluation of social experiments (see Figure 4.1). More so than other sources of selection bias, attrition poses challenges in evaluations. In a clinical intervention, the participants who are more likely to drop out are often those who believe that treatment will no longer be of benefit; therefore, attrition is nonrandom. Relatedly, participants who are more difficult to treat (e.g., who may lead more chaotic lives) are more likely to drop out. Even when groups are randomized, postrandomization attrition may corrupt group balance. In many experimental studies with random assignment, treatment assignment is ignorable, but postassignment dropout produces selection bias because participants in either the treatment or control conditions differentially drop out. Although advances have been made in statistical methods for longitudinal studies, statisticians and researchers are only beginning to develop robust and effective models that correct for attrition bias.

8.1.2 OVERT BIAS VERSUS HIDDEN BIAS

Rosenbaum (2002b) made a careful distinction between overt bias and hidden bias:

An observational study is biased if the treated and control groups differ prior to treatment in ways that matter for the outcomes under study. An overt bias is one that can be seen in the data at hand—for instance, prior to treatment, treated subjects are observed to have lower incomes than controls. A hidden bias is similar but cannot be seen because the required information was not observed or recorded. (p. 71)

Although we often attempt to control for bias by anticipating sources of selection and measuring them as pretreatment covariates, we can never be sure we have thought of everything. Ipso facto, hidden bias is unobserved selection. Unobserved heterogeneity may affect both the dependent variable and the probability of being in the sample. According to Kennedy (2003), the unmeasured nature of hidden bias is crucial because, if it were measured, we could account for it and avoid the bias. Furthermore, because unmeasured variables also affect the probability of a participant being in the sample, we get an unrepresentative (nonrandom) sample; and because unmeasured variables affect the dependent variable, it is possible for this unrepresentative sample to give rise to selection that affects causal inference.

Overt and hidden biases that affect causal inference are controlled in well-implemented randomized experiments. We control—or correct—for potential bias through the mechanism of randomization. It tends to balance data and to make the averages of omitted variables (i.e., the error term of a regression model) equal to zero. This characteristic of balancing data for both the measured variables (i.e., variables available to the analyst but incidentally omitted) and the unmeasured variables (i.e., variables on which data are not collected) is truly the advantage of randomization.

Balancing the data is especially a problem in observational studies. Although researchers control for selection bias through matching, they can adjust only on the observed or measured covariates; thus, selection bias due to unmeasured covariates remains a problem. As such, for most observational studies, what remains unknown is the extent to which matching or other adjustments adequately control for bias and yield estimates of treatment effects that are trustworthy (Rosenbaum, 2002b). To deal with this problem, Rosenbaum developed sensitivity analysis methods that aim to gauge the level of sensitivity of study findings to hidden bias. We discuss these methods in Section 8.3.

8.1.3 CONSEQUENCES OF SELECTION BIAS

Whenever an observational study has selection bias, nonignorable treatment assignment becomes a matter of great concern. In this context, a conventional regression analysis or regression-type model often involves incidental truncation, endogeneity bias, or confoundedness. Throughout this book, we have described the potentially adverse consequences of applying regression models to such data. When used in an attempt to control for selection effects, conventional regression models may produce biased (and often inflated) estimates of treatment effects. Moreover, effect estimates may be inconsistent. Given these undesirable outcomes, how do we correct for selection bias?

8.1.4 STRATEGIES TO CORRECT FOR SELECTION BIAS

A variety of techniques and strategies have been developed for correcting for selection bias. Wooldridge (2002, pp. 551–598) argued that researchers should take selection effects seriously when they cause the error term of a regression model to be correlated with an explanatory variable. Furthermore, he advocated three methods for controlling selection: the maximum likelihood estimation of treatment effects (see Section 4.2), the Heckman two-stage procedure (see Section 4.1), and instrumental variables estimation (see Section 4.3).

We have introduced three additional approaches for adjusting selection bias: the propensity score matching methods, matching estimators, and kernel-based matching estimators. It is important to apply these corrective strategies with caution because each was developed on the basis of a different statistical theory. This section examines the history and development of corrective strategies by linking four methods (the Heckman sample selection model plus the three mentioned above) to the two core traditions (i.e., econometric vs. statistical tradition) in observational studies. Our discussion is guided by the key questions of how well a correction model handles the problem of selection bias and what assumptions are embedded in each method.

In Chapters 1 and 2, we described both the econometric tradition and the statistical tradition of developing methods for observational studies. The econometric tradition stems from Haavelmo's (1943, 1944) linear simultaneous equation model, Quandt's (1958, 1972) switching regression model, and Roy's (1951) choice model between hunting and fishing. According to Maddala (1983) and Manski (2007), Roy's model can be characterized as *outcome optimization*; that is, the model emphasizes that choice behavior might make the observability of a behavioral outcome depend on the value of the outcome. We review this perspective below as a way of illustrating the importance of outcome optimization. Our review follows Maddala (1983, pp. 257–259).

Let Y_{1i} denote the income of the ith individual in hunting and Y_{2i} denote the individual's income in fishing. Individuals make a rational choice by optimizing outcomes, such as income or social status. That is, individual i will choose to be a hunter if and only if $Y_{1i} > Y_{2i}$. Assuming that (Y_{1i}, Y_{2i}) have a joint normal distribution, with means (μ_1, μ_2) and covariance matrix

$$\begin{bmatrix} \sigma_1^2 & \sigma_{12} \\ \sigma_{12} & \sigma_2^2 \end{bmatrix},$$

define

$$u_1 = Y_1 - \mu_1, \; u_2 = Y_2 - \mu_2, \; \sigma^2 = \mathrm{Var}(u_1 - u_2),$$

$$Z = \frac{\mu_1 - \mu_2}{\sigma}, \; \text{and } u = \frac{u_2 - u_1}{\sigma}.$$

The condition $Y_1 > Y_2$ implies $u < Z$. The mean income of hunters is given by

$$E(Y_1 | u < Z) = \mu_1 - \sigma_{1u} \frac{\phi(Z)}{\Phi(Z)},$$

where $\sigma_{1u} = \mathrm{Cov}\,(u_1, u)$, and $\phi(\cdot)$ and $\Phi(\cdot)$ are the standard normal density function and standard normal distribution function, respectively. The mean income of fishermen is given by

$$E(Y_2 | u > Z) = \mu_2 + \sigma_{2u} \frac{\phi(Z)}{1 - \Phi(Z)},$$

where $\sigma_{2u} = \mathrm{Cov}(u_2, u)$. Because

$$\sigma_{1u} = \frac{\sigma_{12} - \sigma_1^2}{\sigma} \; \text{and} \; \sigma_{2u} = \frac{\sigma_2^2 - \sigma_{12}}{\sigma}.$$

we have $\sigma_{2u} - \sigma_{1u} > 0$. Given the above definitions, we now can consider different cases.

Case 1: $\sigma_{1u} < 0$, $\sigma_{2u} > 0$. In this case, the mean income of hunters is greater than μ_1 and the mean income of fishermen is greater than μ_2. Under this condition, those who choose hunting have incomes better than the average income of hunters, and those who choose fishing have incomes better than the average income of fishermen.

Case 2: $\sigma_{1u} < 0$, $\sigma_{2u} < 0$. In this case, the mean income of hunters is greater than μ_1, and the mean income of fishermen is less than μ_2. Under this condition, those who choose hunting are better than average in both hunting and fishing, but they are better in hunting than in fishing. Those who choose fishing are below average in both hunting and fishing, but they are better in fishing than in hunting.

Case 3: $\sigma_{1u} > 0$, $\sigma_{2u} > 0$. This is the reverse of Case 2.

Case 4: $\sigma_{1u} > 0$, $\sigma_{2u} < 0$. This is not possible, given the definitions of σ_{1u} and σ_{2u}.

The above model has a significant impact on all econometric discussions of self-selection and the development of models correcting for self-selection. The key features of the Roy model are (a) that it was based on rational-choice theory, specifically the idea that individuals make a selection by optimizing an outcome, and (b) that the process cannot be assumed to be random, and factors determining the structure of selection should be explicitly specified and modeled.

Maddala (1983) showed that Gronau (1974), Lewis (1974), and Heckman (1974) followed a Roy model when they began their studies examining women in the labor force. In these studies, the observed distribution of wages became a truncated distribution, and the self-selectivity problem became a problem of incidental truncation. It was this pioneering work that instigated discussion of the consequences of self-selectivity and that motivated the development of econometric models to correct for self-selectivity. Although these models differ in methodology, a feature that is shared among all the models is that the impact of selection bias is neither relegated away nor assumed to be random. Rather, extending the choice perspective, it is explicitly used and estimated in the correction model. The more important models are (a) Heckman's two-stage sample selection model and (b) the variant of Heckman's two-stage model—Maddala's treatment effect model using a maximum likelihood estimator.

In contrast, models that correct for selection bias by following the statistical tradition make a crucial assumption. They assume that selection follows a random process. Manski (2007) elaborated on this distinction saying, "Whereas economists have often sought to model treatment selection in nonexperimental settings as conscious choice behavior, statisticians have typically assumed that treatment selection is random conditional on specified covariates. See, for example, Rubin (1974) and Rosenbaum and Rubin (1983)" (p. 151). Assuming random selection offers numerous advantages in the development of correction procedures. However, in practice, researchers need to consider whether such assumptions are realistic and choose a valid approach for correction. Although self-selection might be assumed to be random, other sources of selection (i.e., researcher selection, measurement selection such as rater effects, administrative selection, and attrition-induced selection) cannot be assumed to be random. Corrective methods for these sources of selection bias need to be developed.

Despite these different perspectives (i.e., "selection is a rational choice and should be explicitly modeled" vs. "selection is random conditional on specified covariates"), the two traditions converge on a key feature: both emphasize the importance of controlling for measured covariates that affect selection. Shortly after Heckman (1978, 1979) developed his two-stage estimator that used the predicted probability of receiving treatment, Rosenbaum and Rubin (1983) developed their propensity score matching estimator. Both models share a common feature of using the conditional probability of receiving treatment in correction.

Another difference between the econometric tradition and the statistical tradition lies in the level of restrictiveness of assumptions. Because observational studies analyze potential outcomes or counterfactuals, the developer must impose assumptions on the model to make model parameters identifiable. Thus, a concern or topic of debate is the extent to which model assumptions are realistic.

Based on the above discussion, four key issues in the development of strategies to correct for selection bias emerge: (1) the econometric approach emphasizes the structure of selection and therefore underscores a direct modeling of selection bias; (2) the statistical approach assumes that selection is random conditional on covariates; (3) both approaches emphasize direct control of observed selection by using a conditional probability of receiving treatment; and (4) the two approaches are based on different assumptions and they differ on the restrictiveness of assumptions. It is clear then that assumptions play a crucial role in correction. Users of correction models should be aware of the assumptions embedded in each model, should be willing to check the tenability of assumptions in their data, should choose a correction model that is best suited to the nature of the data and research questions, and should interpret findings with caution.

In Chapters 4 through 7, we reviewed the assumptions embedded in each of the four correction models—the Heckman sample selection model, propensity score matching, matching estimators, and kernel-based matching. A summary of the key assumptions for each model is provided in Table 8.1. Note that these correction models vary by the types of treatment effects estimated. Some models can be used to estimate both the average treatment effect and the average treatment effect for the treated, whereas other models can estimate only one of the two effects. We encourage users of corrective models to consider the assumptions listed in Table 8.1 and condition study findings on the degree to which assumptions are satisfied.

8.2 A Monte Carlo Study Comparing Corrective Models

We conducted a Monte Carlo study to show the importance of checking the tenability of assumptions related to the corrective methods and to compare models under different scenarios of data generation (i.e., different types of selection bias). A Monte Carlo study is a simulation exercise designed to shed light on the small-sample properties of competing estimators for a given estimating problem (Kennedy, 2003). The same objectives can certainly be accomplished by other means, such as using statistical theories to derive the results analytically. That type of analysis underpins the work of the original developers of the corrective methods, and the main findings of those studies have been highlighted in the chapters of this book. We chose to use a Monte Carlo study to compare models because such a simulation approach allows us to examine comparative results in a way that is more intuitive and less technical.

Table 8.1   Key Assumptions and Effects by Correction Model

Correction Model	Effect to Estimate	Key Assumptions
1. Heckman sample selection and Heckit treatment effect model (Chapter 4)	Average treatment effect	1. Normal distribution of the outcome variable; 2. Error terms of the regression equation and the selection equation are bivariate normal and correlated.
2. Propensity score matching (Rosenbaum & Rubin, 1983; Chapter 5)	Average treatment effect	1. Strongly ignorable treatment assignment or $(Y_0, Y_1) \perp W \mid X$; 2. Overlap of conditional probability receiving treatment between treated and control groups.
3. Matching estimators (Chapter 6)	1. Average treatment effect 2. Average treatment effect for the treated 3. Average treatment effect for the control	1. Strongly ignorable treatment assignment or $(Y_0, Y_1) \perp W \mid X$; 2. Overlap of conditional probability receiving treatment between treated and control groups.
4. Kernel-based matching (Chapter 7)	Average treatment effect for the treated	1. A weaker assumption about the strongly ignorable treatment assignment or $(Y_0) \perp W \mid X$; 2. Mean independence or $E(Y_0 \mid W = 1, X) = E(Y_1 \mid W = 0, X)$.

Many Monte Carlo studies have been conducted on the properties of corrective approaches, but most of these studies have been comparisons of properties under different settings within one of the four corrective methods. For instance, Stolzenberg and Relles (1990), Hartman (1991), and Zuehlke and Zeman (1991) conducted Monte Carlo studies that examined various aspects of the Heckman or Heckit models. Kennedy (2003) reviewed these studies and summarized the main findings: relative to subsample ordinary least squares (OLS) and on a mean-square-error criterion, the Heckman procedure does not perform well when the errors do not have a normal distribution, the sample size is small, the amount of censoring is small, the correlation between the errors of the regression and selection equations is small, or the degree of collinearity between the explanatory variables in the regression and selection equations is high. Kennedy warned that the Heckman model does poorly—and even does more harm than good—in the presence of high collinearity.

Zhao (2004) conducted a Monte Carlo study to compare propensity score matching with covariate matching estimators under different conditions. Zhao found that selection bias due only to observables was a strong assumption; however, with a proper data set and if the selection-only-on-observables assumption was justifiable, matching estimators were useful for estimating treatment effects. Furthermore, Zhao found no clear winner among different matching estimators, and that the propensity score matching estimators rely on the balancing property.

Freedman and Berk (2008) conducted data simulations to examine the properties of propensity score weighting. As mentioned earlier, they found that the weighting approach requires correctly specifying a causal model.

In contrast to these studies, few Monte Carlo simulations have been conducted to compare the properties of corrective methods under a fixed setting; this is the objective of our Monte Carlo study. In this study, we compared four models (i.e., the OLS regression, the Heckit treatment effect model using maximum likelihood estimation, the propensity score one-to-one matching in conjunction with a postmatching regression analysis [PSM], and the matching estimators) under a given setting. We ruled out kernel-based matching because it estimates only the average treatment effect for the treated, and the four models listed above were not all designed to estimate this effect. Thus, our Monte Carlo study focused on the average treatment effect.

We simulated two data generation settings (i.e., selection on observables and selection on unobservables) and compared performance across the four models within each setting. Our aim for the Monte Carlo study was to address the following four research questions: (1) Within each setting of selection bias, which model performs the best, and how are the four models ranked in terms of bias and mean-square-error criteria? (2) Within each setting of selection bias and modifying model specifications to simulate the most realistic practice for real-world application, which model performs the best, and how are the four models ranked in terms of bias and mean-square-error criteria? (3) What conclusions, in terms of the sensitivity of model performance to the mechanism of data generation, can we draw from the simulation study? Last, (4) comparing the data generation process defined by a given setting with assumptions embedded in a given model, how important and restrictive are the assumptions?

We emphasize that the Monte Carlo study simulates very limited settings of data generation. There are certainly many other settings—actually an unlimited number of settings.[1] Therefore, the conclusions of our study cannot be generalized to other settings. We did not attempt to compare all four methods in general settings to determine which model was the best. The main purpose of the Monte Carlo study is to show the importance of checking data assumptions, that is, to demonstrate that models have different assumptions and the performances of the models under a common setting of data generation will vary.

8.2.1 DESIGN OF THE MONTE CARLO STUDY

In this subsection, we first present a statistical framework that outlines the process of treatment assignment, and then we show specifications for the two settings of selection bias simulated by the Monte Carlo study. Last, we show the model specifications for each of the four models under these settings and the evaluation criteria we used.

The statistical framework we adopted for this study was drawn from Heckman and Robb (1985, 1986, 1988), which aimed to model the assignment mechanism that generated treatment and control groups.

Let Y_{i1} and Y_{i0} denote potential outcomes for observation i under the conditions of treatment and control, respectively. The potential outcomes Y_{i1} and Y_{i0} can be expressed as deviations from their means:

$$Y_{i0} = E(Y_0) + u_{i0},$$

$$Y_{i1} = E(Y_1) + u_{i1}.$$

Combining these two expressions with the observation rule given by the definition of the treatment assignment dummy variable W_i (i.e., Equation 2.1 of Chapter 2, or $Y_i = W_i Y_{1i} + (1 - W_i) Y_{0i}$), the equation for any Y_i is

$$Y_i = E(Y_0) + W_i[E(Y_1) - E(Y_0)] + u_{i0} + W_i(u_{i1} - u_{i0})$$

$$= E(Y_0) + W_i E(\delta) + u_i, \tag{8.1}$$

where $u_i = u_{i0} + W_i(u_{i1} - u_{i0})$. Equation 8.1 is known as the *structural equation*. For a consistent estimate of the true average treatment effect, W_i and u_i must be uncorrelated.

Consider a supplemental equation, known as the *assignment* or *selection equation*, that determines W_i. Denoting W_i^* as a latent continuous variable, we have

$$W_i^* = Z_i \alpha' + v_i, \tag{8.2}$$

where Z_i is a row vector of values on various exogenous observed variables that affect the assignment process, α is a vector of parameters that typically needs to be estimated, and v_i is an error term that captures unobserved factors that affect assignment. The treatment dummy variable W is determined by the following rule: $W_i = 1$ if $W_i^* > c$, and $W_i = 0$ if $W_i^* < c$ (c is a cutoff value).

Additional covariates X_i may be included in Equation 8.1, and X_i and Z_i may be in common. We can distinguish between two settings in which W_i and the error term u_i in Equation 8.1 can be correlated.

Setting 1: Selection on the observables. When Z_i and u_i are correlated, but u_i and v_i are uncorrelated, we have the condition known as *selection on the observables.* This condition is alternatively known as the *strongly ignorable treatment assignment* assumption (Rosenbaum & Rubin, 1983). Under this condition, a statistical control such as an OLS regression of Equation 8.1 will be unbiased as long as Z_i is sufficiently included in the equation.

Setting 2: Selection on the unobservables. When Z_i and u_i are uncorrelated, but u_i and v_i are correlated, the condition is known as *selection on the unobservables.* This setting simulates hidden selection bias and the violation of the regression assumption about uncorrelated error terms with an independent variable. Under this setting, results of OLS estimation are expected to be biased and inconsistent. In practice, controlling for selection bias due to unobservables is difficult.

The data generation process for each of the two settings is shown in Figure 8.1. In addition to the above specifications, this process further imposes the following conditions: (a) three variables or covariates (x_1, x_2, and x_3) that affect the outcome variable y; (b) z determines treatment assignment w only; and (c) x_3 also affects treatment assignment w.

The above data generation for each setting was implemented by using Stata version 9.0; the syntax files for the data generation and analysis using four models are available on the companion Web page of this book. We encourage readers to replicate the study and to use our syntax as a baseline to generate more settings or to compare additional models.

1. *Specifications of Setting 1 in Stata:* The Stata syntax generates Setting 1 using the following specifications:

$$Y = 100 + .5x_1 + .2x_2 - .05x_3 + .5W + u \qquad (8.3)$$

$$W^* = .5Z + .1x_3 + v,$$

where x_1, x_2, x_3, Z, and u are random variables, normally distributed with a mean vector of (3 2 10 5 0), standard deviation vector (.5 .6 9.5 2 1), and the following symmetric correlation matrix:

$$
r_{(x_1,x_2,x_3,Z,u)} = \begin{bmatrix}
1 & & & & \\
.2 & 1 & & & \\
.3 & 0 & 1 & & \\
0 & 0 & 0 & 1 & \\
0 & 0 & 0 & .4 & 1
\end{bmatrix}.
$$

Setting 1: Selection on the observables

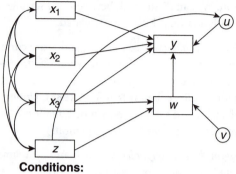

Conditions:

$r_{zu} = .4$, $r_{uv} = .00$.

Setting 2: Selection on the unobservables

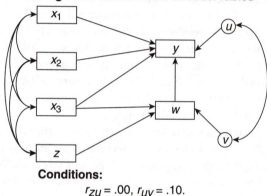

Conditions:

$r_{zu} = .00$, $r_{uv} = .10$.

Figure 8.1   Design of the Monte Carlo Study: Two Settings of Selection Bias

In addition, v is a random variable that is normally distributed with mean zero and variance one; and $W = 1$, if $W^* > \text{Median}(W^*)$, and $W = 0$ otherwise.

The above specifications create a correlation between Z and u of .4, and a correlation between u and v of 0. Thus, the data generation meets the requirements for simulating selection on observables, as shown in Setting 1 in Figure 8.1. The Monte Carlo study generates 10,000 samples for each corrective model with a size of 500 observations per sample. Under this specification, the true average treatment effect in the population is known in advance, that is, $W = .5$, as shown in Equation 8.2.

2. *Specifications of Setting 2 in Stata:* The Stata syntax generates Setting 2 using the following specifications:

$$Y = 100 + .5x_1 + .2x_2 - .05x_3 + .5W + u \qquad (8.4)$$

$$W^* = .5Z + .1x_3 + v$$

$$v = \delta + .15\varepsilon,$$

where x_1, x_2, x_3, Z, u, and ε are random variables, normally distributed with a mean vector of (3 2 10 5 0 0), standard deviation vector (.5 .6 9.5 2 1 1), and the following symmetric correlation matrix:

$$r_{(x_1,x_2,x_3,Z,u,\varepsilon)} = \begin{bmatrix} 1 & & & & & \\ .2 & 1 & & & & \\ .3 & 0 & 1 & & & \\ 0 & 0 & 0 & 1 & & \\ 0 & 0 & 0 & 0 & 1 & \\ 0 & 0 & 0 & 0 & .7 & 1 \end{bmatrix}.$$

In addition, δ is a random variable that is normally distributed with mean zero and variance one; and $W = 1$, if $W^* > \text{Median}(W^*)$, and $W = 0$ otherwise.

The above specifications create a correlation between z and u of 0, and a small correlation between u and v of .1. Thus, the data generation meets the requirements for simulating selection on unobservables as shown in Setting 2 in Figure 8.1. The Monte Carlo study generates 10,000 samples for each corrective model with a size of 500 observations per sample. Under this specification, the true average treatment effect in the population is known in advance, that is, $W = .5$, as shown by Equation 8.3.

3. *Specifications of corrective models in Stata:* The specifications for each of the four corrective models under Setting 1 are shown below.

Model 1.1. OLS regression: $\hat{Y} = \hat{\beta}_0 + \hat{\beta}_1 x_1 + \hat{\beta}_2 x_2 + \hat{\beta}_3 x_3 + \hat{\beta}_4 Z + \hat{\tau} W$.

Model 1.2. PSM: The logistic regression model predicting the conditional probability is

$$\hat{P}(W = 1) = \hat{e}(x) = \frac{1}{1 + e^{-(\hat{\beta}_0 + \hat{\beta}_1 x_1 + \hat{\beta}_2 x_2 + \hat{\beta}_3 x_3 + \hat{\beta}_4 Z)}};$$

the predicted probability from the logistic regression $\hat{e}(x)$ is defined as the estimated propensity score; the PSM procedure then matches each treated case to a control case (i.e., a 1-to-1 match) on the estimated propensity score using nearest neighbor within a caliper of .086 (i.e., a quarter of the standard deviation of the estimated propensity scores); and the postmatching analysis performs the following OLS regression of $\hat{Y} = \hat{\beta}_0 + \hat{\beta}_1 x_1 + \hat{\beta}_2 x_2 + \hat{\beta}_3 x_3 + \hat{\beta}_4 Z + \hat{\tau} W$ based on the matched sample.

Model 1.3. Treatment effect model: The regression equation is $\hat{Y} = \hat{\beta}_0 + \hat{\beta}_1 x_1 + \hat{\beta}_2 x_2 + \hat{\beta}_3 x_3 + \hat{\tau} W$; The selection equation is $W^* = \gamma Z + v$, $W = 1$ if $W^* > \text{Median}(W^*)$, and $W = 0$ otherwise; $\text{Prob}(W = 1 \mid Z) = \Phi(Z\gamma)$ and $\text{Prob}(W = 0 \mid Z) = 1 - \Phi(Z\gamma)$; and the model is estimated by the maximum likelihood estimation.[2]

Model 1.4. Matching estimator: The matching covariates include x_1, x_2, x_3, and Z; the model is estimated by the bias-corrected and robust-variance estimator where the vector norm uses the inverse of the sample variance matrix.

The four corrective models under Setting 2 are the same as those under Setting 1. That is,

Model 2.1. OLS regression: Same as Model 1.1.

Model 2.2. PSM: Same as Model 1.2.

Model 2.3. Treatment effect model: Same Model 1.3.

Model 2.4. Matching estimator: Same as Model 1.4.

In this design, Setting 1 simulates selection on observables. By design, Z is an important variable that determines sample selection, and the key of selection on observables is to control for Z correctly in the analysis model. This need to control for Z is why we specify Z as a covariate affecting selection in the OLS regression, in the logistic regression for the PSM model, in the selection equation for the treatment effect model, and in the matching estimator. In practice, the analyst may not know that Z is important and may inadvertently omit it from the analysis. The omission of Z from analysis creates overt selection bias. To simulate this scenario, we ran another set of models under Setting 1 in which Z was not used in all models. The specifications used for this set of models are shown below.

Model 1.1.1. OLS regression: $\hat{Y} = \hat{\beta}_0 + \hat{\beta}_1 x_1 + \hat{\beta}_2 x_2 + \hat{\beta}_3 x_3 + \hat{\tau} W$

Model 1.2.1. PSM: The logistic regression model predicting the conditional probability is

$$\hat{P}(W = 1) = \hat{e}(x) = \frac{1}{1 + e^{-(\hat{\beta}_0 + \hat{\beta}_1 x_1 + \hat{\beta}_2 x_2 + \hat{\beta}_3 x_3)}};$$

the predicted probability from the logistic regression $\hat{e}(x)$ is defined as the estimated propensity score; the PSM procedure then matches each treated case to a control case (i.e., a 1-to-1 match) on the estimated propensity score using nearest neighbor within a caliper of .06 (i.e., a quarter of one standard deviation of the estimated propensity scores); and the postmatching analysis performs the following OLS regression of $\hat{Y} = \hat{\beta}_0 + \hat{\beta}_1 x_1 + \hat{\beta}_2 x_2 + \hat{\beta}_3 x_3 + \hat{\tau} W$; based on the matched sample. Model 1.3.1. Treatment effect model: The regression equation is $\hat{Y} = \hat{\beta}_0 + \hat{\beta}_1 x_1 + \hat{\beta}_2 x_2 + \hat{\beta}_3 x_3 + \hat{\tau} W$; the selection equation is $W^* = \gamma_1 x_1 + \gamma_2 x_2 + \gamma_3 x_3 + v$, $W = 1$ if $W^* > \text{Median}(W^*)$, and $W = 0$ otherwise; $\text{Prob}(W = 1 \mid X) = \Phi(X\gamma)$

and Prob($W = 0 \mid X$) = $1 - \Phi(X\gamma)$; and the model is estimated by the maximum likelihood method.

Model 1.4.1. Matching estimator: The matching covariates include x_1, x_2, and x_3; the model is estimated by the bias-corrected and robust-variance estimator where the vector norm uses the inverse of the sample variance matrix.

4. *Criteria used to assess model performance.* We used two criteria to assess model performance. One criterion was the estimated bias, which is based on the mean value of the estimated average treatment effect of the 10,000 samples. Because the true treatment effect is known (i.e., 0.5), the mean estimated average treatment effect of the 10,000 samples minus 0.5 provides an estimation of bias for a given model. The second criterion was the estimated *mean square error* (*MSE*), which is estimated by the average of the squared differences between $\hat{\tau}_{\text{MODEL}-n}$ and the true value of treatment effect 0.5:

$$MSE = \sum_{i=1}^{10,000} \left(\hat{\tau}_{\text{MODEL}-n,i} - .5 \right)^2 / 10,000.$$

MSE provides an estimation of the variation of the sampling distribution for the estimated treatment effects; a small *MSE* value indicates low variation.

8.2.2 RESULTS OF THE MONTE CARLO STUDY

Table 8.2 presents the findings of the Monte Carlo study under the two settings. The main findings of model performances under Setting 1 are summarized in the table.

Under Setting 1, that is, the selection on observables, the PSM model performed the best: On average, the PSM model estimated a treatment effect of 0.4875, which was 0.0125 below the true effect (or an underestimation of 2.5%) with a low *MSE* of 0.0152.

The OLS regression also worked reasonably well and was ranked second: On average, OLS regression estimated a treatment effect of 0.5375, which was 0.0375 above the true effect (or an overestimation of 7.5%) with a low *MSE* of 0.012. It is worth noting that OLS works reasonably well in this setting because x_3 and Z are the main variables determining selection, Z and u are correlated, and both source variables x_3 and Z are controlled in the analysis. These conditions are restrictive and may not hold in practice: In a typical application, we may not know that x_3 and Z are the major source of selection; x_3 and Z may not be available or collected; and Z and u may not be correlated.

The matching estimator did not provide acceptable bias correction and was ranked third among the four models: On average, the matching estimator estimated the treatment effect of 0.4531, which was 0.0469 below the true effect

Table 8.2   Results of Monte Carlo Study Comparing Models

Analysis Model	Estimated Average Treatment Effect		MSE	Rank
	Mean of 10,000 Samples	Bias = Mean – True Effect (% Bias)		
Setting 1: Selection on the observables				
Model 1.1: OLS regression	0.5375	+0.0375 (7.5% Over)	0.0120	2
Model 1.2: PSM	0.4875	–0.0125 (2.5% Under)	0.0152	1
Model 1.3: Treatment effect model	1.9285	+1.4285 (285.7% Over)	2.0469	4
Model 1.4: Matching estimator	0.4531	–0.0469 (9.4% Under)	0.0237	3
Setting 2: Selection on the unobservables				
Model 2.1: OLS regression	0.6900	+0.1900 (38.0% Over)	0.0486	4
Model 2.2: PSM	0.6464	+0.1464 (29.3% Over)	0.0395	3
Model 2.3: Treatment effect model	0.5036	+0.0036 (0.7% Over)	0.0005	1
Model 2.4: Matching estimator	0.6377	+0.1377 (27.5% Over)	0.0441	2

(or an underestimation of 9.4%) with a medium-sized MSE of 0.0237. The primary reason for the poor showing of matching estimation of the treatment effect was that Z and u are correlated in this setting. Although it is true that this assumption is also embedded in PSM and OLS regression, these two models seem to provide a more robust response to the violation than the matching estimator. Note that all matching variables used in the matching model (i.e., x_1, x_2, x_3, and Z) were continuous, so that the matching was not exact. Although the matching estimator made efforts to correct for bias by using a least-squares regression to adjust the difference within the matches for the differences in covariate values, under the given setting, matching via vector norm was inferior to matching via a one-dimensional propensity score (i.e., the PSM).

In Setting 1, the variant of Heckman's two-stage model, Maddala's treatment effect model, performed the worst among the four models: On average, the treatment effect model estimated the treatment effect of 1.9285, which was 1.4285 above the true effect (or an overestimation of 285.7%) with a huge MSE of 2.0469. The primary reason for this poor estimation was that in Setting 1, u and v were not correlated, so that the data violated the assumption about correlation of errors between the regression and selection equations. This finding underscores the fact that the assumption about a nonzero correlation of two error terms is crucial to a successful application of the treatment effect model. In addition, this finding confirmed that the Heckman selection model depends strongly not only on the model being correct but also on the tenability of model assumptions; and this requirement is more pronounced than that of OLS regression, a point we made in Chapter 4.

We now summarize the main findings of model performances under Setting 2. In stark contrast to the findings under Setting 1, when u and v are correlated or selection is on unobservables, the Maddala treatment effect model provided excellent estimation of the average treatment effect and was ranked first among the four models: On average, the treatment effect model estimated the treatment effect of 0.5036, which was 0.0036 above the true effect (or an overestimation of 0.7%) with an excellent MSE of 0.0005. No other models under this setting could compete with the treatment effect model. In contrast, the other three models produced unacceptable estimates for the treatment effects: The overestimation ranged from 27.5% for the matching estimator to 38.0% for OLS regression; and the estimated MSE ranged from 0.0395 for the PSM to 0.0486 for OLS regression. Among the three poor-performing models, the PSM and matching estimator did a relatively better job than the OLS. This finding confirms the danger inherent in using OLS regression to correct for selection bias, particularly when hidden selection bias is present. Furthermore, this finding indicates that PSM and the matching estimator are sensitive to hidden selection bias, and they are not robust estimators under that condition.

Under Setting 1, the crucial feature of data generation is understanding (i.e., observing) the source of selection and the use of controls for Z and x_3 in the analysis. This is a wishful condition that is unlikely to occur in practice. Had the analysis omitted the specification of Z or encountered a situation of overt bias, the estimation results would be unacceptable. Table 8.3 presents results of

Table 8.3   Results of Monte Carlo Study Comparing Models Not Controlling for Z Under Setting 1

	Estimated Average Treatment Effect			
Analysis Model	Mean of 10,000 Samples	Bias = Mean − True Effect (% Bias)	MSE	Rank
Setting 1: Selection on the observables				
Model 1.1.1: OLS regression	1.0005	+0.5005 (100.1% Over)	0.2565	3
Model 1.2.1: PSM	0.9903	+0.4903 (98.06% Over)	0.2482	2
Model 1.3.1: Treatment effect model	1.8118	+1.3118 (262.4% Over)	1.8056	4
Model 1.4.1: Matching estimator	0.9433	+0.4433 (88.9% Over)	0.2038	1

the Monte Carlo study under this setting. Indeed, when Z is omitted in the analysis, all models fail to provide unbiased estimation about the treatment effect. The models are ranked as follows: matching estimator performs the best (88.9% overestimation with MSE of 0.2038), PSM is ranked second (98.06% overestimation with MSE of 0.2482), OLS regression is ranked third (100.1% overestimation with MSE of 0.2565), and the treatment effect model performs the worst of the four (262.4% overestimation with MSE of 1.8056).

8.2.3 IMPLICATIONS

We can distill several implications from the findings of the Monte Carlo analyses. First, no single model works well in all scenarios. The "best" results depend on the fit between the assumptions embedded in a model and the process of data generation. A model performing well in one setting may perform poorly in another setting. Thus, the models are not robust against a variety of data situations. It is important to check the tenability of model assumptions in alternative applications, and choose a model that is suited to the nature of the data at hand. With a mismatch of the data structure and the analytic model, it is easy to observe a misleading result.

Second, when information regarding the tenability of model assumptions is not available (e.g., when there is no way of knowing whether the study omits important covariates), findings must be conditioned on a discussion of model assumptions. At a minimum, when disseminating results, the assumptions which underpin estimated treatment effects should be disclosed and the conditions under which estimation may be compromised should be described.

Third, and more specifically, the Maddala treatment effect model relies strongly on the assumption that the errors of the selection and regression equations are correlated, and, when this assumption is violated, the model fails catastrophically to provide an unbiased estimation. But, in contrast to critiques of the Heckman selection model, the Monte Carlo study showed the treatment effect model to be robust against hidden bias. It was the only model (i.e., Model 2.3) that provided accurate estimation of the treatment effect under the setting of selection on unobservables.

Fourth, relative to PSM, OLS regression appears to work well only in very restrictive settings (i.e., Model 1.1) that require the user to have previous knowledge about the main sources of selection bias, to have collected data to measure the main sources of selection bias, and to have correctly used those variables in the model. Had Z been omitted in the analysis, the OLS results would have been marked by extreme overestimation (i.e., Model 1.1.1 overestimated the effect by 100%). Thus, the Monte Carlo study confirms the caveat repeatedly given by experienced observational researchers: OLS regression is not a valid approach to correct for selection bias. The findings suggest that a matching estimator or PSM is preferred in this data situation.

Finally, the Monte Carlo study points again to the challenges imposed by hidden bias. Under the condition of selection on unobservables, or under the condition of selection on observables but when researchers have omitted measured variables from analysis, all models except Model 2.3 provided biased estimates of treatment effect. Therefore, it is important to conduct sensitivity analyses to gauge bias induced by hidden selection. The development of these procedures is the topic of the next section.

8.3 Rosenbaum's Sensitivity Analysis

Hidden bias is essentially a problem created by the omission of important variables in statistical analysis, and omission renders nonrandom the unobserved heterogeneity reflected by an error term in regression equations. Although correcting the problem may involve taking steps such as respecifying an analytic model by using additional variables, collecting additional data, or redesigning a study as a randomized experiment, these strategies can be expensive and time-consuming. It is often preferable to start by conducting sensitivity analyses to estimate the level of bias. This type of analysis seeks to answer this question: How sensitive are findings to hidden bias? Although sensitivity analysis is exploratory, it is an important step in analyses using the models we have described in this book. Rosenbaum and Rubin (1983) and Rosenbaum (2002b, 2005) have recommended that researchers routinely conduct sensitivity analyses in observational studies. In this section, we describe an emerging framework for sensitivity analysis.

8.3.1 THE BASIC IDEA

Rosenbaum (2002b, 2005) provides a succinct and well-organized description of sensitivity analysis. As summarized by Berk (2004), Rosenbaum's approach is simple:

> One manipulates the estimated odds of receiving a particular treatment to see how much the estimated treatment effects may vary. What one wants to find is that the estimated treatment effects are robust to a plausible range of selection biases. (p. 231)

Because sensitivity analysis is so important, we illustrate the basic idea of this method below and explain one procedure (i.e., sensitivity analysis for matched pair studies using the Wilcoxon's signed-rank test) in detail.

According to Rosenbaum (2005),

> A sensitivity analysis in an observational study addresses this possibility: it asks what the unmeasured covariate would have to be like to alter the conclusions of the study. Observational studies vary markedly in their sensitivity to hidden bias: some are sensitive to very small biases, while others are insensitive to quite large biases. (p. 1809)

The original framework for this perspective came from Cornfield et al. (1959), who studied evidence linking smoking to lung cancer. In their effort to sort out conflicting claims, Cornfield and his colleagues derived an inequality for a risk ratio of the probability of death from lung cancer for smokers over the probability of death from lung cancer for nonsmokers. Cornfield et al. argued that to explain the association between smoking and lung cancer seen in a given study, an analyst would need a hidden bias of a particular magnitude in the inequality. If the association was strong, they proposed, then the hidden bias needed to explain it would be large. Therefore, the fundamental task for sensitivity analysis is to derive a range of possible values attributable to hidden bias (i.e., the so-called Γs).

Specifically, suppose that there are two units j and k, and that the two units have the same observed covariates x but possibly different chances of receiving treatment π; that is, $x_{[j]} = x_{[k]}$, but $\pi_{[j]} \neq \pi_{[k]}$. Then, units j and k might be matched to form a matched pair or placed together in the same subclass in an attempt to control overt bias due to x. The odds that units j and k receive the treatment are $\pi_{[j]} / (1 - \pi_{[j]})$ and $\pi_{[k]} / (1 - \pi_{[k]})$, respectively. The odds ratio is

$$\frac{\pi_{[j]}/(1 - \pi_{[j]})}{\pi_{[k]}/(1 - \pi_{[k]})} = \frac{\pi_{[j]}(1 - \pi_{[k]})}{\pi_{[k]}(1 - \pi_{[j]})}.$$

The sensitivity analysis goes further to assume that this odds ratio for units with the same x was at most some number of $\Gamma \geq 1$, that is,

$$\frac{1}{\Gamma} \leq \frac{\pi_{[j]}(1 - \pi_{[k]})}{\pi_{[k]}(1 - \pi_{[j]})} \leq \Gamma \quad \text{for all } j, k \text{ with } x_{[j]} = x_{[k]}.$$

By the above definitions, if Γ were 1, then $\pi_{[j]} = \pi_{[k]}$ whenever $x_{[j]} = x_{[k]}$, so the study would be free of hidden bias. If $\Gamma = 2$, then two units that appear similar, that have the same x, could differ in their odds of receiving the treatment by as much as a factor of 2, so one unit might be twice as likely as the other to receive treatment. As explained by Rosenbaum (2002b),

> In other words, Γ is a measure of the degree of departure from a study that is free of hidden bias. A sensitivity analysis will consider several possible values of

Γ and show how the inferences might change. A study is sensitive if values of Γ close to 1 could lead to inferences that are very different from those obtained assuming the study is free of hidden bias. A study is insensitive if extreme values of Γ are required to alter the inference. (p. 107)

Whereas the original sensitivity analysis of Cornfield et al. (1959) ignored sampling variability (which is hazardous except in extremely large samples), Rosenbaum (2002b) attended to sampling variation by developing methods to compute the bounds on inference quantities, such as p values or confidence intervals. Thus, for each $\Gamma > 1$, we obtain an interval of p values that reflect uncertainty because of hidden bias. "As Γ increases, this interval becomes longer, and eventually it becomes uninformative, including both large and small p values. The point, Γ, at which the interval becomes uninformative is a measure of sensitivity to hidden bias" (Rosenbaum, 2005, p. 1810).

Rosenbaum developed various methods of sensitivity analysis, including the McNemar's test, the Wilcoxon's signed-rank test, and the Hodges-Lehmann point and interval estimates for sensitivity analysis evaluating matched pair studies, sign-score methods for sensitivity analysis evaluating matching with multiple controls, sensitivity analysis for matching with multiple controls when responses are continuous variables, and sensitivity analysis for comparing two unmatched groups. All methods are explained in detail in Rosenbaum (2002b, Chapter 4). A user-developed program in Stata is available to conduct some of these analyses.

8.3.2 ILLUSTRATION OF THE WILCOXON'S SIGNED-RANK TEST FOR SENSITIVITY ANALYSIS OF MATCHED PAIR STUDY

To illustrate sensitivity analysis, we turn to an example originally used by Rosenbaum (2002b). Table 8.4 shows a data set for 33 pairs of children. The exposed group comprised children whose parents worked in a factory that used lead to manufacture batteries (i.e., the treatment group), and the control group comprised children who were matched to the treated children but whose parents were employed in other industries that did not use lead. The study hypothesized that children were exposed to lead that was inadvertently brought home by their parents. Table 8.4 reports the outcome data measured as the micrograms of lead found in a deciliter of each child's blood (i.e., µg/dl). For example, the blood lead level for the treated (i.e., exposed) child in the first pair was 38, and the blood lead level for the control child was 16, and the difference was 22. To remove selection bias, the study matched each treated child to a control child on age and neighborhood of residence. The study found that when controlling for age and neighborhood of residence, the average blood lead level of the treated children was 15.97 µg/dl higher than that of the control children; the Wilcoxon's signed-rank test shows that this treatment

Table 8.4   Example of Sensitivity Analysis: Blood Lead Levels (μg/dl) of
Children Whose Parents Are Exposed to Lead at Their Places
of Work Versus Children Whose Parents Are Unexposed to Lead
at Their Places of Work

Pair	Exposed	Control	Difference
1	38	16	22
2	23	18	5
3	41	18	23
4	18	24	−6
5	37	19	18
6	36	11	25
7	23	10	13
8	62	15	47
9	31	16	15
10	34	18	16
11	24	18	6
12	14	13	1
13	21	19	2
14	17	10	7
15	16	16	0
16	20	16	4
17	15	24	−9
18	10	13	−3
19	45	9	36
20	39	14	25
21	22	21	1
22	35	19	16
23	49	7	42
24	48	18	30
25	44	19	25
26	35	12	23
27	43	11	32
28	39	22	17
29	34	25	9
30	13	16	−3
31	73	13	60
32	25	11	14
33	27	13	14

SOURCE: Rosenbaum (2002b, p. 82). Reprinted with kind permission of Springer Science + Business Media.

effect was statistically significant at the .0001 level. Even though the study used matching on two covariates to remove selection bias, to what extent was this study finding sensitive to hidden bias?

The sensitivity analysis for the matched pair study using Wilcoxon's signed-rank test involves the following steps.

Step 1: Compute ranked absolute differences d_s. This step includes the following procedures. Take the absolute value of differences, sort the data in an ascending order of the absolute differences, and create d_s that ranks the absolute value of differences and adjusts for ties. The results of Step 1 are shown in Table 8.5. Note that the first data line in this table shows the information for Pair 15, rather than Pair 1 as in Table 8.4. This is because Pair 15's absolute value of difference is 0, and after the sorting procedure, Pair 15 appears on the first line because it has the lowest absolute difference value in the sample. Note how the column of d_s is determined: d_s was first determined on the basis of the pair's rank and then adjusted in value for tied cases. For instance, among the first four cases, the second and third cases are tied. So the d_s value for each case is the average rank or $d_s = (2 + 3)/2 = 2.5$. Other tied pairs include Pairs 18 and 30, Pairs 4 and 11, Pairs 17 and 29, Pairs 32 and 33, Pairs 10 and 22, and Pairs 3 and 26; the d_s values for these pairs are all average ranks.

Step 2: Compute the Wilcoxon signed-rank statistic for the outcome difference between treated and control groups. Table 8.6 shows results of the procedures under Step 2. First, we calculate two variables, c_{s1} and c_{s2}, which are comparisons of outcome values between the exposed child and the control child in each pair: Let $c_{s1} = 1$ if "the outcome of the exposed child" > "the outcome of the control"; $c_{s1} = 0$ otherwise; and $c_{s1} = 0$, if "the outcome of the exposed child" = "the outcome of the control." Similarly, let $c_{s2} = 1$ if "the outcome of the control" > "the outcome of the exposed child"; $c_{s2} = 0$ otherwise; and $c_{s2} = 0$, if "the outcome of the exposed child" = "the outcome of the control." The computations of the two variables c_{s1} and c_{s2} are shown in columns (G) and (H). Next, we need to computed $\sum_{i=1}^{2} c_{si}Z_{si}$ where $Z_{s1} = 1$ for the treated case, and $Z_{s2} = 0$ for the control case. Results of this computation are shown in column (I). Next, we need to compute $d_s \sum_{i=1}^{2} c_{si}Z_{si}$ which is a product of d_s and $\sum_{i=1}^{2} c_{si}Z_{si}$ (i.e., column $[J]$). Finally, we take the sum of column (J) to obtain the Wilcoxon signed-rank statistic for the outcome difference between groups, which equals 527.

Step 3: Compute needed statistics for obtaining the one-sided significance level for the standardized deviate when $\Gamma = 1$. Table 8.7 shows the results of the procedures related to Step 3. Note that for the purpose of clarity, we deleted columns (A) to

(Text continued on page 308)

Table 8.5   Exhibit of Step 1: Take the Absolute Value of Differences, Sort the Data in an Ascending Order of the Absolute Differences, and Create d_s That Ranks the Absolute Value of Differences and Adjusts for Ties

Pair	Exposed	Control	Difference	Absolute Value of Difference	d_s	
15	16	16	0	0	1	
12	14	13	1	1	2.5	
21	22	21	1	1	2.5	Among the first 4 cases, the 2nd and 3rd cases are tied. So the d_s value for each case is the average rank. That is, $d_s = (2+3)/2 = 2.5$.
13	21	19	2	2	4	
18	10	13	-3	3	5.5	
30	13	16	-3	3	5.5	
16	20	16	4	4	7	
2	23	18	5	5	8	
4	18	24	-6	6	9.5	
11	24	18	6	6	9.5	
14	17	10	7	7	11	
17	15	24	-9	9	12.5	
29	34	25	9	9	12.5	
7	23	10	13	13	14	
32	25	11	14	14	15.5	
33	27	13	14	14	15.5	
9	31	16	15	15	17	
10	34	18	16	16	18.5	
22	35	19	16	16	18.5	
28	39	22	17	17	20	
5	37	19	18	18	21	
1	38	16	22	22	22	
3	41	18	23	23	23.5	
26	35	12	23	23	23.5	
6	36	11	25	25	26	
20	39	14	25	25	26	
25	44	19	25	25	26	
24	48	18	30	30	28	
27	43	11	32	32	29	
19	45	9	36	36	30	
23	49	7	42	42	31	
8	62	15	47	47	32	
31	73	13	60	60	33	

Table 8.6 Exhibit of Step 2: Calculate the Wilcoxon's Signed-Rank Statistic for the Differences in the Outcome Variable Between Treated and Control Groups

Let $c_{s1} = 1$ if "Exposed" > "Control"; $c_{s1} = 0$ otherwise; and $c_{s1} = 0$, if "Exposed" = "Control."

Let $c_{s2} = 1$ if "Control" > "Exposed"; $c_{s2} = 0$ otherwise; and $c_{s2} = 0$, if "Exposed" = "Control."

$= (c_{s1}*1) + (c_{s2}*0)$;
$(I) = (G)*1 + (H)*0$

$(J) = (F)*(I)$

Pair	Exposed	Control	Difference	Absolute Value of Difference	d_s	c_{s1}	c_{s2}	$\sum_{i=1}^{2} c_s^i Z_s^i$	$d_s \sum_{i=1}^{2} c_s^i Z_s^i$
(A)	(B)	(C)	(D)	(E)	(F)	(G)	(H)	(I)	(J)
15	16	16	0	0	1	0	0	0	0
12	14	13	1	1	2.5	1	0	1	2.5
21	22	21	1	1	2.5	1	0	1	2.5
13	21	19	2	2	4	1	0	1	4
18	10	13	−3	3	5.5	0	1	0	0
30	13	16	−3	3	5.5	0	1	0	0
16	20	16	4	4	7	1	0	1	7
2	23	18	5	5	8	1	0	1	8
4	18	24	−6	6	9.5	0	1	0	0
11	24	18	6	6	9.5	1	0	1	9.5
14	17	10	7	7	11	1	0	1	11
17	15	24	−9	9	12.5	0	1	0	0
29	34	25	9	9	12.5	1	0	1	12.5
7	23	10	13	13	14	1	0	1	14
32	25	11	14	14	15.5	1	0	1	15.5
33	27	13	14	14	15.5	1	0	1	15.5
9	31	16	15	15	17	1	0	1	17

(Continued)

303

Table 8.6 (Continued)

Pair	Exposed	Control	Difference	Absolute Value of Difference	d_s	c_{s1}	c_{s2}	$\sum_{i=1}^2 c_s^i Z_s^i$	$d_s \sum_{i=1}^2 c_s^i Z_s^i$
(A)	(B)	(C)	(D)	(E)	(F)	(G)	(H)	(I)	(J)
10	34	18	16	16	18.5	1	0	1	18.5
22	35	19	16	16	18.5	1	0	1	18.5
28	39	22	17	17	20	1	0	1	20
5	37	19	18	18	21	1	0	1	21
1	38	16	22	22	22	1	0	1	22
3	41	18	23	23	23.5	1	0	1	23.5
26	35	12	23	23	23.5	1	0	1	23.5
6	36	11	25	25	26	1	0	1	26
20	39	14	25	25	26	1	0	1	26
25	44	19	25	25	26	1	0	1	26
24	48	18	30	30	28	1	0	1	28
27	43	11	32	32	29	1	0	1	29
19	45	9	36	36	30	1	0	1	30
23	49	7	42	42	31	1	0	1	31
8	62	15	47	47	32	1	0	1	32
31	73	13	60	60	33	1	0	1	33
									527

Let $c_{s1} = 1$ if "Exposed" > "Control", $c_{s1} = 0$ otherwise; and $c_{s1} = 0$, if "Exposed" = "Control."

Let $c_{s2} = 1$ if "Control" > "Exposed", $c_{s2} = 0$ otherwise; and $c_{s2} = 0$, if "Exposed" = "Control."

$= (c_{s1} * 1) + (c_{s2} * 0)$;
$(I) = (G) * 1 + (H) * 0$

$(J) = (F) * (I)$

Take the sum of column (J) to obtain 527. This is the Wilcoxon signed rank statistic for the differences in the outcome between treated and control groups.

304

Table 8.7 Exhibit of Step 3: Calculate Statistics Necessary for Obtaining the One-Sided Significance Level for the Standardized Deviate When $\Gamma = 1$

$P_s^+ = 0$, if $c_{s1} = c_{s2} = 0$

$P_s^+ = 1$, if $c_{s1} = c_{s2} = 1$

$P_s^+ = \dfrac{\Gamma}{1+\Gamma} = \dfrac{1}{2} = .5$, if $c_{s1} \neq c_{s2}$

$P_s^- = 0$, if $c_{s1} = c_{s2} = 0$

$P_s^- = 1$, if $c_{s1} = c_{s2} = 1$

$P_s^+ = \dfrac{\Gamma}{1+\Gamma} = \dfrac{1}{2} = .5$, if $c_{s1} \neq c_{s2}$

When $\Gamma = 1$:

d_s	c_{s1}	c_{s2}	$\sum_{i=1}^{2} c_{si} Z_{si}$	$d_s \sum_{i=1}^{2} c_{si} Z_{si}$	P_s^+	P_s^-	$d_s P_s^+$	d_s^2	$d_s^2 P_s^+ (1-P_s^+)$
(F)	(G)	(H)	(I)	(J)	(K)	(L)	(M) = (K)*(F)	(N) = (F)*(F)	(O) = (N)*(K)*[1−(K)]
1	0	0	0	0	0	0	0.0000	1.0000	0.0000
2.5	1	0	1	2.5	0.5	0.5	1.2500	6.2500	1.5625
2.5	1	0	1	2.5	0.5	0.5	1.2500	6.2500	1.5625
4	1	0	1	4	0.5	0.5	2.0000	16.0000	4.0000
5.5	0	1	0	0	0.5	0.5	2.7500	30.2500	7.5625
5.5	0	1	0	0	0.5	0.5	2.7500	30.2500	7.5625
7	1	0	1	7	0.5	0.5	3.5000	49.0000	12.2500
8	1	0	1	8	0.5	0.5	4.0000	64.0000	16.0000
9.5	0	1	0	0	0.5	0.5	4.7500	90.2500	22.5625

(Continued)

305

Table 8.7 (Continued)

$P_s^+ = 0$, if $c_{s1} = c_{s2} = 0$
$P_s^+ = 1$, if $c_{s1} = c_{s2} = 1$
$P_s^+ = \dfrac{\Gamma}{1+\Gamma} = \dfrac{1}{2} = .5$, if $c_{s1} \neq c_{s2}$

$P_s^- = 0$, if $c_{s1} = c_{s2} = 0$
$P_s^- = 1$, if $c_{s1} = c_{s2} = 1$
$P_s^- = \dfrac{\Gamma}{1+\Gamma} = \dfrac{1}{2} = .5$, if $c_{s1} \neq c_{s2}$

When $\Gamma = 1$:

d_s	c_{s1}	c_{s2}	$\sum_{i=1}^2 c_{si} Z_{si}$	$d_s \sum_{i=1}^2 c_{si} Z_{si}$	P_s^+	P_s^-	$d_s P^+$	d_s^2	$d_s^2 P_s^+ (1 - P_s^+)$
(F)	(G)	(H)	(I)	(J)	(K)	(L)	$(M) = (K)\star(F)$	$(N) = (F)\star(F)$	$(O) = (N)\star(K)\star[1-(K)]$
9.5	1	0	1	9.5	0.5	0.5	4.7500	90.2500	22.5625
11	1	0	1	11	0.5	0.5	5.5000	121.0000	30.2500
12.5	0	1	0	0	0.5	0.5	6.2500	156.2500	39.0625
12.5	1	0	1	12.5	0.5	0.5	6.2500	156.2500	39.0625
14	1	0	1	14	0.5	0.5	7.0000	196.0000	49.0000
15.5	1	0	1	15.5	0.5	0.5	7.7500	240.2500	60.0625
15.5	1	0	1	15.5	0.5	0.5	7.7500	240.2500	60.0625
17	1	0	1	17	0.5	0.5	8.5000	289.0000	72.2500
18.5	1	0	1	18.5	0.5	0.5	9.2500	342.2500	85.5625
18.5	1	0	1	18.5	0.5	0.5	9.2500	342.2500	85.5625
20	1	0	1	20	0.5	0.5	10.0000	400.0000	100.0000
21	1	0	1	21	0.5	0.5	10.5000	441.0000	110.2500
22	1	0	1	22	0.5	0.5	11.0000	484.0000	121.0000

23.5	1	0	23.5	0.5	11.7500	552.2500	138.0625
23.5	1	0	23.5	0.5	11.7500	552.2500	138.0625
26	1	0	26	0.5	13.0000	676.0000	169.0000
26	1	0	26	0.5	13.0000	676.0000	169.0000
26	1	0	26	0.5	13.0000	676.0000	169.0000
28	1	0	28	0.5	14.0000	784.0000	196.0000
29	1	0	29	0.5	14.5000	841.0000	210.2500
30	1	0	30	0.5	15.0000	900.0000	225.0000
31	1	0	31	0.5	15.5000	961.0000	240.2500
32	1	0	32	0.5	16.0000	1024.0000	256.0000
33	1	0	33	0.5	16.5000	1089.0000	272.2500
			527		280.0000		3130.6250

Take the sum of column (M). This is $E(T^+)$, or the expectation of the signed rank statistic under the null hypothesis of no treatment effect.

Take the sum of column (O). This is $\mathrm{Var}(T^+)$, or

$$\mathrm{Var}(T^+) = \sum_{s=1}^{s} d_s^2 P_s^+ (1 - P_s^+).$$

(E) from this table. In Step 3, we first need to calculate P_s^+ and P_s^- using the following rules:

$$P_s^+ = 0, \quad \text{if } c_{s1} = c_{s2} = 0; \qquad\qquad P_s^- = 0, \quad \text{if } c_{s1} = c_{s2} = 0;$$
$$P_s^+ = 1, \quad \text{if } c_{s1} = c_{s2} = 1; \qquad\qquad P_s^- = 1, \quad \text{if } c_{s1} = c_{s2} = 1;$$
$$P_s^+ = \frac{\Gamma}{1+\Gamma} = \frac{1}{2} = .5, \quad \text{if } c_{s1} \neq c_{s2}; \qquad P_s^- = \frac{1}{1+\Gamma} = \frac{1}{2} = .5, \quad \text{if } c_{s1} \neq c_{s2}$$

Next, by multiplying d_s with P_s^+ we obtained column (M); taking the sum of column (M), we obtained $E(T^+) = 280$, which is the expectation for the signed-rank statistic under the null hypothesis of no treatment effect. Our next calculation is to compute the variance of signed-rank statistic under the null hypothesis of no treatment effect or $\text{Var}(T^+) = \sum_{s=1}^{S} d_s^2 P_s^+ (1 - P_s^+)$. The last two columns (N) and (O) show the calculation of the variance. After we create column (O), we take the sum of column (O) to obtain the variance, that is, $\text{Var}(T^+) = \sum_{s=1}^{S} d_s^2 P_s^+ (1 - P_s^+)$. Finally, using the Wilcoxon signed-rank statistic for the outcome difference between groups of 527, $E(T^+) = 280$, and $\text{Var}(T^+) = 3130.625$, we can calculate the standard deviate as $(527 - 280)/\sqrt{3130.63} = 4.414$. This statistic follows a standard normal distribution. By checking a table of standard normal distribution or using the normal distribution function of a spreadsheet, we obtain an approximate one-sided significance level of $p < .0001$.[3] The lower and upper bounds of the p value for the standard deviate when $\Gamma = 1$ are the same; that is, the minimum and maximum p values are both less than .0001.

Step 4: Compute needed statistics for obtaining the one-sided significance levels for the standardized deviates (i.e., the lower and upper bounds of p values) when $\Gamma = 2$ and when $\Gamma = $ other values. If the study were free of hidden bias, we would find strong evidence that a parent's occupational exposure to lead increased the level of lead in their children's blood. The sensitivity analysis asks how this conclusion might be changed by hidden biases of various magnitudes. Therefore, we need to compute the one-sided significance levels for standardized deviates (i.e., the lower and upper bounds of p value) under other values of Γ. Table 8.8 shows the calculation of the p values when $\Gamma = 2$. The calculation of the needed statistics for $\Gamma = 2$ is similar to the procedure of Step 3 when $\Gamma = 1$. The difference is that when $\Gamma > 1$, we also need to calculate $E(T^-)$ and $\text{Var}(T^-)$, which are the expectation and variance needed for computing the lower bound of the p value. In this step, as shown in Table 8.8, the calculation of $E(T^+)$ and $\text{Var}(T^+)$ follows the same procedure we followed for $\Gamma = 1$. The table shows that $E(T^+) = 373.3333$ and $\text{Var}(T^+) = 2782.7778$. Note that the column (N), $d_s P^-$, is an addition to this table, and is the product of d_s and P^- (i.e., column (N) = column (L) multiplied by column (F)). Taking the sum of column (N), we obtained $E(T^-) = 186.6667$. The variance for the lower bound statistic $\text{Var}(T^-)$ is the same as $\text{Var}(T^+)$, so $\text{Var}(T^-) = 2782.7778$. Given these statistics, we can calculate the deviates as follows: The deviate for the upper bound is $(527 - 373.33)/\sqrt{2782.78} = 2.91$; and the deviate for the lower bound is $(527 - 186.67)/\sqrt{2782.78} = 6.45$. Checking a table of standard normal distribution or using the normal distribution function of a spreadsheet, we find that the p value associated with 2.91 is .0018, and the p value associated with 6.45 is less than .0001. Because this interval of p values does not approach a nonsignificant cutoff value of .05, a higher value of Γ can be used for further testing. We can then replicate Step 4 for other Γ values, such as $\Gamma = 3$, $\Gamma = 4$. To ease our exposition, we jump to the calculation of the lower and upper bounds of p values when $\Gamma = 4.25$. It is at this value of Γ that the interval becomes uninformative. Table 8.9 presents results of the calculation of needed statistics for the p values when $\Gamma = 4.25$. As shown in the table, when $\Gamma = 4.25$, we have $E(T^+) = 453.3333$, $\text{Var}(T^+) = 1930.9070$, $E(T^-) = 106.6667$. Therefore, the deviate for the upper bound is $(527 - 453.33)/\sqrt{1930.91} = 1.676$; and the deviate for the lower bound is $(527 - 106.67)/\sqrt{1930.91} = 9.566$. Checking a table of standard normal distribution or using the normal distribution function of a spreadsheet, we find that the p value associated with 1.676 is .0468, and the p value associated with 9.566 is less than .0001.

Table 8.10 shows the range of significance levels for the test statistic when G is equal to various values. As depicted, G = 4.25 is the value at which the significance interval becomes uninformative. Rosenbaum (200b) interpreted this finding in the following way:

> The table shows that to explain away the observed association between parental exposure to lead and child's lead level, a hidden bias or unobserved covariate would need to increase the odds of exposure by more than a factor of $\Gamma = 4.25$. (p. 115)

Thus, based on this study, it appears that there is an association between parental occupational exposure to lead and child blood lead level. Moreover, the study finding is robust against hidden bias.

8.4 Overview of the Stata Program *rbounds*

Few software programs are available to conduct the kind of sensitivity analyses developed by Rosenbaum. But the Stata user-developed program *rbounds* (Gangl, 2007) allows users to perform sensitivity analyses for matched pair studies using

(Text continued on page 316)

Table 8.8 Exhibit of Step 4: Calculate Needed Statistics for Obtaining the One-Sided Significance Levels for the Standardized Deviates (i.e., the Lower and Upper Bounds of p Value) When $\Gamma = 2$

Box for P_s^+:

$P_s^+ = 0$, if $c_{s1} = c_{s2} = 0$,

$P_s^+ = 1$, if $c_{s1} = c_{s2} = 1$

$P_s^+ = \dfrac{\Gamma}{1+\Gamma} = \dfrac{2}{3} = .666667$, if $c_{s1} \neq c_{s2}$

Box for P_s^-:

$P_s^- = 0$, if $c_{s1} = c_{s2} = 0$

$P_s^- = 1$, if $c_{s1} = c_{s2} = 1$

$P_s^- = \dfrac{\Gamma}{1+\Gamma} = \dfrac{1}{3} = .333333$, if $c_{s1} \neq c_{s2}$

When $\Gamma = 2$:

d_s	c_{s1}	c_{s2}	$\sum_{i=1}^2 c_{si} Z_{si}$	$d_s \sum_{i=1}^2 c_{si} Z_{si}$	P_s^+	P_s^-	$d_s P^+$	$d_s P^-$	d_s^2	$d_s^2 P_s^+ (1 - P_s^+)$
(F)	( )	(H)	(I)	(J)	(K)	(L)	(M) = (K)*(F)	(N) = (L)*(F)	(O) = (F)*(F)	(P) = (O)*(K)*[1-(K)]
1	0	0	0	0	0.0000	0.0000	0.0000	0.0000	1.0000	0.0000
2.5	1	0	1	2.5	0.6667	0.3333	1.6667	0.8333	6.2500	1.3889
2.5	1	0	1	2.5	0.6667	0.3333	1.6667	0.8333	6.2500	1.3889
4	1	0	1	4	0.6667	0.3333	2.6667	1.3333	16.0000	3.5556
5.5	0	1	0	0	0.6667	0.3333	3.6667	1.8333	30.2500	6.7222
5.5	0	1	0	0	0.6667	0.3333	3.6667	1.8333	30.2500	6.7222
7	1	0	1	7	0.6667	0.3333	4.6667	2.3333	49.0000	10.8889
8	1	0	1	8	0.6667	0.3333	5.3333	2.6667	64.0000	14.2222

9.5	0	1	0	0.6667	0.3333	6.3333	3.1667	90.2500	20.0556
9.5	1	0	9.5	0.6667	0.3333	6.3333	3.1667	90.2500	20.0556
11	1	0	11	0.6667	0.3333	7.3333	3.6667	121.0000	26.8889
12.5	0	1	0	0.6667	0.3333	8.3333	4.1667	156.2500	34.7222
12.5	1	0	12.5	0.6667	0.3333	8.3333	4.1667	156.2500	34.7222
14	1	0	14	0.6667	0.3333	9.3333	4.6667	196.0000	43.5556
15.5	1	0	15.5	0.6667	0.3333	10.3333	5.1667	240.2500	53.3889
15.5	1	0	15.5	0.6667	0.3333	10.3333	5.1667	240.2500	53.3889
17	1	0	17	0.6667	0.3333	11.3333	5.6667	289.0000	64.2222
18.5	1	0	18.5	0.6667	0.3333	12.3333	6.1667	342.2500	76.0556
18.5	1	0	18.5	0.6667	0.3333	12.3333	6.1667	342.2500	76.0556
20	1	0	20	0.6667	0.3333	13.3333	6.6667	400.0000	88.8889
21	1	0	21	0.6667	0.3333	14.0000	7.0000	441.0000	98.0000
22	1	0	22	0.6667	0.3333	14.6667	7.3333	484.0000	107.5556
23.5	1	0	23.5	0.6667	0.3333	15.6667	7.8333	552.2500	122.7222
23.5	1	0	23.5	0.6667	0.3333	15.6667	7.8333	552.2500	122.7222
26	1	0	26	0.6667	0.3333	17.3333	8.6667	676.0000	150.2222
26	1	0	26	0.6667	0.3333	17.3333	8.6667	676.0000	150.2222
26	1	0	26	0.6667	0.3333	17.3333	8.6667	676.0000	150.2222

(Continued)

Table 8.8 (Continued)

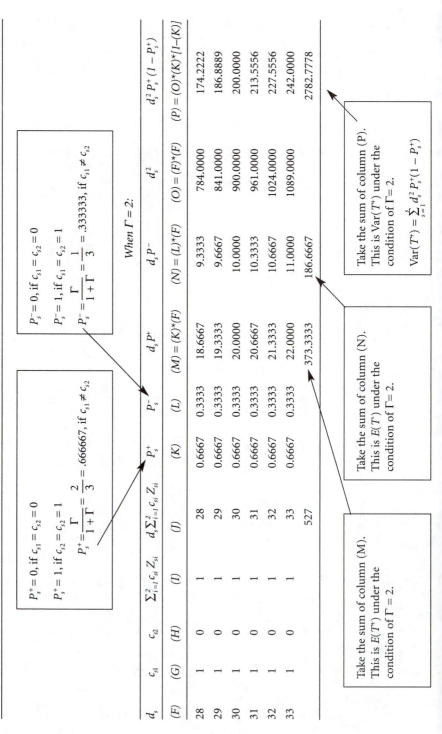

$P_s^+ = 0$, if $c_{s1} = c_{s2} = 0$

$P_s^+ = 1$, if $c_{s2} = c_{s2} = 1$

$P_s^+ = \dfrac{\Gamma}{1+\Gamma} = \dfrac{2}{3} = .666667$, if $c_{s1} \neq c_{s2}$

$P_s^- = 0$, if $c_{s1} = c_{s2} = 0$

$P_s^- = 1$, if $c_{s1} = c_{s2} = 1$

$P_s^- = \dfrac{\Gamma}{1+\Gamma} = \dfrac{1}{3} = .333333$, if $c_{s1} \neq c_{s2}$

When $\Gamma = 2$:

d_s	c_{s1}	c_{s2}	$\sum_{i=1}^2 c_{si} Z_{si}$	$d_s \sum_{i=1}^2 c_{si} Z_{si}$	P_s^+	P_s^-	$d_s P_s^+$	$d_s P_s^-$	d_s^2	$d_s^2 P_s^+(1 - P_s^+)$
(F)	(G)	(H)	(I)	(I)	(K)	(L)	(M) = (K)*(F)	(N) = (L)*(F)	(O) = (F)*(F)	(P) = (O)*(K)*[1−(K)]
28	1	0	1	28	0.6667	0.3333	18.6667	9.3333	784.0000	174.2222
29	1	0	1	29	0.6667	0.3333	19.3333	9.6667	841.0000	186.8889
30	1	0	1	30	0.6667	0.3333	20.0000	10.0000	900.0000	200.0000
31	1	0	1	31	0.6667	0.3333	20.6667	10.3333	961.0000	213.5556
32	1	0	1	32	0.6667	0.3333	21.3333	10.6667	1024.0000	227.5556
33	1	0	1	33	0.6667	0.3333	22.0000	11.0000	1089.0000	242.0000
				527			373.3333	186.6667		2782.7778

Take the sum of column (M). This is $E(T^+)$ under the condition of $\Gamma = 2$.

Take the sum of column (N). This is $E(T^-)$ under the condition of $\Gamma = 2$.

Take the sum of column (P). This is $\mathrm{Var}(T^+)$ under the condition of $\Gamma = 2$.

$$\mathrm{Var}(T^+) = \sum_{s=1}^s d_s^2 P_s^+(1 - P_s^+)$$

312

Table 8.9 Exhibit of Step 4: Calculate Needed Statistics for Obtaining the One-Sided Significance Levels for the Standardized Deviates (i.e., the Lower and Upper Bounds of p Value) When $\Gamma = 4.25$

$$P_s^+ = 0, \text{ if } c_{s1} = c_{s2} = 0$$
$$P_s^+ = 1, \text{ if } c_{s1} = c_{s2} = 1$$
$$P_s^+ = \frac{\Gamma}{1+\Gamma} = \frac{4.25}{5.25} = .809524, \text{ if } c_{s1} \neq c_{s2}$$

$$P_s^- = 0, \text{ if } c_{s1} = c_{s2} = 0$$
$$P_s^- = 1, \text{ if } c_{s1} = c_{s2} = 1$$
$$P_s^- = \frac{1}{1+\Gamma} = \frac{1}{5.25} = .190476, \text{ if } c_{s1} \neq c_{s2}$$

When $\Gamma = 4.25$:

d_s	c_{s1}	c_{s2}	$\sum_{i=1}^2 c_i Z_{si}$	$d_s \sum_{i=1}^2 c_{si} Z_{si}$	P_s^+	P_s^-	$d_s P^+$	$d_s P^-$	d_s^2	$d_s^2 P_s^+(1 - P_s^+)$
(F)	(G)	(H)	(I)	(J)	(K)	(L)	$(M) = (K)^\star(F)$	$(N) = (L)^\star(F)$	$(O) = (F)^\star(F)$	$(P) = (O)^\star(K)^\star[1-(K)]$
1	0	0	0	0	0.0000	0.0000	0.0000	0.0000	1.0000	0.0000
2.5	1	0	1	2.5	0.8095	0.1905	2.0238	0.4762	6.2500	0.9637
2.5	1	0	1	2.5	0.8095	0.1905	2.0238	0.4762	6.2500	0.9637
4	1	0	1	4	0.8095	0.1905	3.2381	0.7619	16.0000	2.4671
5.5	0	1	0	0	0.8095	0.1905	4.4524	1.0476	30.2500	4.6644
5.5	0	1	0	0	0.8095	0.1905	4.4524	1.0476	30.2500	4.6644
7	1	0	1	7	0.8095	0.1905	5.6667	1.3333	49.0000	7.5556
8	1	0	1	8	0.8095	0.1905	6.4762	1.5238	64.0000	9.8685

(Continued)

313

Table 8.9 (Continued)

$P_s^+ = 0$, if $c_{s1} = c_{s2} = 0$

$P_s^+ = 1$, if $c_{s1} = c_{s2} = 1$

$P_s^+ = \dfrac{\Gamma}{1+\Gamma} = \dfrac{4.25}{5.25} = .809524$, if $c_{s1} \neq c_{s2}$

$P_s^- = 0$, if $c_{s1} = c_{s2} = 0$

$P_s^- = 1$, if $c_{s1} = c_{s2} = 1$

$P_s^- = \dfrac{1}{5.25} = .190476$, if $c_{s1} \neq c_{s2}$

When $\Gamma = 4.25$:

d_s	c_{s1}	c_{s2}	$\sum_{i=1}^2 c_{si} Z_{si}$	$d_s \sum_{i=1}^2 c_{si} Z_{si}$	P_s^+	P_s^-	$d_s P^+$	$d_s P^-$	d_s^2	$d_s^2 P_s^+ (1-P_s^+)$
(F)	(G)	(H)	(I)	(J)	(K)	(L)	(M) = (K)*(F)	(N) = (L)*(F)	(O) = (F)*(F)	(P) = (O)*(K)*[1−(K)]
9.5	0	1	0	0	0.8095	0.1905	7.6905	1.8095	90.2500	13.9161
9.5	1	0	1	9.5	0.8095	0.1905	7.6905	1.8095	90.2500	13.9161
11	1	0	1	11	0.8095	0.1905	8.9048	2.0952	121.0000	18.6576
12.5	0	1	0	0	0.8095	0.1905	10.1190	2.3810	156.2500	24.0930
12.5	1	0	1	12.5	0.8095	0.1905	10.1190	2.3810	156.2500	24.0930
14	1	0	1	14	0.8095	0.1905	11.3333	2.6667	196.0000	30.2222
15.5	1	0	1	15.5	0.8095	0.1905	12.5476	2.9524	240.2500	37.0454
15.5	1	0	1	15.5	0.8095	0.1905	12.5476	2.9524	240.2500	37.0454
17	1	0	1	17	0.8095	0.1905	13.7619	3.2381	289.0000	44.5624
18.5	1	0	1	18.5	0.8095	0.1905	14.9762	3.5238	342.2500	52.7732

1	0	1	18.5	0.8095	0.1905	14.9762	3.5238	342.2500	52.7732
1	0	1	20	0.8095	0.1905	16.1905	3.8095	400.0000	61.6780
1	0	1	21	0.8095	0.1905	17.0000	4.0000	441.0000	68.0000
1	0	1	22	0.8095	0.1905	17.8095	4.1905	484.0000	74.6304
1	0	1	23.5	0.8095	0.1905	19.0238	4.4762	552.2500	85.1542
1	0	1	23.5	0.8095	0.1905	19.0238	4.4762	552.2500	85.1542
1	0	1	26	0.8095	0.1905	21.0476	4.9524	676.0000	104.2358
1	0	1	26	0.8095	0.1905	21.0476	4.9524	676.0000	104.2358
1	0	1	26	0.8095	0.1905	21.0476	4.9524	676.0000	104.2358
1	0	1	28	0.8095	0.1905	22.6667	5.3333	784.0000	120.8889
1	0	1	29	0.8095	0.1905	23.4762	5.5238	841.0000	129.6780
1	0	1	30	0.8095	0.1905	24.2857	5.7143	900.0000	138.7755
1	0	1	31	0.8095	0.1905	25.0952	5.9048	961.0000	148.1814
1	0	1	32	0.8095	0.1905	25.9048	6.0952	1024.0000	157.8957
1	0	1	33	0.8095	0.1905	26.7143	6.2857	1089.0000	167.9184
			527			453.3333	106.6667		1930.9070

Take the sum of column (M). This is $E(T^+)$ under the condition of $\Gamma = 4.25$.

Take the sum of column (N). This is $E(T^-)$ under the condition of $\Gamma = 4.25$.

Take the sum of column (P). This is $\mathrm{Var}(T^+)$ under the condition of $\Gamma = 4.25$.

$$\mathrm{Var}(T^+) = \sum_{s=1}^{s} d_s^2 \, P_s^+ (1 - P_s^+)$$

Table 8.10   Results of the Sensitivity Analysis for Blood Lead Levels of Children: Range of Significance Levels for the Signed-Rank Statistic

Γ	Minimum	Maximum
1	< .0001	< .0001
2	< .0001	.0018
3	< .0001	.0136
4	< .0001	.0388
4.25	< .0001	.0468
5	< .0001	.0740

SOURCE: Rosenbaum (2002b, p. 115). Reprinted with kind permission of Springer Science + Business Media.

the Wilcoxon's signed-rank test as well as the Hodges-Lehmann point and interval estimates. As a user-developed program, *rbounds* is not included in the regular Stata package, but the program can be downloaded from the Internet. To locate this software on the Internet, Stata users can use the *findit* command followed by *rbounds* (i.e., *findit rbounds*), and then follow the online instructions to download and install the program. After installing *rbounds*, users can access the basic instructions for running the program by going to the program's *help* file.

The *rbounds* program is started with the following syntax:

rbounds *varname, gamma(numlist)*

In this command, *varname* is the outcome difference between the treated group and the control group. Before running the *rbounds* program, users should organize the data files at the pair level, so that each data line represents pair information. Specifically, the data should look like the data shown in Table 8.4. The data file should contain the following two variables: pair and difference, with difference being a *varname* the user specifies in the command. *gamma(numlist)* is the only required key word. It specifies the values of Γ for the sensitivity analysis. Users specify particular Γ values in the parentheses. By running the command, *rbounds* returns the minimum and maximum values of the p value using the Wilcoxon's signed-rank test, the minimum and maximum values of the Hodges-Lehmann point estimate, and the lower and upper bounds of the 95% confidence interval (i.e., the default) of the Hodges-Lehmann interval estimate.

8.5 Examples

8.5.1 SENSITIVITY ANALYSIS OF THE EFFECTS OF LEAD EXPOSURE

Table 8.11 exhibits the syntax and output of running *rbounds* using the sensitivity analysis for the study of the effects of lead exposure (see Section 8.3.2).

In the syntax, *dlead* is the variable showing outcome difference between a treated participant and a control for each pair, *gamma(1 (1) 4 4.25 5 6)* is a shortcut specification that forces the Γ value to start at 1, with increments of 1 (i.e., $\Gamma = 2, 3$) up to the value of $\Gamma = 4$, and then to use the listed Γ values 4.25, 5, and 6. In this case, (1) specified within the parentheses tells the program that the Γ value increases with an increment of 1. Thus, the specification is equivalent to *gamma(1 2 3 4 4.25 5 6)*.

Essentially, the output provides the minimum and maximum values of various inference quantities. The maximum and minimum *p* values using the Wilcoxon's signed-rank test are shown in the columns labeled as *sig+* and *sig–*; following the *sig–* column the output lists the Hodges-Lehmann point and interval estimates. Results of the output for *sig+* and *sig–* are identical to Table 8.10.

8.5.2 SENSITIVITY ANALYSIS FOR THE STUDY USING PAIR MATCHING

In Chapter 5, we conducted a regression analysis of difference scores based on a matched pair sample following pair matching (see Section 5.9.4). The study found that in 1997, children who used AFDC had an average letter-word

Table 8.11 Exhibit of Stata *rbounds* Syntax and Output (Example 8.5.1)

```
cd "D:\Sage\ch8"
use rbtest, clear
rbounds dlead, gamma(1 (1) 4 4.25 5 6)

(Output)
Rosenbaum bounds for dlead (N = 33 matched pairs)
```

Gamma	sig+	sig-	t-hat+	t-hat-	CI+	CI-
1	5.1e-06	5.1e-06	15	15	9.5	20.5
2	.00179	5.5e-11	10.25	19.5	4.5	27.5
3	.013615	6.7e-16	8	23	1	32.5
4	.03879	0	6.5	25	-1	37
4.25	.046825	0	6	25	-1.5	38.5
5	.073991	0	5	26.5	-3	42
6	.11502	0	4	28	-6	48

```
*  gamma  - log odds of differential assignment due to unobserved factors
   sig+   - upper bound significance level
   sig-   - lower bound significance level
   t-hat+ - upper bound Hodges-Lehmann point estimate
   t-hat- - lower bound Hodges-Lehmann point estimate
   CI+    - upper bound confidence interval (a=  .95)
   CI-    - lower bound confidence interval (a=  .95)
```

identification score that was 3.17 points lower than that of children who never used AFDC ($p < .05$). Furthermore, the study adjusted selection based on six observed variables (i.e., the ratio of family income to poverty threshold in 1996, caregiver's education in 1997, caregiver's history of using welfare, child's race, child's age in 1997, and child's gender). We now examine the same data in an attempt to determine to what extent this study is sensitive to hidden selection bias.

Results of the sensitivity analysis are shown in Table 8.12. Using the Wilcoxon's signed-rank test, the sensitivity analysis shows that the study becomes sensitive to hidden bias at $\Gamma = 1.43$. Because 1.43 is a small value, we can conclude that the study is very sensitive to hidden bias, and therefore, further analysis that controls for additional biases is warranted.

We can further compare the level of sensitivity to bias of the AFDC study with that of other studies. Table 8.13 presents results of sensitivity analyses for four observational studies (Rosenbaum, 2005, table 4). Among the four studies, Study 2 is the least sensitive to hidden bias; the study of the effects of diethylstilbestrol becomes sensitive at about $\Gamma = 7$. In contrast, Study 4 is the most sensitive to hidden bias; the study of the effects of coffee becomes sensitive at about $\Gamma = 1.3$. Rosenbaum (2005) provided the following explanation of the implications of the sensitivity analyses for these studies:

> A small bias could explain away the effects of coffee, but only an enormous bias could explain away the effects of diethylstilbestrol. The lead exposure study, although quite insensitive to hidden bias, is about halfway between these two other studies, and is slightly more sensitive to hidden bias than the study of the effects of smoking. (p. 1812)

Compared with these four studies, our study of the effect of welfare use on academic achievement is slightly better (i.e., less sensitive to unobserved bias)

Table 8.12   Results of the Sensitivity Analysis for the Study of Children's Letter-Word Identification Score: Range of Significance Levels for the Signed-Rank Statistic (Example 8.5.2)

Γ	Minimum	Maximum
1	< .00001	< .00001
1.3	< .00001	.0096
1.42	< .00001	.0420
1.43	< .00001	.0466
1.44	< .00001	.0515
1.45	< .00001	.0567
1.5	< .00001	.0887
2	< .00001	.7347

Table 8.13   Sensitivity to Hidden Bias in Four Observational Studies

Treatment	$\Gamma = 1$	$(\Gamma, Max\ p\ Value)$
Smoking/lung cancer (Study 1)	< .0001	(5, .03)
Diethylstilbestrol/vaginal cancer (Study 2)	< .0001	(7, .054)
Lead/blood lead (Study 3)	< .0001	(4.25, .047)
Coffee/myocardial infarction (Study 4)	.0038	(1.3, .056)

SOURCE: Rosenbaum (2005, table 4). Reprinted with permission from John Wiley & Sons.

Study 1: Hammond (1964).

Study 2: Herbst, Ulfelder, and Poskanzer (1971).

Study 3: Morton, Saah, Silberg, Owens, Roberts, and Saah (1982).

Study 4: Jick, Miettinen, Neff, Jick, Miettinen, Neff, Shapiro, Heinonen, and Slone (1973).

than the study of the effects of coffee, but it is more sensitive to hidden bias than the three other studies.

8.6 Conclusions

Selection bias is the most challenging analytic problem in observational studies. Although corrective approaches have been developed, a valid application of these approaches requires broad knowledge and skill. Properly using corrective models involves (a) having a thorough understanding of the sources of selection bias, (b) conducting a careful investigation of existing data and literature to identify all possible covariates that might affect selection and be used as covariates in a correction effort, (c) developing an understanding of the fit between the data generation process and the assumptions in correction models, (d) providing a cautious interpretation of study findings that is conditioned on the tenability of assumptions, and (e) conducting a sensitivity analysis to gauge the level of sensitivity of findings to hidden bias.

Notes

1. For instance, sample size in our study was fixed at 500. We could change this value to see model performance under different settings of sample size. This feature is

helpful if the researcher needs to assess model properties when sample size is small. In fact, each fixed value used in the data generation can be changed, which allows a comparison of a set of scenarios for the parameter. We did not do this and fixed our settings at two because we attempted to accomplish a narrowly defined objective for this study.

2. Note that the selection equation includes Z only. This is not exactly equivalent to the propensity score matching model in which the logistic regression employs x_1, x_2, x_3, and Z. We tried the model specifying x_1, x_2, x_3, and Z in the selection equation, but the model did not converge. For comparative purposes, the current model captures the main features of Setting 1 and is the best possible model we can specify.

3. To find the p value, consult a table of the standard normal distribution (i.e., a Z table) or use an Excel function to obtain the p value by typing "=1-NORMSDIST(4.414)" in a cell. In this case, Excel returns a p value of .0000050739.

9

Concluding Remarks

I n this chapter, we conclude by making a few remarks on criticisms of observational studies, on the debate regarding the approximation of randomization using bias-correcting methods, and on directions for future development. Section 9.1 describes common pitfalls in observational studies. Section 9.2 highlights both the debate and the criticism of approximating experiments using propensity score approaches. Section 9.3 reviews advances in modeling causality that are methodologically different from those we have described. Here we discuss James Robins's marginal structural models and Judea Pearl's directed acyclic graphs (DAGs). Last, Section 9.4 speculates on future developments.

9.1 Common Pitfalls in Observational Studies: A Checklist for Critical Review

In evaluation and intervention research, it is common to think of designs as having differential capacity for inferring causal relationships between programs and observed outcomes. At the top of what is called the "evidentiary hierarchy" sits meta-analyses of randomized controlled trials (RCTs), which are viewed as superior to single RCTs. In turn, RCTs are viewed as superior to studies in which participants are not randomly assigned to treatment and control conditions. These goodness criteria for ranking research designs are used often in assessing proposals and in valuing the importance of findings.

Researchers have generally agreed that inferential methods based on the assumption of a randomized assignment mechanism are superior to other approaches. Indeed, Rubin (2008) recently argued,

> The existence of these assignment-based methods, and their success in practice, documents that the model for the assignment mechanism is more fundamental for inference for causal effects than a model for the science. These methods lead to concepts such as unbiased estimation and asymptotic confidence intervals

(due to Neyman), and *p*-values or significance levels for sharp null hypotheses (due to Fisher), all defined by the distribution of statistics (e.g., the difference of treatment and control sample means) induced by the assignment mechanism. In some contexts, such as the U.S. Food and Drug Administration's approval of a new drug, such assignment mechanism-based analyses are considered the gold standard for confirmatory inferences. (pp. 814–815)

In statistical analysis as opposed to research design, goodness criteria are less clear. We often argue that the method must fit the research question and that assumptions must always be met as a test of statistical conclusion validity (Shadish et al., 2002). In a rapidly developing field such as propensity score analysis, criteria may be murky because we are just beginning to understand the sensitivity of models to the assumptions on which they rest. As in propensity score analysis, we often have choices in the selection of statistical methods and our choices should fit the data situation. Toward a better understanding of when and how to use the four methods described in previous chapters, we list below 18 pitfalls that can trip up evaluators when they use propensity score methods.

1. *Mismatch of the research question, the design, and the analytic method:* Perhaps the most obvious pitfall of all involves a mismatch of the research question, the design, and the statistical method. When a study uses observational data to address research questions that are clearly related to causality (e.g., evaluating the effectiveness of a service or treatment), the research question and the design are mismatched. When causal attributions are to be made, control group designs with randomization are preferred. To be sure, some quasi-experimental designs—for example, regression-discontinuity designs— permit relatively strong causal inference. If a strong design is not used or if the design is compromised, bias-correcting methods should be used.

In the absence of a strong design and/or a propensity score approach (or other types of correction models), addressing research questions related to causal inference is ill-advised. We have explained why: In observational studies where treatment assignment is nonignorable, the treated and comparison groups are formed naturally, are prone to numerous selection biases, and may be imbalanced on observed and unobserved covariates. The treated and comparison groups cannot be compared analytically using common covariance adjustments such as ordinary least squares (OLS) regression or other analyses that do not explicitly control for selection. In these situations, propensity score methods may be used conditionally.

2. *Randomization failure:* A randomized experiment may fail in practice because the conditions required to implement randomization were either infeasible or simply not met. Although there are a variety of ways in which randomization can fail, it typically fails when the rules for assigning participants

to the treated and control conditions are inadvertently violated or purposively thwarted. Because it relies on probability theory, random assignment can also be implemented correctly but produce imbalanced groups because of an insufficient sample size. Group-randomized designs appear particularly vulnerable to selection when a small number of units—say, schools—is available. Finally, though it is technically not a failure of randomization, random assignment can be compromised by a variety of postrandomization effects in which participants react to assignment to treatment or control conditions. These include, for example, "spillover" or experimental contamination effects, in which participants in control conditions become aware of the elements of services provided to participants in treatment conditions.

Randomization failure becomes a research pitfall when data are analyzed as if the design were not compromised. To avoid this common mistake, sample balance must be assessed as a test of whether randomization has worked as planned. When a balance problem is found, remedial measures, such as use of a correction method developed for observational studies to adjust for bias, must be considered.

3. *Insufficient selection information:* Assuming now that the researcher intends to use a propensity score or other type of correction analysis, what pitfalls occur in the context of implementing a bias-correction strategy? It is at this point that early decisions in research design may affect the degree to which statistical methods can be used to adjust for imbalance. If covariates to explain potential selection were not included in the measurement model, insufficient information may be available for the selection equation or matching. At the design stage, selection biases observed in previous studies should be considered and used as a basis for the adoption of measures and instruments. Because unobserved bias is a more serious problem than observed bias, measures of potential hidden selection effects should be incorporated in data collection. We recommend that reviewing unmeasured variables affecting selection bias in prior studies and proposing to collect such data in a new study should be a routine element of proposal development and critically reviewing whether or not such elements are included in grant proposals should be a criterion for awarding funds.

4. *Insufficient pre- and postadjustment analysis:* In this case, the researcher may have properly measured potential selection effects and used a propensity score or other type of correction analysis, but analyses are not conducted to assess the effect of the adjustment strategy. The degree to which a statistical adjustment produces balance—that is, the success of correction using a model—must be demonstrated in reports. Specifically, a study fails to present sufficient information if it does not report pre- and postadjustment balances on covariates.

5. *Failure to evaluate assumptions related to the Heckit model:* Turning now to specific models, this pitfall occurs when a study uses the Heckit treatment effect model but does not provide information on the degree to which data meet the assumptions embedded in the model. Specifically, use of the Heckit model requires discussion of the (normal) distribution of the outcome variable, the (nonzero) correlation of the error terms of the regression and selection equations, the level of collinearity of independent variables included in both equations, and the size of the sample. The sample must be sufficiently large to permit the use of the maximum likelihood estimator.

6. *Failure to evaluate assumptions related to propensity score matching:* This problem arises when a study employs propensity score matching but does not fully explain whether data meet the assumptions embedded in the model. When propensity score matching is used, the strongly ignorable treatment assignment and overlap assumptions in both the pre- and postmatching samples must be discussed explicitly and evidence of overt selection must be provided.

7. *Insufficient information on logistic regression:* In the same vein, the process used in estimating the selection equation should be described. An adequately described study will include description of the specifications of the logistic regression used to predict the propensity scores. For example, a study might be flawed because it has insufficient information about the rationale for choosing the conditioning variables and the specification of the functional forms of predictor variables. An adequate study should include discussion of alternative or competing conditioning models and description of the procedure used to derive the final logistic regression equation.

8. *Failure to show justification of caliper size:* In studies that use 1-to-1 nearest neighbor matching within calipers, the caliper size must be justified. The failure to justify the caliper size comes about through using only one caliper size without explaining why one size is adequate, failing to discuss the potential limitation of inexact matching or incomplete matching that is produced by a given caliper size, and failing to provide justification for using (or not using) other types of matching procedures, such as Mahalanobis metric matching. Relatedly, a substantial reduction of sample size after caliper-based matching is often troublesome and requires scrutiny because conclusions based on subsamples may differ from those based on original samples.

9. *Failure to discuss limitations of greedy matching:* Greedy matching has a number of limitations, and, in our view, they should always be discussed. Principal among the limitations of greedy matching is the assumption that the propensity scores of the treated and control groups overlap. In addition, estimated scores must be patterned so as to provide a common support region that

is of sufficient size for the method to work. Finally, findings based on greedy matching should be assessed against additional analyses using different approaches, such as optimal matching.

10. *Insufficient information on optimal matching procedures:* Use of optimal matching always involves a set of decisions. These include decisions about forcing a 1-to-1 pair matching, forcing a pair matching with a constant ratio of treated and control participants, specification of a minimum and maximum number of controls for each treated participant in a variable matching, and specification of a matching structure in a full matching. Together and separately, these decisions affect both the level of bias reduction and efficiency. A detailed description of these decisions ensures appropriate interpretation of study results, and it enhances replication.

11. *Erroneous selection of analytic procedures following optimal matching:* Sometimes the researcher may simply choose a wrong analytic procedure following optimal matching. Based on the matched sample following an optimal matching, it is not uncommon for analysts to err by using OLS regression with a dummy treatment variable instead of conducting regression adjustment using difference scores, outcome analysis with the Hodges-Lehmann aligned rank test, or other types of analyses (see Section 5.5).

12. *Failure to evaluate assumptions related to matching estimators:* Like other procedures, matching estimators are based on clear assumptions, which must be explored as a condition of appropriate use. These include the strongly ignorable treatment assignment and overlap assumptions.

13. *Failure to correct for bias in matching estimators:* When using matching estimators, a correction may be insufficient and additional applications may improve efficiency. For instance, when two or more continuous covariates are used, the researcher should not assume that matching is exact. With continuous covariates, the analysis needs to include bias-corrected matching that involves an additional regression adjustment (see Section 6.2.2). Additionally, when the treatment effect is not constant, researchers should employ a variance estimator allowing for heteroscedasticity (see Section 6.2.4).

14. *Failure to evaluate assumptions related to kernel-based matching:* So too, in kernel-based matching, the data must meet the assumptions embedded in the model. Specifically, the conditional and mean independence assumptions with regard to the treatment assignment must be evaluated. Kernel-based matching estimates an average treatment effect for the treated. This should not be confused with, or compared with, the sample (or population) average treatment effect.

15. *Failure to estimate outcomes under a variety of bandwidths:* In kernel-based matching, treatment effects are estimated within bandwidths. To our knowledge, there are no unequivocal criteria for the selection of the proper bandwidth. Thus, the robustness of findings should always be tested under different specifications of bandwidth values. Failure to do so leaves findings vulnerable to the speculation that the use of alternative bandwidths would have produced nontrivially different findings.

16. *Lack of trimming when local linear regression is used:* When local linear regression matching is used, matching treated participants is sometimes challenging at the region where controls are sparse. We have recommended the use of different trimming schemes to determine the robustness of findings. Failure to using trimming to assess the stability of findings leaves open the possibility that findings are related to marginal matches.

17. *Failure to note concerns regarding bootstrapping:* As we have indicated, kernel-based matching with bootstrapping continues to be controversial. Studies that use bootstrapping should warn reviewers and readers that findings derived from such a procedure may be prone to errors. This is a rapidly developing area, and, as more findings come out, we will post information on significance testing in the kernel-based matching to our Web site.

18. *Insufficient cross-validation:* The promising methods we have described are changing quickly, as researchers develop and revise algorithms and code. Our Monte Carlo analyses suggest that findings are sensitive to different data situations. So there is much work to be done. From a practice perspective, when it is difficult to test assumptions, cross-validation of findings becomes imperative. This last pitfall occurs when a researcher draws conclusions from an observational study by using one correction method but fails to provide a warning note that the findings have not been cross-validated by other methods. Cross-validation strengthens inference and generalization.

9.2 Approximating Experiments With Propensity Score Approaches

Over the past 30 years, methods have evolved in complexity as researchers have recognized the need to develop more efficient approaches for assessing the effects of social and health policies, including both federal and state programs that arise from legislative initiatives. As shown in this book, particularly useful advances have been made in the development of robust methods for observational studies. The criticism and reformulation of the classical experimental approach symbolize a shift in evaluation methods. Although it has been more than 30 years since Heckman (1978) published his first correction method, debate about correction methods is lively, and it has fueled the development of new approaches.

Proponents of the new methods posit that it is possible to develop robust and efficient analytic methods that approximate randomization. Furthermore, proponents argue that nonexperimental approaches should replace conventional covariance adjustment methods and that these bias-correction approaches provide a reasonable estimate of treatment effects when randomization fails or is impossible.

9.2.1 CRITICISM OF PROPENSITY SCORE METHODS

Not surprisingly, this is a debatable perspective, and some critics do not hold such an optimistic view of statistical advances. In general, opponents question the assumptions made by correction methods, and they are skeptical that the conditions of real-world application can meet these assumptions. For instance, using earnings data from a controlled experiment of the effects of a mandatory welfare-to-work program, Michalopoulos et al. (2004) compared findings from randomization with findings generated using nonexperimental analytic approaches. Their analyses followed a methodology originated by LaLonde (1986) and Fraker and Maynard (1987) to compare "true" experimental impact estimates with those obtained from nonexperimental approaches. Michalopoulos and his colleagues (2004) reported that the nonexperimental estimators all exhibited significant bias. They found that the smallest bias occurred when in-state comparison groups were used and short-term outcomes were examined, in which case comparison group earnings were on average 7% of true control group earnings. However, for longer-term outcomes the bias grew—almost doubling—and it grew even more when out-of-state comparison groups were yoked. Based on this finding, Michalopoulos and colleagues concluded that propensity scores correct less well for studies in which the treated and nontreated groups are not exposed to the same ecological influences. In addition, Michalopoulos and colleagues observed that OLS regression often yielded the same estimates as matching estimators.

In 2004, Agodini and Dynarski conducted a similar study. With data from a randomized experiment designed to prevent school dropout, they used nearest neighbor matching on propensity scores to compare results of matching with results obtained under true randomization. The trial examined the effect of a prevention intervention on several subsequent student outcomes, and the comparison groups were drawn from two sources: (1) control group members in a separate but related experiment in different areas and (2) the national sample of the National Educational Longitudinal Survey. As summarized by Moffitt (2004), Agodini and Dynarski's (2004) results suggest that the propensity score matching estimators perform poorly. Indeed, the bias was the same whether a more geographically distant comparison group drawn from the national survey was used or the comparison group from the experiment was used. The authors conclude that significant selection on unobservables was probably present and compromised the matching procedure.

Debate also occurs among proponents and developers of the nonexperimental approaches. The most prominent debate is that between two schools of researchers, each of whom follows a different tradition in developing correction methods. Throughout this book, we have shown that disagreements between econometricians and statisticians center on the restrictiveness of distributional assumptions made in estimators, the tenability of assumptions in real application settings, the extent to which researchers can assume that the selection process is random, the extent to which researchers should not assume that the selection process is random and should make efforts to model the structure of selection, and the ability to control for hidden selection or unobserved heterogeneity.

9.2.2 CRITICISM OF SENSITIVITY ANALYSIS (Γ)

Although Rosenbaum's approach to sensitivity analysis is considered to be methodologically and mathematically elegant, Robins (2002) expressed skepticism about its usefulness in practice. Robins argued that Rosenbaum's model would be useful only if experts could provide a plausible and logically coherent range for the value of the sensitivity parameter Γ, which measures the potential magnitude of hidden bias. To test this, Robins defined a measure of hidden bias to be *paradoxical* if its magnitude increases as the analyst decreases the amount of hidden bias by measuring some of the unmeasured confounders. Based on this definition, Robins proved that Rosenbaum's Γ fit the criteria of a paradoxical measure. He argued that sensitivity analysis based on a *paradoxical measure of hidden bias* may be scientifically useless because, without prolonged and careful training, users might reach misleading, logically incoherent conclusions.

The debate is continuing as we write. Although we offer no judgment on the issues in dispute, it is our hope that our cautions and caveats will encourage readers to monitor the discussion, to explore the conditions under which correction models work, to discuss transparently the limitations embedded in observational studies, and to exercise critical thinking in using correction methods.

9.2.3 GROUP RANDOMIZED TRIALS

Substantial progress in addressing evaluation challenges has also been made in the area of study design. However, even these efforts have sparked debate. One increasingly common design innovation is called *group randomization*, which aims to solve the problem of randomization failure at the individual level. We review this method below and show the usefulness of the correction methods developed for observational data for analyzing data generated by group randomization.

In the past 15 years, group randomization has gained a footing as an alternative to individual randomization in social behavioral sciences research (for a review, see Bloom, 2004). At its core, the idea of the method is quite simple: Random assignment to treatment or control conditions is done on the group level rather than on the individual level. Specifically, instead of randomly assigning individuals to either the treatment or the control conditions, evaluators randomly assign groups (e.g., hospitals, schools) into study conditions. Thus, all the individuals within a given study group (e.g., all patients in a hospital or all students in a school) are assigned to the same condition (i.e., receive treatment or no treatment). According to Bloom, group randomization is useful when (a) the effects of a program have the potential to "spill over" from treatment participants to nonparticipants, (b) the most efficient delivery of program services is through targeting specific locations, (c) the program is designed to address a spatially concentrated problem or situation, (d) using a place-based group randomization reduces political opposition to randomization, and (e) maintaining the integrity of the experiment requires the physical separation of the treatment group from the control group.

Although group randomization has important practical applications in educational, health, social welfare, and other settings, Murray, Pals, Blitstein, Alfano, and Lehman (2008) found that the data from these designs are often incorrectly analyzed. To identify group-randomized trials, Murray and his colleagues used a set of key words to search the peer-reviewed literature on cancer prevention and control. Their investigation identified group randomization designs in 75 articles that were published in 41 journals. Among these studies, many of the researchers who used group randomization did not adequately attend to the analytic challenges raised in the design. Because individuals within the unit of randomization may have correlated outcomes (e.g., the scores of patients within the same hospital or treated by the same doctor may be correlated), calculations based on sampling variability must take the intracluster correlation into account. Ignoring this correlation will produce standard errors that are too small and increase the potential for Type I errors.

Compounding matters, group-randomized trials often have low power and, as we have mentioned, run the risk of failed randomization. Insufficient sample size at the group level (e.g., a small number of schools in a school-based trial) may cause failure of randomization and makes covariates between study conditions (i.e., treatment vs. control conditions) imbalanced. In such a situation, the study data may remain imbalanced at the individual level. Whenever imbalances of covariates occur, the study design cannot be treated as a randomized experiment. It is important to control for selection bias in the data analysis, and the correction methods described in this book should be considered. For example, we have shown how to use the Heckit treatment effect model to determine the treatment effect of a program that employed a group randomization design (see Section 4.5.2) and an efficacy subset analysis of the same data using matching estimators (see Section 6.4.2). These data suggest that correction methods designed for observational studies may be useful when group randomization does not produce ideal balances of covariates between study conditions.

Whenever applicable, we have reviewed the limitations of propensity score approaches in this text. We agree with critics that even randomized clinical trials are imperfect ways of determining the results of treatment for every member of a population—treated or not. Nor can propensity score methods provide definitive answers to questions of treatment effectiveness. Multiple methods for estimating program effects are indicated for use within and across studies. Researchers using propensity score methods should be cautious because the limitations, as we currently understand them, are not trivial and findings vary markedly when assumptions are violated. Interpretation of study results should be constrained by an understanding of the limits of data and analytical methods. Comparisons between propensity score limitations and those of experimental and "traditional" correlational methods warrant further exploration.

9.3 Other Advances in Modeling Causality

Methods aimed at correcting for bias in observational studies are developing rapidly. We have introduced only four of these correction methods, and we have briefly described other methods that can be used to accomplish the same goal of data balancing (see Section 2.5.2). We have chosen not to describe other methods that differ substantially in methodology from the four methods that are the focus of this text. However, two of these methodologically different approaches are especially important and warrant consideration: (1) the marginal structural model and (2) causal analysis using DAGs.

Marginal Structure Models. In a series of publications, James Robins developed analytic methods known as *marginal structural models* that are appropriate for drawing causal inferences from complex observational and randomized studies with time-varying exposure or treatment (Robins, 1999a, 1999b; Robins, Hernn, & Brumback, 2000). To a large extent, these methods are based on the estimation of the parameters for a new class of causal models—structural nested models—using a new class of estimators. The conventional approach to the estimation of the effect of a time-varying treatment or exposure on time to disease has been to model the hazard incidence of failure at time t as a function of past treatment history using a time-dependent Cox proportional hazards model. However, Robins showed that this conventional approach is biased. In contrast, the marginal structural models allow an analyst to make adjustments for the effects of concurrent nonrandomized treatments or nonrandom noncompliance that occur in many randomized clinical trials. For instance, when researchers need to clarify differences between association and causation, the inverse probability of treatment weighted (IPTW) estimation of a marginal structural model is particularly useful. The IPTW estimation consistently estimates the causal effect of a time-dependent treatment when all relevant confounding factors have been measured. To deal with hidden biases or effects of unmeasured variables, Robins developed a sensitivity analysis to adjust inferences concerning the causal effect of treatment as a function of the magnitude of confounding due to unmeasured variables (Robins, 1999b).

Directed Acyclic Graphs. Judea Pearl (2000) and others (Glymour & Cooper, 1999; Spirtes, Glymour, & Scheines, 1993) developed a formal framework to determine which of many conditional distributions could be estimated from data using DAGs. A DAG is a conventional path diagram with a number of formal mathematical properties attached. Pearl (2000) argued, "A causal structure of a set of variables *V* is a directed acyclic graph (DAG) in which each node corresponds to a distinct element of *V*, and each link represents a direct functional relationship among the corresponding variables" (p. 44).

Pearl's framework focuses on the "inferred causation" inherent in the concept of a latent structure. Two latent structures are equivalent, if they imply the same conditional distributions. Each latent structure in a set of such structures is "minimal" if it can produce the same conditional distributions and only those. A structure is consistent with the data if it reproduces the observed conditional distributions. According to Berk (2004), Pearl's contribution is to provide tools for winnowing down a set of potential causal models and then determining for a subset whether they speak with one voice about a particular causal relationship.

9.4 Directions for Future Development

Given advances such as those developed by Robins and Pearl, the growth of propensity score analysis methods, and the fertility of debates in the field, it is difficult to predict what the future may hold. However, we think that the following three directions are evident and are likely to contribute substantially to the design of evaluation methods for observational studies.

The first direction is the need to refine and expand analytical methods for researchers in the social behavioral and health sciences. Historically, the analytic methods for observational studies were developed and used primarily by statisticians and econometricians. However, when applying these methods to a broader range of research problems in the social behavioral and health sciences, issues that were not critical to econometricians and statisticians have arisen and warrant consideration. These include the need to do the following:

- Develop a framework for power analysis for the correction methods. Although social behavioral scientists have used Cohen's (1988) framework for power analysis, that framework does not provide estimators of sample size for propensity score models. The estimation of sample sizes is more complicated in propensity score analysis because most models require overlapping of propensity scores. Incorporating into the estimation the effect of sample reduction due to the common support region problem is an added complexity.
- Develop a standardized or metric-free measure of effect size (i.e., Cohen's d). The d_{x_m} statistic developed by Haviland et al. (2007) to measure covariance imbalance is similar to Cohen's d for post-optimal-matching analysis (see Sections 5.5.3 and 5.5.4). Similar measures should be developed for other types of propensity score analyses.
- Develop approaches to control for covariates that are differentially correlated with outcomes and treatment assignment. One of the three limitations of propensity score matching is that the approach cannot distinguish among effects of covariates that are differentially related to treatment assignment and observed outcomes (Rubin, 1997). In social behavioral research, covariates are often correlated differentially with outcomes and treatment assignment (i.e., covariates may be correlated with treatment assignment but not with the outcome; alternatively, covariates may be correlated more strongly with outcomes than with treatment assignment). To develop these approaches, Heckman, Ichimura, and Todd's (1998) work on *separability* (i.e., dividing the variables that determine outcomes into observables and unobservables) and *exclusion restriction* (i.e., isolating covariates that determine outcomes and program participation into two sets of variables T and Z, where the T variables determine outcomes, and the Z variables determine program participation) is promising.
- Develop correction models that control for measurement errors. One of the more challenging problems in social behavioral and health research is the non-random measurement error produced by multiple raters, particularly in longitudinal studies (see Section 8.1.1).
- Develop correction models that account for attrition selection in longitudinal studies. As discussed in Chapter 8, attrition cannot be assumed to be random and requires explicit modeling efforts.
- Improve correction models that control for clustering effect. As discussed in Chapter 6, the developers of matching estimators are aware of the limitation of not controlling for clustering in their method and are working to improve the matching estimators along this line.

The second direction is to develop approaches that address challenges outlined in Heckman's (2005) framework for the "scientific model of causality" (see Section 2.8). Heckman weighed the implicit assumptions underlying four widely used methods of causal inference: (1) matching, (2) control functions,

(3) the instrumental variables method, and (4) the method of DAGs. In his framework, he emphasized the need in policy research to forecast the impact of interventions in new environments, to identify parameters (causal or otherwise) from hypothetical population data, and to develop estimators evaluating different types of treatment effects (i.e., average treatment effect, treatment effect for the treated, treatment effect for the untreated, marginal treatment effect, and local average treatment effect).

Finally, the third direction is to develop effective methods based on Rosenbaum's framework. We have cited Rosenbaum extensively in this book. As a statistician, Rosenbaum developed his framework based on the randomization inference in completely randomized experiments. He then extended the covariance adjustment assuming randomization to observational studies free of hidden bias, and finally to observational studies with hidden bias. For a detailed discussion of Rosenbaum's framework, readers are referred to a special issue of *Statistical Science* (2002, Vol. 17, No. 3), which presents an interesting dialogue between Rosenbaum and several prominent researchers in the field, including Angrist and Imbens (2002), Robins (2002), and Hill (2002). Rosenbaum has also made significant contributions to the development of optimal matching and sensitivity analysis. In the context of heated debate and ongoing disagreement, Rosenbaum's work sets a new standard for the analysis of observational studies. The refinements and advancements latent in the Rosenbaum framework serve as a springboard for the future development of methods for observational studies.

References

Abadie, A., Drukker, D., Herr, J. L., & Imbens, G. W. (2004). Implementing matching estimators for average treatment effects in Stata. *Stata Journal, 4,* 290–311.

Abadie, A., & Imbens, G. W. (2002). *Simple and bias-corrected matching estimators* [Technical report]. Department of Economics, University of California, Berkeley. Retrieved August 8, 2008, from http://ideas.repec.org/p/nbr/nberte/0283.html

Abadie, A., & Imbens, G. W. (2006). Large sample properties of matching estimators for average treatment effects. *Econometrica, 74,* 235–267.

Achenbach, T. M. (1991). *Integrative guide for the 1991 CBCL/4–18, YSR, and TRF profiles.* Burlington: University of Vermont, Department of Psychiatry.

Agodini, R., & Dynarski, M. (2004). Are experiments the only option? A look at dropout prevention programs. *Review of Economics and Statistics, 86,* 180–194.

Altonji, J. G., Elder, T. E., & Taber, C. R. (2005). Selection on observed and unobserved variables: Assessing the effectiveness of Catholic schools. *Journal of Political Economy, 113,* 151–184.

Angrist, J. D., & Imbens, G. W. (2002). Comment on "Covariance adjustment in randomized experiments and observational studies." *Statistical Science, 17*(3), 304–307.

Angrist, J. D., Imbens, G. W., & Rubin, D. G. (1996). Identification of causal effects using instrumental variables. *Journal of the American Statistical Association, 91,* 444–472.

Barnow, B. S., Cain, G. G., & Goldberger, A. S. (1980). Issues in the analysis of selectivity bias. In E. Stromsdorfer & G. Farkas (Eds.), *Education studies* (Vol. 5, pp. 42–59). San Francisco: Sage.

Barth, R. P., Gibbons, C., & Guo, S. (2006). Substance abuse treatment and the recurrence of maltreatment among caregivers with children living at home: A propensity score analysis. *Journal of Substance Abuse Treatment, 30,* 93–104.

Barth, R. P., Lee, C. K., Wildfire, J., & Guo, S. (2006). A comparison of the governmental costs of long-term foster care and adoption. *Social Service Review, 80*(1), 127–158.

Barth, R. R., Greeson, J. K., Guo, S., & Green, B. (2007). Outcomes for youth receiving intensive in-home therapy or residential care: A comparison using propensity scores. *American Journal of Orthopsychiatry, 77,* 497–505.

Berk, R. A. (2004). *Regression analysis: A constructive critique.* Thousand Oaks, CA: Sage.

Bloom, H. S. (2004). *Randomizing groups to evaluate place-based programs.* Retrieved August 31, 2008, from www.wtgrantfoundation.org/usr_doc/RSChapter4Final.pdf

Brooks-Gunn, J., & Duncan, G. J. (1997). The effects of poverty on children. *Future of Children, 7,* 55–70.

Campbell, D. T. (1957). Factors relevant to the validity of experiments in social settings. *Psychological Bulletin, 54,* 297–312.

Cochran, W. G. (1965). The planning of observational studies of human populations (with discussion). *Journal of the Royal Statistical Society, Series A, 128,* 134–155.

Cochran, W. G. (1968). The effectiveness of adjustment by subclassification in removing bias in observational studies. *Biometrics, 24,* 295–313.

Cochran, W. G., & Rubin, D. B. (1973). Controlling bias in observational studies: A review. *Sankya, Series A, 35,* 417–446.

Cohen, J. (1988). *Statistical power analysis for the behavioral sciences* (2nd ed.). Hillsdale, NJ: Lawrence Erlbaum.

Corcoran, M., & Adams, T. (1997). Race, sex, and the intergenerational transmission of poverty. In G. J. Duncan & J. Brooks-Gunn (Eds.), *Consequences of growing up poor* (pp. 461–517). New York: Russell Sage Foundation.

Cornfield, J., Haenszel, W., Hammond, E., Lilienfeld, A., Shimkin, M., & Wynder, E. (1959). Smoking and lung cancer: Recent evidence and a discussion of some questions. *Journal of the National Cancer Institute, 22,* 173–203.

Courtney, M. E. (2000). Research needed to improve the prospects for children in out-of-home placement. *Children and Youth Services Review, 22*(9/10), 743–761.

Cox, D. R. (1958). *Planning of experiments.* New York: Wiley.

Cronbach, L. J., Gleser, G. C., Nanda, H., & Rajaratnam, N. (1972). *The dependability of behavioral measurements: Theory of generalizability of sources and profiles.* New York: Wiley.

D'Agostino, R. B., Jr. (1998). Tutorial in biostatistics: Propensity score methods for bias reduction in the comparison of a treatment to a non-randomized control group. *Statistics in Medicine, 17,* 2265–2281.

Dehejia, R., & Wahba, S. (1999). Causal effects in nonexperimental studies: Reevaluating the evaluation of training programs. *Journal of the American Statistical Association, 94,* 1053–1062.

Derigs, U. (1988). Solving non-bipartite matching problems via shortest path techniques. *Annals of Operations Research, 13,* 225–261.

Duncan, G. J., Brooks-Gunn, J., & Klebanov, P. K. (1994). Economic deprivation and early childhood development. *Child Development, 65,* 296–318.

Duncan, G. J., Brooks-Gunn, J., Yeung, W. J., & Smith, J. R. (1998). How much does childhood poverty affect the life chances of children? *American Sociological Review, 63,* 406–423.

Earle, C. C., Tsai, J. S., Gelber, R. D., Weinstein, M. C., Neumann, P. J., & Weeks, J. C. (2001). Effectiveness of chemotherapy for advanced lung cancer in the elderly: Instrumental variable and propensity analysis. *Journal of Clinical Oncology, 19,* 1064–1070.

Edwards, L. N. (1978). An empirical analysis of compulsory schooling legislation 1940–1960. *Journal of Law and Economics, 21,* 203–222.

Eichler, M., & Lechner, M. (2002). An evaluation of public employment programmes in the East German state of Sachsen-Anhalt. *Labour Economics, 9,* 143–186.

English, D., Marshall, D., Brummel, S., & Coghan, L. (1998). *Decision-making in child protective services: A study of effectiveness. Final Report.* Olympia, WA: Department of Social and Health Services.

Fan, J. (1992). Design adaptive nonparametric regression. *Journal of the American Statistical Association, 87,* 998–1004.

Fan, J. (1993). Local linear regression smoothers and their minimax efficiencies. *Annals of Statistics, 21,* 196–216.

Fisher, R. A. (1925). *Statistical methods for research workers*. Edinburgh, UK: Oliver & Boyd.

Fisher, R. A. (1935/1971). *The design of experiments*. Edinburgh, UK: Oliver & Boyd.

Foster, E. M., & Furstenberg, F. F., Jr. (1998). Most disadvantaged children: Who are they and where do they live? *Journal of Poverty, 2*, 23–47.

Foster, E. M., & Furstenberg, F. F., Jr. (1999). The most disadvantaged children: Trends over time. *Social Service Review, 73*, 560–578.

Fox, J. (2000). *Nonparametric simple regression: Smoothing scatterplots*. Thousand Oaks, CA: Sage.

Fraker, T., & Maynard, R. (1987). The adequacy of comparison group designs for evaluations of employment-related programs. *Journal of Human Resources, 22*, 194–227.

Freedman, D. A., & Berk, R. A. (2008). Weighting regressions by propensity scores. *Evaluation Review, 32*, 392–409.

Friedman, J. (2002). Stochastic gradient boosting. *Computational Statistics and Data Analysis, 38*, 367–378.

Friedman, J., Hastie, T., & Tibshirani, R. (2000). Additive logistic regression: A statistical view of boosting (with discussion). *Annals of Statistics, 28*, 337–374.

Frölich, M. (2004). Finite-sample properties of propensity-score matching and weighting estimators. *The Review of Economics and Statistics, 86*, 77–90.

Galati, J. C., Royston, P., & Carlin, J. B. (2009). *MIMSTACK: Stata module to stack multiply-imputed datasets into format required by mim*. Retrieved February 27, 2009, from http://ideas.repec.org/c/boc/bocode/s456826.html

Gangl, M. (2007). *RBOUNDS: Stata module to perform Rosenbaum sensitivity analysis for average treatment effects on the treated*. Retrieved August 1, 2007, from fmwww.bc.edu/RePEc/bocode/r

Glymour, C., & Cooper, G. (Eds.). (1999). *Computation, causation, and discovery*. Cambridge: MIT Press.

Greene, W. H. (1981). Sample selection bias as specification error: Comment. *Econometrica, 49*, 795–798.

Greene, W. H. (1995). *LIMDEP, version 7.0: User's manual*. Bellport, NY: Econometric Software.

Greene, W. H. (2003). *Econometric analysis* (5th ed.). Upper Saddle River, NJ: Prentice Hall.

Gronau, R. (1974). Wage comparisons: A selectivity bias. *Journal of Political Economy, 82*, 1119–1143.

Gum, P. A., Thamilarasan, M., Watanabe, J., Blackstone, E. H., & Lauer, M. S. (2001). Aspirin use and all-cause mortality among patients being evaluated for known or suspected coronary artery disease: A propensity analysis. *Journal of the American Medical Association, 286*, 1187–1194.

Guo, S. (2005). Analyzing grouped data with hierarchical linear modeling. *Children and Youth Services Review, 27*, 637–652.

Guo, S. (2008a). *The Stata hodgesl program*. Available in the companion Web page of this book.

Guo, S. (2008b). *The Stata imbalance program*. Available in the companion Web page of this book.

Guo, S., Barth, R. P., & Gibbons, C. (2006). Propensity score matching strategies for evaluating substance abuse services for child welfare clients. *Children and Youth Services Review, 28*, 357–383.

Guo, S., & Hussey, D. L. (1999). Analyzing longitudinal rating data: A three-level hierarchical linear model. *Social Work Research, 23,* 258–269.

Guo, S., & Hussey, D. L. (2004). Nonprobability sampling in social work research: Dilemmas, consequences, and strategies. *Journal of Social Service Research, 30,* 1–18.

Guo, S., & Lee, J. (2008). *Optimal propensity score matching and its applications to social work evaluations and research.* Unpublished working paper. School of Social Work, University of North Carolina, Chapel Hill.

Guo, S., & Wildfire, J. (2005, June 9). *Quasi-experimental strategies when randomization is not feasible: Propensity score matching.* Paper presented at the Children's Bureau Annual Conference on the IV-E Waiver Demonstration Project, Washington, DC.

Haavelmo, T. (1943). The statistical implications of a system of simultaneous equations. *Econometrica, 11,* 1–12.

Haavelmo, T. (1944). The probability approach in econometrics. *Econometrica, 12,* 1–115.

Hahn, J. (1998). On the role of the propensity score in efficient semiparametric estimation of average treatment effects. *Econometrica, 66,* 315–331.

Hammond, E. C. (1964). Smoking in relation to mortality and morbidity: Findings in first thirty-four months of follow-up in a prospective study started in 1959. *Journal of the National Cancer Institute, 32,* 1161–1188.

Hansen, B. B. (2004). Full matching in an observational study of coaching for the SAT. *Journal of the American Statistical Association, 99,* 609–618.

Hansen, B. B. (2007). Optmatch: Flexible, optimal matching for observational studies. *R News, 7,* 19–24.

Hansen, B. B., & Klopfer, S. O. (2006). Optimal full matching and related designs via network flows. *Journal of Computational and Graphical Statistics, 15,* 1–19.

Hardle, W. (1990). *Applied nonparametric regression.* Cambridge, UK: Cambridge University Press.

Hartman, R. S. (1991). A Monte Carlo analysis of alternative estimators in models involving selectivity. *Journal of Business and Economic Statistics, 9,* 41–49.

Haviland, A., Nagin, D. S., & Rosenbaum, P. R. (2007). Combining propensity score matching and group-based trajectory analysis in an observational study. *Psychological Methods, 12,* 247–267.

Heckman, J. J. (1974). Shadow prices, market wages, and labor supply. *Econometrica, 42,* 679–694.

Heckman, J. J. (1976). Simultaneous equations model with continuous and discrete endogenous variables and structural shifts. In S. M. Goldfeld & R. E. Quandt (Eds.), *Studies in non-linear estimation* (pp. 235–272). Cambridge, MA: Ballinger.

Heckman, J. J. (1978). Dummy endogenous variables in a simultaneous equations system. *Econometrica, 46,* 931–960.

Heckman, J. J. (1979). Sample selection bias as a specification error. *Econometrica, 47,* 153–161.

Heckman, J. J. (1992). Randomization and social policy evaluation. In C. Manski & I. Garfinkel (Eds.), *Evaluating welfare and training programs* (pp. 201–230). Cambridge, MA: Harvard University Press.

Heckman, J. J. (1996). Comment on "Identification of causal effects using instrumental variables" by Angrist, Imbens, & Rubin. *Journal of the American Statistical Association, 91,* 459–462.

Heckman, J. J. (1997). Instrumental variables: A study of implicit behavioral assumptions used in making program evaluations. *Journal of Human Resources, 32,* 441–462.

Heckman, J. J. (2005). The scientific model of causality. *Sociological Methodology, 35,* 1–97.

Heckman, J. J., Ichimura, H., Smith, J., & Todd, P. E. (1998). Characterizing selection bias using experimental data. *Econometrica, 66,* 1017–1098.

Heckman, J. J., Ichimura, H., & Todd, P. E. (1997). Matching as an econometric evaluation estimator: Evidence from evaluating a job training programme. *Review of Economic Studies, 64,* 605–654.

Heckman, J. J., Ichimura, H., & Todd, P. E. (1998). Matching as an econometric evaluation estimator. *Review of Economic Studies, 65,* 261–294.

Heckman, J. J., LaLonde, R. J., & Smith, J. A. (1999). The economics and econometrics of active labor market programs. In O. Ashenfelter & D. Card (Eds.), *Handbook of labor economics* (Vol. 3, pp. 1865–2097). New York: Elsevier.

Heckman, J. J., & Robb, R. (1985). Alternative methods for evaluating the impact of interventions. In J. Heckman & B. Singer (Eds.), *Longitudinal analysis of labor market data* (pp. 156–245). Cambridge, UK: Cambridge University Press.

Heckman, J. J., & Robb, R. (1986). Alternative methods for solving the problem of selection bias in evaluating the impact of treatments on outcomes. In H. Wainer (Ed.), *Drawing inferences from self-selected samples* (pp. 63–113). New York: Springer-Verlag.

Heckman, J. J., & Robb, R. (1988). The value of longitudinal data for solving the problem of selection bias in evaluating the impact of treatment on outcomes. In G. Duncan & G. Kalton (Eds.), *Panel surveys* (pp. 512–538). New York: Wiley.

Heckman, J., & Smith, J. (1995). Assessing the case for social experiments. *Journal of Economic Perspectives, 9,* 85–110.

Heckman, J., & Smith, J. (1998). Evaluating the welfare state. *Frisch centenary econometric monograph series.* Cambridge, UK: Cambridge University Press.

Heckman, J., Smith, J., & Clements, N. (1997). Making the most out of social experiments: Accounting for heterogeneity in programme impacts. *Review of Economic Studies, 64,* 487–536.

Heckman, J. J., & Vytlacil, E. J. (1999). Local instrumental variables and latent variable models for identifying and bounding treatment effects. *Proceedings of the National Academy of Sciences, 96,* 4730–4734.

Heckman, J. J., & Vytlacil, E. J. (2005). Structural equations, treatment effects, and econometric policy evaluation. *Econometrica, 73,* 669–738.

Helmreich, J. E., & Pruzek, R. M. (2008). *The PSA graphics package.* Retrieved October 30, 2008, from www.r-project.org

Herbst, A., Ulfelder, H., & Poskanzer, D. (1971). Adenocarcinoma of the vagina: Association of maternal stilbestrol therapy with tumor appearance in young women. *New England Journal of Medicine, 284,* 878–881.

Hill, J. (2002). Comment on "Covariance adjustment in randomized experiments and observational studies." *Statistical Science, 17,* 307–309.

Hirano, K., & Imbens, G. W. (2001). Estimation of causal effects using propensity score weighting: An application to data on right heart catheterization. *Health Services & Outcomes Research Methodology, 2,* 259–278.

Hirano, K., Imbens, G. W., & Ridder, G. (2003). Efficient estimation of average treatment effects using the estimated propensity score. *Econometrica, 71,* 1161–1189.

Ho, D., Imai, K., King, G., & Stuart, E. (2004). *Matching as nonparametric preprocessing for improving parametric causal inference.* Retrieved October 20, 2008, from http://gking.harvard.edu/files/abs/matchp-abs.shtml

Hodges, J., & Lehmann, E. (1962). Rank methods for combination of independent experiments in the analysis of variance. *Annals of Mathematical Statistics, 33,* 482–497.

Hofferth, S., Stafford, F. P., Yeung, W. J., Duncan, G. J., Hill, M. S., Lepkowski, J., et al. (2001). *Panel study of income dynamics, 1968–1999: Supplemental files (computer file), ICPSR version*. Ann Arbor: University of Michigan Survey Research Center.

Holland, P. (1986). Statistics and causal inference (with discussion). *Journal of the American Statistical Association, 81*, 945–970.

Horvitz, D., & Thompson, D. (1952). A generalization of sampling without replacement from a finite population. *Journal of the American Statistical Association, 47*, 663–685.

Hosmer, D. W., & Lemeshow, S. (1989). *Applied logistic regression*. New York: Wiley.

Hume, D. (1748/1959). *An enquiry concerning human understanding*. LaSalle, IL: Open Court Press.

Iacus, S. M., King, G., & Porro, G. (2008). *Matching for causal inference without balance checking*. Retrieved October 30, 2008, from http://gking.harvard.edu/files/abs/cem-abs.shtml

Imai, K., & Van Dyk, D. A. (2004). Causal inference with general treatment regimes: Generalizing the propensity score. *Journal of the American Statistical Association, 99*, 854–866.

Imbens, G. W. (2000). The role of the propensity score in estimating dose-response functions. *Biometrika, 87*, 706–710.

Imbens, G. W. (2004). Nonparametric estimation of average treatment effects under exogeneity: A review. *Review of Economics and Statistics, 86*, 4–29.

Jick, H., Miettinen, O., Neff, R., Jick, H., Miettinen, O. S., Neff, R. K., et al. (1973). Coffee and myocardial infarction. *New England Journal of Medicine, 289*, 63–77.

Joffe, M. M., & Rosenbaum, P. R. (1999). Invited commentary: Propensity scores. *American Journal of Epidemiology, 150*, 327–333.

Jones, A. S., D'Agostino, R. B., Gondolf, E. W., & Heckert, A. (2004). Assessing the effect of batterer program completion on reassault using propensity scores. *Journal of Interpersonal Violence, 19*, 1002–1020.

Kang, J. D. Y., & Schafer, J. L. (2007). Demystifying double robustness: A comparison of alternative strategies for estimating a population mean from incomplete data. *Statistical Science, 22*, 523–539.

Keele, L. J. (2008). *The RBOUNDS package*. Retrieved October 30, 2008, from www.r-project.org

Kempthorne, O. (1952). *The design and analysis of experiments*. New York: Wiley.

Kennedy, P. (2003). *A guide to econometrics* (5th ed.). Cambridge: MIT Press.

King, G., & Zeng, L. (2006). The dangers of extreme counterfactuals. *Political Analysis, 14*(2), 131–159. Retrieved October 30, 2008, from http://gking.harvard.edu

King, G., & Zeng, L. (2007). When can history be our guide? The pitfalls of counterfactual inference. *International Studies Quarterly, 51*(1), 183–210. Retrieved October 30, 2008, from http://gking.harvard.edu

Knight, D. K., Logan, S. M., & Simpson, D. D. (2001). Predictors of program completion for women in residential substance abuse treatment. *American Journal of Drug and Alcohol Abuse, 27*, 1–18.

Kutner, M. H., Nachtsheim, C. J., & Neter, J. (2004). *Applied linear regression models* (4th ed.). New York: McGraw-Hill/Irwin.

LaLonde, R. J. (1986). Evaluating the econometric evaluations of training programs with experimental data. *American Economic Review, 76*, 604–620.

Landes, W. (1968). The economics of fair employment laws. *Journal of Political Economy, 76*, 507–552.

Lazarsfeld, P. F. (1959). Problems in methodology. In R. K. Merton, L. Broom, & L. S. Cottrell Jr. (Eds.), *Sociology today: Problems and prospects* (Vol. 1, pp. 39–72). New York: Basic Books.

Lechner, M. (1999). Earnings and employment effects of continuous off-the-job training in East Germany after unification. *Journal of Business and Economic Statistics, 17,* 74–90.

Lechner, M. (2000). An evaluation of public sector sponsored continuous vocational training programs in East Germany. *Journal of Human Resources, 35,* 347–375.

Lehmann, E. L. (2006). *Nonparametrics: Statistical methods based on ranks* (Rev. ed.). New York: Springer.

Leuven, E., & Sianesi, B. (2003). *PSMATCH2 (version 3.0.0): Stata module to perform full Mahalanobis and propensity score matching, common support graphing, and covariate imbalance testing.* Retrieved August 22, 2008, from http://ideas.repec.org/c/boc/bocode/s432001.html

Lewis, D. (1973). *Counterfactuals.* Cambridge, MA: Harvard University Press.

Lewis, D. (1986). *Philosophical papers* (Vol. 2). New York: Oxford University Press.

Lewis, H. G. (1974). Comments on selectivity biases in wage comparisons. *Journal of Political Economy, 82,* 1145–1155.

Littell, J. H. (2001). Client participation and outcomes of intensive family preservation services. *Social Work Research, 25*(2), 103–113.

Littell, J. H. (2005). Lessons from a systematic review of effects of multisystemic therapy. *Children and Youth Services Review, 27,* 445–463.

Little, R. A., & Rubin, D. B. (2002). *Statistical analysis with missing data* (2nd ed.). New York: Wiley.

Lochman, J. E., Boxmeyer, C., Powell, N., Roth, D. L., & Windle, M. (2006). Masked intervention effects: Analytic methods for addressing low dosage of intervention. *New Directions for Evaluation, 110,* 19–32.

Long, J. S. (1997). *Regression models for categorical and limited dependent variables.* Thousand Oaks, CA: Sage.

Lu, B., Zanutto, E., Hornik, R., & Rosenbaum, P. R. (2001). Matching with doses in an observational study of a media campaign against drug abuse. *Journal of the American Statistical Association, 96,* 1245–1253.

Maddala, G. S. (1983). *Limited-dependent and qualitative variables in econometrics.* Cambridge, UK: Cambridge University Press.

Magura, S., & Laudet, A. B. (1996). Parental substance abuse and child maltreatment: Review and implications for intervention. *Children and Youth Services Review, 3,* 193–220.

Manning, W. G., Duan, N., & Rogers, W. H. (1987). Monte-Carlo evidence on the choice between sample selection and 2-part models. *Journal of Econometrics, 35,* 59–82.

Manski, C. F. (2007). *Identification for prediction and decision.* Cambridge, MA: Harvard University Press.

Mantel, N. (1963). Chi-square tests with one degree of freedom: Extensions of the Mantel-Haenszel procedure. *Journal of the American Statistical Association, 58,* 690–700.

Mantel, N., & Haenszel, W. (1959). Statistical aspects of retrospective studies of disease. *Journal of the National Cancer Institute, 22,* 719–748.

Maxwell, S. E., & Delaney, H. D. (1990). *Designing experiments and analyzing data: A model comparison perspective.* Pacific Grove, CA: Brooks/Cole.

McCaffrey, D. F., Ridgeway, G., & Morral, A. R. (2004). Propensity score estimation with boosted regression for evaluating causal effects in observational studies. *Psychological Methods, 9,* 403–425.

McCullagh, P. (1980). Regression models for ordinal data. *Journal of the Royal Statistical Society, Series B, 42,* 109–142.

McCullagh, P., & Nelder, J. (1989). *Generalized linear models* (2nd ed.). London: Chapman & Hall.

McMahon, T. J., Winkel, J. D., Suchman, N. E., & Luthar, S. S. (2002). Drug dependence, parenting responsibilities, and treatment history: Why doesn't mom go for help? *Drug and Alcohol Dependence, 65,* 105–114.

McNemar, Q. (1947). Note on the sampling error of the differences between correlated proportions or percentage. *Psychometrika, 12,* 153–157.

Mease, D., Wyner, A. J., & Buja, A. (2007). Boosted classification trees and class probability/quantile estimation. *Journal of Machine Learning Research, 8,* 409–439.

Michalopoulos, C., Bloom, H. S., & Hill, C. J. (2004). Can propensity-score methods match the findings from a random assignment evaluation of mandatory welfare-to-work programs? *Review of Economics and Statistics, 86,* 156–179.

Mill, J. S. (1843). *System of logic* (Vol. 1). London: John Parker.

Miller, A. (1990). *Subset selection in regression.* London: Chapman & Hall.

Ming, K., & Rosenbaum, P. R. (2001). A note on optimal matching with variable controls using the assignment algorithm. *Journal of Computational and Graphical Statistics, 10,* 455–463.

Moffitt, R. A. (2004). Introduction to the symposium on the econometrics of matching. *Review of Economics and Statistics, 86,* 1–3.

Morgan, S. L. (2001). Counterfactuals, causal effect, heterogeneity, and the Catholic school effect on learning. *Sociology of Education, 74,* 341–374.

Morton, D., Saah, A., Silberg, S., Owens, W., Roberts, M., & Saah, M. (1982). Lead absorption in children of employees in a lead-related industry. *American Journal of Epidemiology, 115,* 549–555.

Murray, D. M., Pals, S. L., Blitstein, J. L., Alfano, C. M., & Lehman, J. (2008). Design and analysis of group-randomized trials in cancer: A review of current practices. *Journal of the National Cancer Institute, 100,* 483–491.

Neyman, J. S. (1923). Statistical problems in agricultural experiments. *Journal of the Royal Statistical Society, Series B, 2,* 107–180.

Nobel Prize Review Committee. (2000). *The Sveriges Riksbank Prize in economic sciences in memory of Alfred Nobel 2000.* Retrieved August 8, 2008, from http://nobelprize .org/nobel_prizes/economics/laureates/2000

Normand, S. T., Landrum, M. B., Guadagnoli, E., Ayanian, J. Z., Ryan, T. J., Cleary, P. D., et al. (2001). Validating recommendations for coronary angiography following acute myocardial infarction in the elderly: A matched analysis using propensity scores. *Journal of Clinical Epidemiology, 54,* 387–398.

NSCAW Research Group. (2002). Methodological lessons from the National Survey of Child and Adolescent Well-being: The first three years of the USA's first national probability study of children and families investigated for abuse and neglect. *Children and Youth Services Review, 24,* 513–541.

Nunnally, J. C., & Bernstein, I. H. (1994). *Psychometric theory* (3rd ed.). New York: McGraw-Hill.

Obenchain, B. (2007). *The USPS package.* Retrieved October 30, 2008, from www.r-project.org

Parsons, L. S. (2001). *Reducing bias in a propensity score matched-pair sample using greedy matching techniques [SAS SUGI paper 214-26].* Proceedings of the 26th annual SAS Users' Group International Conference, Cary, NC: SAS Institute. Retrieved August 22, 2008, from www2.sas.com/proceedings/sugi26/p214-26.pdf

Pearl, J. (2000). *Causality: Models, researching, and inference.* Cambridge, UK: Cambridge University Press.

Perkins, S. M., Tu, W., Underhill, M. G., Zhou, X., & Murray, M. D. (2000). The use of propensity scores in pharmacoepidemiologic research. *Pharmacoepidemiology and Drug Safety, 9,* 93–101.

Quandt, R. E. (1958). The estimation of the parameters of a linear regression system obeying two separate regimes. *Journal of the American Statistical Association, 53,* 873–880.

Quandt, R. E. (1972). A new approach to estimating switching regressions. *Journal of the American Statistical Association, 67,* 306–310.

R Foundation for Statistical Computing. (2008). *R version 2.6.2.* Retrieved August 8, 2008, from http://cran.stat.ucla.edu/

Ridgeway, G. (1999). The state of boosting. *Computing Science and Statistics, 31,* 172–181.

Robins, J. M. (1999a). Association, causation, and marginal structural models. *Synthese, 121,* 151–179.

Robins, J. M. (1999b). Marginal structural models versus structural nested models as tools for causal inference. In M. E. Halloran & D. Berry (Eds.), *Statistical models in epidemiology: The environment and clinical trials* (pp. 95–134). New York: Springer-Verlag.

Robins, J. M. (2002). Comment on "Covariance adjustment in randomized experiments and observational studies." *Statistical Science, 17,* 309–321.

Robins, J. M., Hernn, M., & Brumback, B. (2000). Marginal structural models and causal inference in epidemiology. *Epidemiology, 11,* 550–560.

Robins, J. M., & Ronitzky, A. (1995). Semiparametric efficiency in multivariate regression models with missing data. *Journal of the American Statistical Association, 90,* 122–129.

Rosenbaum, P. R. (1987). Model-based direct adjustment. *Journal of the American Statistical Association, 82,* 387–394.

Rosenbaum, P. R. (2002a). Covariance adjustment in randomized experiments and observational studies. *Statistical Science, 17,* 286–304.

Rosenbaum, P. R. (2002b). *Observational studies* (2nd ed.). New York: Springer.

Rosenbaum, P. R. (2005). Sensitivity analysis in observational studies. In B. S. Everitt & D. C. Howell (Eds.), *Encyclopedia of statistics in behavioral science* (pp. 1809–1814). New York: Wiley.

Rosenbaum, P. R., Ross, R. N., & Silber, J. H. (2007). Minimum distance matched sampling with fine balance in an observational study of treatment for ovarian cancer. *Journal of the American Statistical Association, 102,* 75–83.

Rosenbaum, P. R., & Rubin, D. B. (1983). The central role of the propensity score in observational studies for causal effects. *Biometrika, 70,* 41–55.

Rosenbaum, P. R., & Rubin, D. B. (1984). Reducing bias in observational studies using subclassification on the propensity score. *Journal of the American Statistical Association, 79,* 516–524.

Rosenbaum, P. R., & Rubin, D. B. (1985). Constructing a control group using multivariate matched sampling methods that incorporate the propensity score. *American Statistician, 39,* 33–38.

Rossi, P. H., & Freeman, H. E. (1989). *Evaluation: A systematic approach* (4th ed.). Newbury Park, CA: Sage.

Roy, A. (1951). Some thoughts on the distribution of earnings. *Oxford Economic Papers, 3,* 135–146.

Rubin, D. B. (1974). Estimating causal effects of treatments in randomized and non-randomized studies. *Journal of Educational Psychology, 66,* 688–701.

Rubin, D. B. (1976). Matching methods that are equal percent bias reducing: Some examples. *Biometrics, 32,* 109–120.

Rubin, D. B. (1977). Assignment to treatment groups on the basis of a covariate. *Journal of Educational Statistics, 2,* 1–26.

Rubin, D. B. (1978). Bayesian inference for causal effects: The role of randomization. *Annals of Statistics, 6,* 34–58.

Rubin, D. B. (1979). Using multivariate matched sampling and regression adjustment to control bias in observational studies. *Journal of the American Statistical Association, 74,* 318–328.

Rubin, D. B. (1980a). Discussion of "Randomization analysis of experimental data in the Fisher randomization test" by Basu. *Journal of the American Statistical Association, 75,* 591–593.

Rubin, D. B. (1980b). Percent bias reduction using Mahalanobis metric matching. *Biometrics, 36,* 293–298.

Rubin, D. B. (1986). Which ifs have causal answers? *Journal of the American Statistical Association, 81,* 961–962.

Rubin, D. B. (1996). Multiple imputation after 18+ years. *Journal of the American Statistical Association, 91,* 473–489, 507–515, 515–517.

Rubin, D. B. (1997). Estimating causal effects from large data sets using propensity scores. *Annals of Internal Medicine, 127,* 757–763.

Rubin, D. B. (2008). For objective causal inference, design trumps analysis. *Annals of Applied Statistics, 2,* 808–840.

Schafer, J. L. (1997). *Analysis of incomplete multivariate data.* Boca Raton, FL: Chapman Hall/CRC.

Schonlau, M. (2007). *Boost module for Stata.* Retrieved August 22, 2008, from www.stata-journal.com/software/sj5-3

Sekhon, J. S. (2007). Multivariate and propensity score matching software with automated balance optimization. *Journal of Statistical Software.* http://sekhon.berkeley.edu/papers/MatchingJSS.pdf

Shadish, W. R., Cook, T. D., & Campbell, D. T. (2002). *Experimental and quasi-experimental designs for generalized causal inference.* Boston: Houghton Mifflin.

Shavelson, R. J., & Webb, N. M. (1991). *Generalizability theory: A primer.* Newbury Park, CA: Sage.

Smith, H. L. (1997). Matching with multiple controls to estimate treatment effects in observational studies. *Sociological Methodology, 27,* 325–353.

Smith, J. A., & Todd, P. E. (2005). Does matching overcome LaLonde's critique of non-experimental estimators? *Journal of Econometrics, 125,* 305–353.

Smith, P. K., & Yeung, W. J. (1998). Childhood welfare receipt and the implications of welfare reform. *Social Service Review, 72,* 1–16.

Sobel, M. E. (1996). An introduction to causal inference. *Sociological Methods & Research, 24,* 353–379.

Sobel, M. E. (2005). Discussion: "The scientific model of causality." *Sociological Methodology, 35,* 99–133.

Sosin, M. R. (2002). Outcomes and sample selection: The case of a homelessness and substance abuse intervention. *British Journal of Mathematical and Statistical Psychology, 55,* 63–91.

Spirtes, P., Glymour, C., & Scheines, R. (1993). *Causation, prediction, and search.* New York: Springer-Verlag.

StataCorp. (2003). *Stata release 8: [R].* College Station, TX: Stata Corporation.

StataCorp. (2008). *Stata release 10: [R].* College Station, TX: Stata Corporation.

Stolzenberg, R. M., & Relles, D. A. (1990). Theory testing in a world of constrained research design. *Sociology Methods & Research, 18,* 395–415.

Thurstone, L. (1930). *The fundamentals of statistics.* New York: Macmillan.

Toomet, O., & Henningsen, A. (2008). Sample selection models in R: Package sample selection. *Journal of Statistical Software, 27(7).* Retrieved October 30, 2008, from www.jstatsoft.org

UCLA Academic Technology Services. (2008). *FAQ: What are pseudo R-squareds?* Retrieved April 28, 2008, from www.ats.ucla.edu/stat/mult_pkg/faq/general/Psuedo_RSquareds.htm

U.S. Department of Health and Human Services. (1999). *Blending perspectives and building common ground: A report to Congress on substance abuse and child protection.* Retrieved August 22, 2008, from http://aspe.dhhs.gov/hsp/subabuse99/subabuse.htm

Weigensberg, E. C., Barth, R. P., & Guo, S. (2009). Family group decision making: A propensity score analysis to evaluate child and family services at baseline and after 36-months. *Children and Youth Services Review, 31,* 383–390.

Weisner, C., Jennifer, M., Tam, T., & Moore, C. (2001). Factors affecting the initiation of substance abuse treatment in managed care. *Addiction, 96,* 705–716.

Wilcoxon, F. (1945). Individual comparisons by ranking methods. *Biometrics, 1,* 80–83.

Winship, C., & Morgan, S. L. (1999). The estimation of causal effects from observational data. *Annual Review of Sociology, 25,* 659–707.

Wolock, I., & Horowitz, B. (1981). Child maltreatment and material deprivation among AFDC recipient families. *Social Service Review, 53,* 175–194.

Wooldridge, J. M. (2002). *Econometric analysis of cross section and panel data.* Cambridge: MIT Press.

Yoshikawa, H., Maguson, K. A., Bos, J. M., & Hsueh, J. (2003). Effects of earnings-supplement policies on adult economic and middle-childhood outcomes differ for the "hardest to employ." *Child Development, 74,* 1500–1521.

Zhao, Z. (2004). Using matching to estimate treatment effects: Data requirements, matching metrics, and Monte Carlo evidence. *Review of Economics and Statistics, 86,* 91–107.

Zuehlke, T. W., & Zeman, A. R. (1991). A comparison of two-stage estimators of censored regression models. *Review of Economics and Statistics, 73,* 185–188.

Author Index

Subject Index

About the Authors

Shenyang Guo, PhD, is a Professor at the University of North Carolina (UNC) at Chapel Hill. He has done postdoctoral work at Brown University and held research associate or faculty appointments at the University of Michigan, Case Western Reserve University, the University of Tennessee, and the University of North Carolina. He is the author of numerous research reports in child welfare, child mental health services, welfare, and health care. He has expertise in applying advanced statistical models to solving social welfare problems and has taught graduate courses that address survival analysis, hierarchical linear modeling, growth curve modeling, and program evaluation. He has given many invited workshops on statistical methods—including event history analysis and propensity score matching—to NIH Summer Institute, Children's Bureau, and Society of Social Work and Research conferences. He is the director of the Applied Statistical Working Group at UNC. He led the data analysis planning for the National Survey of Child and Adolescent Well-Being (NSCAW) longitudinal analysis and has developed analytic strategies that address issues of weighting, clustering, growth modeling, and propensity score analysis. He is also directing the analysis of data from the Making Choices Project, a prevention trial funded by the National Institute on Drug Abuse and the Institute of Education Sciences. He has published many articles that include methodological works on the analysis of longitudinal data, multivariate failure time data, program evaluation, and multilevel modeling. He is on the editorial board of *Social Service Review* and a frequent guest reviewer for journals seeking a critique of advanced methodological analyses. He has an MA in economics from Fudan University and a PhD in sociology from the University of Michigan.

Mark W. Fraser, PhD, holds the John A. Tate Distinguished Professorship for Children in Need at the School of Social Work, University of North Carolina at Chapel Hill, where he serves as Associate Dean for Research. He has written numerous chapters and articles on risk and resilience, child behavior, child and family services, and research methods. With colleagues, he is the coauthor or editor of eight books. These include *Families in Crisis*, a study of intensive family-centered services, and *Evaluating Family-Based Services*, a text on methods for family research. In *Risk and Resilience in Childhood*, he and his colleagues describe resilience-based perspectives for child maltreatment, school dropout, substance abuse, violence, unwanted pregnancy, and other social problems. In *Making Choices*, he and his coauthors outline a program to help children build enduring social relationships with peers and adults. In *The Context of Youth Violence*, he explores violence from the perspective of resilience, risk, and protection, and in *Intervention With Children and Adolescents*, he and his colleagues review advances in intervention knowledge for social and health problems. His award-winning text, *Social Policy for Children and Families*, reviews the bases for public policy in child welfare, juvenile justice, mental health, developmental disabilities, and health. His most recent book, *Intervention Research: Developing Social Programs*, describes five steps in the design and development of evidence-based programs.